MULTIVARIATE DENSITY ESTIMATION

MULTIVARIATE DENSITY ESTIMATION

Theory, Practice, and Visualization

Second Edition

DAVID W. SCOTT
Rice University
Houston, Texas

Library of Congress Cataloging-in-Publication Data:

Scott, David W., 1950–
  Multivariate density estimation : theory, practice, and visualization / David W. Scott. – Second edition.
    pages   cm
  Includes bibliographical references and index.
  ISBN 978-0-471-69755-8 (cloth)
1. Estimation theory.   2. Multivariate analysis.   I. Title.
  QA276.8.S28 2014
  519.5′35–dc23

                                                        2014043897

Set in 10/12pts Times Lt Std by SPi Publisher Services, Pondicherry, India

Printed in the United States of America

10 9 8 7 6 5 4 3 2 1

1   2015

*To Jean, Hilary,
Elizabeth, Warren,
and my parents, John
and Nancy Scott*

CONTENTS

9 Other Applications

APPENDIX A Computer Graphics in $\Re^3$

PREFACE TO SECOND EDITION

The past 25 years have seen confirmation of the importance of density estimation and nonparametric methods in modern data analysis, in this era of "big data." This updated version retains its focus on fostering an intuitive understanding of the underlying methodology and supporting theory. I have sought to retain as much of the original material as possible and, in particular, the point of view of its development from the histogram. In every chapter, new material has been added to highlight challenges presented by massive datasets, or to clarify theoretical opportunities and new algorithms. However, no claim to comprehensive coverage is professed.

I have benefitted greatly from interactions with a number of gifted doctoral students who worked in this field—Lynette Factor, Donna Nezames, Rod Jee, Ferdie Wang, Michael Minnotte, Steve Sain, Keith Baggerly, John Salch, Will Wojciechowski, H.-G. Sung, Alena Oetting, Galen Papkov, Eric Chi, Jonathan Lane, Justin Silver, Jaime Ramos, and Yeshaya Adler—their work is represented here. In addition, contributions were made by many students taking my courses. I would also like to thank my colleagues and collaborators, especially my co-advisor Jim Thompson and my frequent co-authors George Terrell (VPI), Bill Szewczyk (DoD) and Masahiko Sagae (Kanazawa University). They have made the lifetime of learning, teaching, and discovery especially delightful and satisfying. I especially wish to acknowledge the able help of Robert Kosar in assembling the final versions of the color figures and reviewing new material.

Not a few mistakes have been corrected. For example, the constant in the expression for the asymptotic mean integrated squared error for the multivariate histogram in Theorem 3.5 is now correct. The content of Tables 3.6 and 3.7 has been modified accordingly, and the effect of dimension on sample size is seen to be even more dramatic in the corrected version. Any mistakes remain the responsibility of the author, who would appreciate hearing of such. All will be recorded in an appropriate repository.

Steve Quigley of John Wiley & Sons was infinitely patient awaiting this second edition until his retirement, and Kathryn Sharples completed the project. Steve made a freshly minted LaTeX version available as a starting point. All figures in S-Plus have been re-engineered into R. Figures in color or using color have been transformed to gray scale for the printed version, but the original figures will also be available in the same repository. In the original edition, I also neglected to properly acknowledge the generous support of the ARO (DAAL-03-88-G-0074 through my colleague James Thompson) and the ONR (N00014-90-J-1176).

As with the original edition, this revision would not have been possible without the tireless and enthusiastic support of my wife, Jean, and family. Thanks for everything.

Houston, Texas   DAVID W. SCOTT
August, 2014

PREFACE TO FIRST EDITION

With the revolution in computing in recent years, access to data of unprecedented complexity has become commonplace. More variables are being measured, and the sheer volume of data is growing. At the same time, advancements in the performance of graphical workstations have given new power to the data analyst. With these changes has come an increasing demand for tools that can detect and summarize the multivariate structure in difficult data. Density estimation is now recognized as a tool useful with univariate and bivariate data; my purpose is to demonstrate that it is also a powerful tool in higher dimensions, with particular emphasis on trivariate and quadrivariate data. I have written this book for the reader interested in the theoretical aspects of nonparametric estimation as well as for the reader interested in the application of these methods to multivariate data. It is my hope that the book can serve as an introductory textbook and also as a general reference.

I have chosen to introduce major ideas in the context of the classical histogram, which remains the most widely applied and most intuitive nonparametric estimator. I have found it instructive to develop the links between the histogram and more statistically efficient methods. This approach greatly simplifies the treatment of advanced estimators, as much of the novelty of the theoretical context has been moved to the familiar histogram setting.

The nonparametric world is more complex than its parametric counterpart. I have selected material that is representative of the broad spectrum of theoretical results available, with an eye on the potential user, based on my assessments of usefulness, prevalence, and tutorial value. Theory particularly relevant to application or understanding is covered, but a loose standard of rigor is adopted in order to emphasize the methodological and application topics. Rather than present a cookbook of techniques, I have adopted a hierarchical approach that emphasizes the similarities among the different estimators. I have tried to present new ideas and practical advice, together with numerous examples and problems, with a graphical emphasis.

Visualization is a key aspect of effective multivariate nonparametric analysis, and I have attempted to provide a wide array of graphic illustrations. All of the figures in this book were composed using S, S-PLUS, Exponent Graphics from IMSL, and Mathematica. The color plates were derived from S-based software. The color graphics with transparency were composed by displaying the S output using the MinneView program developed at the Minnesota Geometry Project and printed on hardware under development by the 3M Corporation. I have not included a great deal of computer code. A collection of software, primarily Fortran-based with interfaces to the S language, is available by electronic mail at scottdw@rice.edu. Comments and other feedback are welcomed.

I would like to thank many colleagues for their generous support over the past 20 years, particularly Jim Thompson, Richard Tapia, and Tony Gorry. I have especially drawn on my collaboration with George Terrell, and I gratefully acknowledge his major contributions and influence in this book. The initial support for the high-dimensional graphics came from Richard Heydorn of NASA. This work has been generously supported by the Office of Naval Research under grant N00014-90-J-1176 as well as the Army Research Office. Allan Wilks collaborated on the creation of many of the color figures while we were visiting the Geometry Project, directed by Al Marden and assisted by Charlie Gunn, at the Minnesota Supercomputer Center.

I have taught much of this material in graduate courses not only at Rice but also during a summer course in 1985 at Stanford and during an ASA short course in 1986 in Chicago with Bernard Silverman. Previous Rice students Lynette Factor, Donna Nezames, Rod Jee, and Ferdie Wang all made contributions through their theses. I am especially grateful for the able assistance given during the final phases of preparation by Tim Dunne and Keith Baggerly, as well as Steve Sain, Monnie McGee, and Michael Minnotte. Many colleagues have influenced this work, including Edward Wegman, Dan Carr, Grace Wahba, Wolfgang Härdle, Matthew Wand, Simon Sheather, Steve Marron, Peter Hall, Robert Launer, Yasuo Amemiya, Nils Hjort, Linda Davis, Bernhard Flury, Will Gersch, Charles Taylor, Imke Janssen, Steve Boswell, I.J. Good, Iain Johnstone, Ingram Olkin, Jerry Friedman, David Donoho, Leo Breiman, Naomi Altman, Mark Matthews, Tim Hesterberg, Hal Stern, Michael Trosset, Richard Byrd, John Bennett, Heinz-Peter Schmidt, Manny Parzen, and Michael Tarter. Finally, this book could not have been written without the patience and encouragement of my family.

Houston, Texas   DAVID W. SCOTT
February, 1992

1

REPRESENTATION AND GEOMETRY OF MULTIVARIATE DATA

A complete analysis of multidimensional data requires the application of an array of statistical tools—parametric, nonparametric, and graphical. Parametric analysis is the most powerful. Nonparametric analysis is the most flexible. And graphical analysis provides the vehicle for discovering the unexpected.

This chapter introduces some graphical tools for visualizing structure in multidimensional data. One set of tools focuses on depicting the data points themselves, while another set of tools relies on displaying of functions estimated from those points. Visualization and contouring of functions in more than two dimensions is introduced. Some mathematical aspects of the geometry of higher dimensions are reviewed. These results have consequences for nonparametric data analysis.

1.1 INTRODUCTION

Classical linear multivariate statistical models rely primarily on analysis of the covariance matrix. So powerful are these techniques that analysis is almost routine for datasets with hundreds of variables. While the theoretical basis of parametric models lies with the multivariate normal density, these models are applied in practice to many kinds of data. Parametric studies provide neat inferential summaries and parsimonious representation of the data.

For many problems second-order information is inadequate. Advanced modeling or simple variable transformations may provide a solution. When no simple

Multivariate Density Estimation, First Edition. David W. Scott.
© 2015 John Wiley & Sons, Inc. Published 2015 by John Wiley & Sons, Inc.

parametric model is forthcoming, many researchers have opted for fully "unparametric" methods that may be loosely collected under the heading of exploratory data analysis. Such analyses are highly graphical; but in a complex non-normal setting, a graph may provide a more concise representation than a parametric model, because a parametric model of adequate complexity may involve hundreds of parameters.

There are some significant differences between parametric and nonparametric modeling. The focus on optimality in parametric modeling does not translate well to the nonparametric world. For example, the histogram might be proved to be an inadmissible estimator, but that theoretical fact should not be taken to suggest histograms should not be used. Quite to the contrary, some methods that are theoretically superior are almost never used in practice. The reason is that the ordering of algorithms is not absolute, but is dependent not only on the unknown density but also on the sample size. Thus the histogram is generally superior for small samples regardless of its asymptotic properties. The exploratory school is at the other extreme, rejecting probabilistic models, whose existence provides the framework for defining optimality.

In this book, an intermediate point of view is adopted regarding statistical efficacy. No nonparametric estimate is considered wrong; only different components of the solution are emphasized. Much effort will be devoted to the data-based calibration problem, but nonparametric estimates can be reasonably calibrated in practice without too much difficulty. The "curse of optimality" might suggest that this is an illogical point of view. However, if the notion that optimality is all important is adopted, then the focus becomes matching the theoretical properties of an estimator to the assumed properties of the density function. Is it a gross inefficiency to use a procedure that requires only two continuous derivatives when the curve in fact has six continuous derivatives? This attitude may have some formal basis but should be discouraged as too heavy-handed for nonparametric thinking. A more relaxed attitude is required. Furthermore, many "optimal" nonparametric procedures are unstable in a manner that slightly inefficient procedures are not. In practice, when faced with the application of a procedure that requires six derivatives, or some other assumption that cannot be proved in practice, it is more important to be able to recognize the signs of estimator failure than to worry too much about assumptions. Detecting failure at the level of a discontinuous fourth derivative is a bit extreme, but certainly the effects of simple discontinuities should be well understood. Thus only for the purposes of illustration are the best assumptions given.

The notions of efficiency and admissibility are related to the choice of a criterion, which can only imperfectly measure the quality of a nonparametric estimate. Unlike optimal parametric estimates that are useful for many purposes, nonparametric estimates must be optimized for each application. The extra work is justified by the extra flexibility. As the choice of criterion is imperfect, so then is the notion of a single optimal estimator. This attitude reflects not sloppy thinking, but rather the imperfect relationship between the practical and theoretical aspects of our methods. Too rigid a point of view leads one to a minimax view of the world where nonparametric methods should be abandoned because there exist difficult problems.

Visualization is an important component of nonparametric data analysis. *Data visualization* is the focus of exploratory methods, ranging from simple scatterplots to sophisticated dynamic interactive displays. *Function visualization* is a significant component of nonparametric function estimation, and can draw on the relevant literature in the fields of scientific visualization and computer graphics. The focus of multivariate data analysis on points and scatterplots has meant that the full impact of scientific visualization has not yet been realized. With the new emphasis on smooth functions estimated nonparametrically, the fruits of visualization will be attained. Banchoff (1986) has been a pioneer in the visualization of higher dimensional mathematical surfaces. Curiously, the surfaces of interest to mathematicians contain singularities and discontinuities, all producing striking pictures when projected to the plane. In statistics, visualization of the smooth density surface in four, five, and six dimensions cannot rely on projection, as projections of smooth surfaces to the plane show nothing. Instead, the emphasis is on contouring in three dimensions and slicing of surfaces beyond. The focus on three and four dimensions is natural because one and two are so well understood. Beyond four dimensions, the ability to explore surfaces carefully decreases rapidly due to the curse of dimensionality. Fortunately, statistical data seldom display structure in more than five dimensions, so guided projection to those dimensions may be adequate. It is these threshold dimensions from three to five that are and deserve to be the focus of our visualization efforts.

There is a natural flow among the parametric, exploratory, and nonparametric procedures that represents a rational approach to statistical data analysis. Begin with a fully exploratory point of view in order to obtain an overview of the data. If a probabilistic structure is present, estimate that structure nonparametrically and explore it visually. Finally, if a linear model appears adequate, adopt a fully parametric approach. Each step conceptually represents a willingness to more strongly *smooth* the raw data, finally reducing the dimension of the solution to a handful of interesting parameters. With the assumption of normality, the mind's eye can easily imagine the d-dimensional egg-shaped elliptical data clusters. Some statisticians may prefer to work in the reverse order, progressing to exploratory methodology as a diagnostic tool for evaluating the adequacy of a parametric model fit.

There are many excellent references that complement and expand on this subject. In exploratory data analysis, references include Tukey (1977), Tukey and Tukey (1981), Cleveland and McGill (1988), and Wang (1978).

In density estimation, the classic texts of Tapia and Thompson (1978), Wertz (1978), and Thompson and Tapia (1990) first indicated the power of the nonparametric approach for univariate and bivariate data. Silverman (1986) has provided a further look at applications in this setting. Prakasa Rao (1983) has provided a theoretical survey with a lengthy bibliography. Other texts are more specialized, some focusing on regression (Müller, 1988; Härdle, 1990), some on a specific error criterion (Devroye and Györfi, 1985; Devroye, 1987), and some on particular solution classes such as splines (Eubank, 1988; Wahba, 1990). A discussion of additive models may be found in Hastie and Tibshirani (1990).

1.2 HISTORICAL PERSPECTIVE

One of the roots of modern statistical thought can be traced to the empirical discovery of correlation by Galton in 1886 (Stigler, 1986). Galton's ideas quickly reached Karl Pearson. Although best remembered for his methodological contributions such as goodness-of-fit tests, frequency curves, and biometry, Pearson was a strong proponent of the geometrical representation of statistics. In a series of lectures a century ago in November 1891 at Gresham College in London, Pearson spoke on a wide-ranging set of topics (Pearson, 1938). He discussed the foundations of the science of pure statistics and its many divisions. He discussed the collection of observations. He described the classification and representation of data using both numerical and geometrical descriptors. Finally, he emphasized statistical methodology and discovery of statistical laws. The syllabus for his lecture of November 11, 1891, includes this cryptic note:

> Erroneous opinion that Geometry is only a means of popular representation: *it is a fundamental method of investigating and analysing statistical material.* (his italics)

In that lecture Pearson described 10 methods of geometrical data representation. The most familiar is a representation "by columns," which he called the "histogram." (Pearson is usually given credit for coining the word "histogram" later in a 1894 paper.) Other familiar-sounding names include "diagrams," "chartograms," "topograms," and "stereograms." Unfamiliar names include "stigmograms," "euthygrams," "epipedograms," "radiograms," and "hormograms."

Beginning 21 years later, Fisher advanced the numerically descriptive portion of statistics with the method of maximum likelihood, from which he progressed on to the analysis of variance and other contributions that focused on the optimal use of data in parametric modeling and inference. In *Statistical Methods for Research Workers*, Fisher (1932) devotes a chapter titled "Diagrams" to graphical tools. He begins the chapter with this statement:

> The preliminary examination of most data is facilitated by the use of diagrams. Diagrams prove nothing, but bring outstanding features readily to the eye; they are therefore no substitute for such critical tests as may be applied to the data, but are valuable in suggesting such tests, and in explaining the conclusions founded upon them.

An emphasis on optimization and the efficiency of statistical procedures has been a hallmark of mathematical statistics ever since. Ironically, Fisher was criticized by mathematical statisticians for relying too heavily upon geometrical arguments in proofs of his results.

Modern statistics has experienced a strong resurgence of geometrical and graphical statistics in the form of exploratory data analysis (Tukey, 1977). Given the parametric emphasis on optimization, the more relaxed philosophy of exploratory data analysis has been refreshing. The revolution has been fueled by the low cost of graphical workstations and microcomputers. These machines have enabled current work on *statistics in motion* (Scott, 1990), that is, the use of animation and kinematic display for visualization of data structure, statistical analysis, and algorithm performance. No longer are static displays sufficient for comprehensive analysis.

All of these events were anticipated by Pearsonand his visionary statistical computing laboratory. In his lecture of April 14, 1891, titled "The Geometry of Motion," he spoke of the "ultimate elements of sensations we represent as motions in space and time." In 1918, after his many efforts during World War I, he reminisced about the excitement created by wartime work of his statistical laboratory:

> The work has been so urgent and of such value that the Ministry of Munitions has placed eight to ten computers and draughtsmen at my disposal ... (Pearson, 1938, p. 165).

These workers produced hundreds of statistical graphs, ranging from detailed maps of worker availability across England (chartograms) to figures for sighting antiaircraft guns (diagrams). The use of stereograms allowed for representation of data with three variables. His "computers," of course, were not electronic but human. Later, Fisher would be frustrated because Pearson would not agree to allocate his "computers" to the task of tabulating percentiles of the t-distribution. But Pearson's capabilities for producing high-quality graphics were far superior to those of most modern statisticians prior to 1980. Given Pearson's joint interests in graphics and kinematics, it is tantalizing to speculate on how he would have utilized modern computers.

1.3 GRAPHICAL DISPLAY OF MULTIVARIATE DATA POINTS

The modern challenge in data analysis is to be able to cope with whatever complexities may be intrinsic to the data. The data may, for example, be strongly non-normal, fall onto a nonlinear subspace, exhibit multiple modes, or be asymmetric. Dealing with these features becomes exponentially more difficult as the dimensionality of the data increases, a phenomenon known as the *curse of dimensionality*. In fact, datasets with hundreds of variables and millions of observations are routinely compiled that exhibit all of these features. Examples abound in such diverse fields as remote sensing, the US Census, geological exploration, speech recognition, and medical research. The expense of collecting and managing these large datasets is often so great that no funds are left for serious data analysis. The role of statistics is clear, but too often no statisticians are involved in large projects and no creative statistical thinking is applied. The goal of statistical data analysis is to extract the maximum information from the data, and to present a product that is as accurate and as useful as possible.

1.3.1 Multivariate Scatter Diagrams

The presentation of multivariate data is often accomplished in tabular form, particularly for small datasets with named or labeled objects. For example, Table B.1 contains economic data spanning the depression years of the 1930s, and Table B.2 contains information on a selected sample of American universities. It is easy enough to scan an individual column in these tables, to make comparisons of library size, for example, and to draw conclusions *one variable at a time* (see Tufte (1983) and Wang (1978)). However, variable-by-variable examination of multivariate data can be overwhelming and tiring, and cannot reveal any relationships among the variables. Looking at all pairwise scatterplots provides an improvement (Chambers et al., 1983). Data on four variables of three species of *Iris* are displayed in Figure 1.1. (A listing of the Fisher–Anderson *Iris* data, one of the few familiar four-dimensional datasets, may be found in several references and is provided with the S package (Becker et al., 1988)). What multivariate structure is apparent from this figure? The *setosa* variety does not overlap the other two varieties. The *versicolor* and *virginica* varieties are not as well separated, although a close examination reveals that they are almost nonoverlapping. If the 150 observations were unlabeled and plotted with the same symbol, it is likely that only two clusters would be observed. Even if it were known *a priori* that there were three clusters, it would still be unlikely that all three clusters would be properly identified. These alternative presentations reflect the two related problems of discrimination and clustering, respectively.

If the observations from different categories overlap substantially or have different sample sizes, scatter diagrams become much more difficult to interpret properly. The data in Figure 1.2 come from a study of 371 males suffering from chest pain (Scott et al., 1978): 320 had demonstrated coronary artery disease (occlusion or narrowing of the heart's own arteries) while 51 had none (see Table B.3). The blood fat concentrations of plasma cholesterol and triglyceride are predictive of heart disease, although the correlation is low. It is difficult to estimate the predictive power of these variables in this setting solely from the scatter diagram. A nonparametric analysis will reveal some interesting nonlinear interactions (see Chapters 5 and 9).

An easily overlooked practical aspect of scatter diagrams is illustrated by these data, which are integer valued. To avoid problems of overplotting, the data have been *jittered* or *blurred* (Chambers et al., 1983); that is, uniform $U(-0.5, 0.5)$ noise is

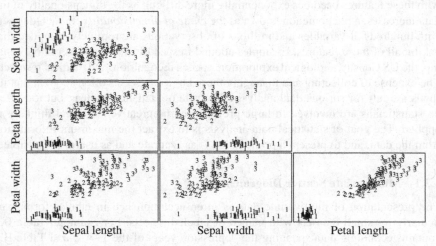

FIGURE 1.1 Pairwise scatter diagrams of the *Iris* data with the three species labeled. 1, setosa; 2, versicolor; 3, virginica.

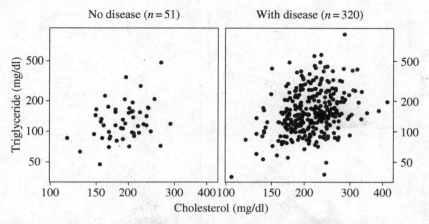

FIGURE 1.2　Scatter diagrams of blood lipid concentrations for 320 diseased and 51 nondiseased males.

added to each element of the original data. This trick should be regularly employed for data recorded with three or fewer significant digits (with an appropriate range on the added uniform noise). Jittering reduces visual miscues that result from the vertical and horizontal synchronization of regularly spaced data.

The visual perception system can easily be overwhelmed if the number of points is more than several thousand. Figure 1.3 displays three pairwise scatterplots derived from measurements taken in 1977 by the Landsat remote sensing system over a 5 mile by 6 mile agricultural region in North Dakota with $n = 22,932 = 117 \times 196$ *pixels* or picture elements, each corresponding to an area approximately 1.1 acres in size (Scott and Thompson, 1983; Scott and Jee, 1984). The Landsat instrument measures the intensity of light in four spectral bands reflected from the surface of the earth. A principal components transformation gives two variables that are commonly referred to as the "brightness" and "greenness" of each pixel. Every pixel is measured at regular intervals of approximately 3 weeks. During the summer of 1977, six useful replications were obtained, giving 24 measurements on each pixel. Using an agronometric growth model for crops, Badhwar et al. (1982) nonlinearly transformed this 24-dimensional data to three dimensions. Badhwar described these synthetic variables, (x_1, x_2, x_3), as (1) the calendar time at which peak greenness is observed, (2) the length of crop ripening, and (3) the peak greenness value, respectively. The scatter diagrams in Figure 1.3 have also been enhanced by jittering, as the raw data are integers between $(0, 255)$. The use of integers allows compression to eight bits of computer memory. Only structure in the boundary and tails is readily seen. The overplotting problem is apparent and the blackened areas include over 95% of the data. Other techniques to enhance scatter diagrams are needed to see structure in the bulk of the data cloud, such as plotting random subsets (see Tukey and Tukey (1981)).

Pairwise scatter diagrams lack one important property necessary for identifying more than two-dimensional features—strong interplot linkage among the plots. In

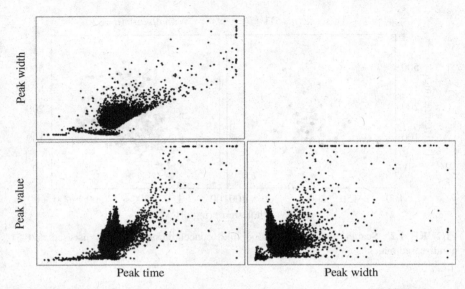

FIGURE 1.3   Pairwise scatter diagram of transformed Landsat data from 22,932 pixels over a 5 by 6 nautical mile region. The range on all the axes is (0, 255).

principle, it should be possible to locate the same point in each figure, assuming the data are free of ties. But it is not practical to do so for samples of any size. For quadrivariate data, Diaconis and Friedman (1983) proposed drawing lines between corresponding points in the scatterplots of (x_1, x_2) and (x_3, x_4) (see Problem 1.2). But a more powerful dynamic technique that takes full advantage of computer graphics has been developed by several research groups (McDonald, 1982; Becker and Cleveland, 1987; see the many references in Cleveland and McGill, 1988). The method is called *brushing* or *painting* a scatterplot matrix. Using a pointing device such as a mouse, a subset of the points in one scatter diagram is selected and the corresponding points are simultaneously highlighted in the other scatter diagrams. Conceptually, a subset of points in $\Re^d$ is tagged, for example, by painting the points red or making the points blink synchronously, and that characteristic is inherited by the linked points in all the "linked" graphs, including not only scatterplots but also histograms and regression plots as well. The *Iris* example in Figure 1.1 illustrates the flavor of brushing with three tags. Usually the color of points is changed rather than the symbol type. Brushing is an excellent tool for identifying outliers and following well-defined clusters. It is well-suited for conditioning on some variable, for example, $1 < x_3 < 3$.

These ideas are illustrated in Figure 1.4 for the PRIM4 dataset (Friedman and Tukey, 1974; the data summarize 500 high-energy particle physics scattering experiments) provided in the S language. Using the brushing tool in S-PLUS (1990), the left cluster in the 1–2 scatterplot was brushed, and then the left cluster in the 2–4 scatterplot was brushed with a different symbol. Try to imagine linking the clusters throughout the scatterplot matrix without any highlighting.

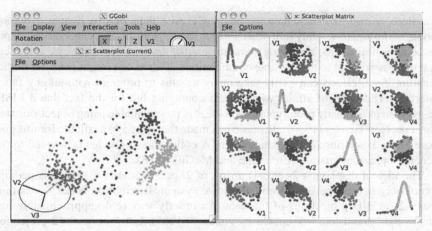

FIGURE 1.4 Pairwise scatterplots of the transformed PRIM4s data using the ggobi visualization system. Two clumps of points are highlighted by brushing.

There are limitations to the brushing technique. The number of pairwise scatterplots is $\binom{d}{2}$, so viewing more than 5 or 10 variables at once is impractical. Furthermore, the physical size of each scatter diagram is reduced as more variables are added, so that fewer distinct data points can be plotted. If there are more than a few variables, the eye cannot follow many of the dynamic changes in the pattern of points during brushing, except with the simplest of structure. It is, however, an open question as to the number of dimensions of structure that can be perceived by this method of linkage. Brushing remains an important and well-used tool that has proven successful in real data analysis.

If a 2-D array of bivariate scatter diagrams is useful, then why not construct a 3-D array of *trivariate* scatter diagrams? Navigating the collection of $\binom{d}{3}$ trivariate scatterplots is difficult even with modest values of d. But a single 3-D scatterplot can easily be rotated in real time with significant perceptual gain compared to three bivariate diagrams in the scatterplot matrix. Many statistical packages now provide this capability. The program MacSpin (Donoho et al., 1988) was the first widely used software of this type. The top middle panel in Figure 1.4 displays a particular orientation of a rotating 3-D scatterplot. The kinds of structure available in 3-D data are more complex (and hence more interesting) than in 2-D data. Furthermore, the overplotting problem is reduced as more data points can be resolved in a rotating 3-D scatterplot than in a static 2-D view (although this is resolution dependent—a 2-D view printed by a laser device can display significantly more points than is possible on a computer monitor). Density information is still relatively difficult to perceive, however, and the sample size definitely influences perception.

Beyond three dimensions, many novel ideas are being pursued (see Tukey and Tukey (1981)). Six-dimensional data could be viewed with two rotating 3-D scatter diagrams linked by brushing. Carr and Nicholson (1988) have actively pursued using stereography as an alternative and adjunct to rotation. Some workers report that stereo viewing of static data can be more precise than viewing dynamic rotation alone. Unfortunately, many individuals suffer from color blindness and various depth perception limitations, rendering some techniques useless. Nevertheless, it is clear that there is no limit to the possible combinations of ideas one might consider implementing. Such efforts can easily take many months to program without any fancy interface. This state of affairs would be discouraging but for the fact that a LISP-based system for easily prototyping such ideas is now available using object-oriented concepts (see Tierney (1990)). RStudio has made the *shiny* app available for this purpose as well: see http://shiny.rstudio.com. A collection of articles is devoted to the general topic of animation (Cleveland and McGill, 1988).

The idea of displaying 2- or 3-D arrays of 2- or 3-D scatter diagrams is perhaps too closely tied to the Euclidean coordinate system. It might be better to examine many 2- or 3-D projections of the data. An orderly way to do approximately just that is the "grand tour" discussed by Asimov (1985). Let P be a $d \times 2$ projection matrix, which takes the d-dimensional data down to a plane. The author proposed examining a sequence of scatterplots obtained by a smoothly changing sequence of projection matrices. The resulting kinematic display shows the n data points moving in a continuous (and sometimes seemingly random) fashion. It may be hoped that most interesting projections will be displayed at some point during the first several minutes of the grand tour, although for even 10 variables several hours may be required (Huber, 1985).

Special attention should be drawn to representing multivariate data in the bivariate scatter diagram with points replaced by *glyphs*, which are special symbols whose shapes are determined by the remaining data variables $(x_3, \ldots, x_d)$. Figure 1.5 displays the *Iris* data in such a form following Carr et al. (1986). The length and angle of the glyph are determined by the sepal length and width, respectively. Careful examination of the glyphs shows that there is no gap in 4-D between the *versicolor* and *virginica* species, as the angles and lengths of the glyphs are similar near the boundary.

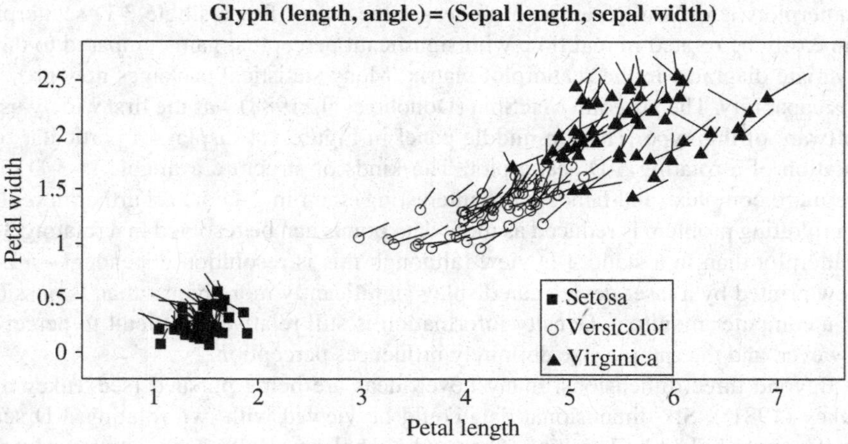

Glyph (length, angle) = (Sepal length, sepal width)

FIGURE 1.5   Glyph scatter diagram of the *Iris* data.

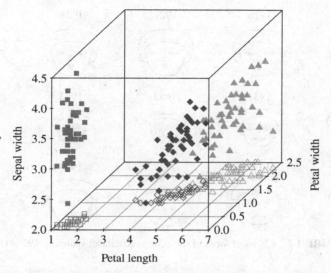

FIGURE 1.6 A three-dimensional scatter diagram of the Fisher–Anderson *Iris* data, omitting the sepal length variable. From left to right, the 50 points for each of the three varieties of *setosa*, *versicolor*, and *virginica* are distinguished by symbol type (square, diamond, triangle), respectively. The symbol is required to indicate the presence of three clusters rather than only two. The same basic picture results from any choice of three variables from the full set of four variables.

A second glyph representation shown in Figure 1.6 is a 3-D scatterplot omitting sepal length, one of the four variables. This figure clearly depicts the structure in these data. Plotting glyphs in 3-D scatter diagrams with stereography is a more powerful visual tool (Carr and Nicholson, 1988). The glyph technique does not treat variables "symmetrically" and all variable–glyph combinations could be considered. This complaint affects most multivariate procedures (with a few exceptions).

All of these techniques are an outgrowth of a powerful system devised to analyze data in up to nine dimensions called PRIM-9 (Fisherkeller et al., 1974; reprinted in Cleveland and McGill, 1988). The PRIM-9 system contained many of the capabilities of current systems. The letters are an acronym for "Picturing, Rotation, Isolation, and Masking." The latter two serve to identify and select subsets of the multivariate data. The "picturing" feature was implemented by pressing two buttons that cycled through all of the $\binom{9}{2}$ pairwise scatter diagrams in current coordinates. An IBM 360 mainframe was specially modified to drive the custom display system.

1.3.2 Chernoff Faces

Chernoff (1973) proposed a special glyph that associates variables to facial features, such as the size and shape of the eyes, nose, mouth, hair, ears, chin, and facial outline. Certainly, humans are able to discriminate among nearly identical faces very well. Chernoff has suggested that most other multivariate point methods "seem to be

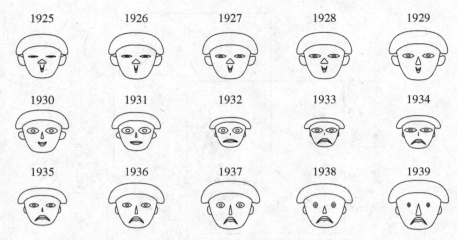

FIGURE 1.7   Chernoff faces of the economic dataset spanning 1925–1939.

less valuable in producing an emotional response" (Wang, 1978, p. 6).Whether an emotional response is desired is debatable. Chernoff faces for the time series dataset in Table B.1 are displayed in Figure 1.7. (The variable–feature associations are listed in the table.) By carefully studying an individual facial feature such as the smile over the sequence of all the faces, simple trends can be recognized. But it is the overall multivariate impression that makes Chernoff faces so powerful. Variables should be carefully assigned to features. For example, Chernoff faces of the colleges' data in Table B.2 might logically assign variables relating to the library to the eyes rather than to the mouth (see Problem 1.3). Such subjective judgments should not prejudice our use of this procedure.

   One early application not in a statistics journal was constructed by Hiebert-Dodd (1982), who had examined the performance of several optimization algorithms on a suite of test problems. She reported that several referees felt this method of presentation was too frivolous. Comparing the endless tables in the paper as it appeared to the Chernoff faces displayed in the original technical report, one might easily conclude the referees were too cautious. On the other hand, when Rice University administrators were shown Chernoff faces of the colleges' dataset, they were quite open to its suggestions and enjoyed the exercise. The practical fact is that repetitious viewing of large tables of data is tedious and haphazard, and broad-brush displays such as faces can significantly improve data digestion. Several researchers have noted that Chernoff faces contain redundant information because of symmetry. Flury and Riedwyl (1981) have proposed using asymmetrical faces, as did Turner and Tidmore (1980), although Chernoff has stated he believes the additional gain does not justify such nonrealistic figures.

1.3.3   Andrews' Curves and Parallel Coordinate Curves

Three intriguing proposals display not the data points themselves but rather a unique curve determined by the data vector $\mathbf{x}$. Andrews (1972) proposed representing

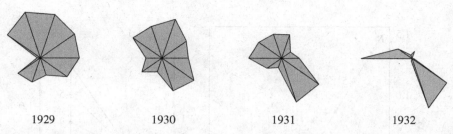

1929   1930   1931   1932

FIGURE 1.8 Star diagram for 4 years of the economic dataset shown in Figure 1.7.

high-dimensional data by replacing each point in $\Re^d$ with a curve $s(t)$ for $|t| < \pi$, where

$$s(t \mid x_1,\ldots,x_d) = \frac{x_1}{\sqrt{2}} + x_2 \sin t + x_3 \cos t + x_4 \sin 2t + x_5 \cos 2t + \cdots,$$

the so-called *Fourier series representation*. This mapping provides the first "complete" continuous view of high-dimensional points on the plane, because, in principle, the original multivariate data point can be recovered from this curve. Clearly, an Andrews' curve is dominated by the variables placed on the low-frequency terms, so care should be taken to put the most interesting variables early in the expansion (see Problem 1.4).

A simple graphical device that treats the d variables symmetrically is the star diagram, which is discussed by Fienberg (1979). The d axes are drawn as spokes on a wheel. The coordinate data values are plotted on those axes and connected as shown in Figure 1.8.

Another novel multivariate approach that treats variables in a symmetric fashion is the *parallel coordinates plot*, introduced by Inselberg (1985) in a mathematical setting and extended by Wegman (1990) to the analysis of stochastic data. Cartesian coordinates are abandoned in favor of d axes drawn parallel and equally spaced. Each multivariate point $\mathbf{x} \in \Re^d$ is plotted as a piecewise linear curve connecting the d points on the parallel axes. For reasons shown by Inselberg and Wegman, there are advantages to simply drawing piecewise linear line segments, rather than a smoother line such as a spline. The disadvantage of this choice is that points that have identical values in any coordinate dimension cannot be distinguished in parallel coordinates. However, with this choice a duality may be deduced between points and lines in Euclidean and parallel coordinates. In the left frame of Figure 1.9, six points that fall on a straight line with negative slope are plotted. The right frame shows those same points in parallel coordinates. Thus a scatter diagram of highly correlated normal points displays a nearly common point of intersection in parallel coordinates. However, if the correlation is positive, that point is not "between" the parallel axes (see Problem 1.6). The location of the point where the lines all intersect can be used to recover the equation of the line back in Euclidean coordinates (see Problem 1.8).

A variety of other properties with potential applications are explored by Inselberg and Wegman. One result is a graphical means of deciding if a point $\mathbf{x} \in \Re^d$ is on the

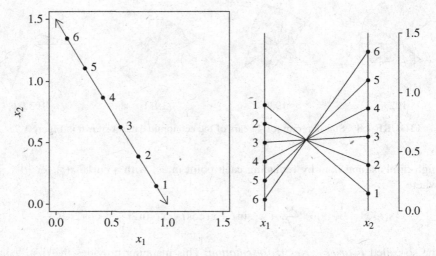

FIGURE 1.9 Example of duality of points and lines between Euclidean and parallel coordinates. The points are labeled 1 to 6 in both coordinate systems.

inside or the outside of a convex closed hypersurface. If all the points on the hypersurface are plotted in parallel coordinates, then a well-defined geometrical outline will appear on the plane. If a portion of the line segments defining the point **x** in parallel coordinates fall outside the outline, then **x** is not inside the hypersurface, and vice versa. One of the more fascinating extensions developed by Wegman is a grand tour of all variables displayed in parallel coordinates. The advantage of parallel coordinates is that all d of the rotating variables are visible simultaneously, whereas in the usual presentation, only two of the grand tour variables are visible in a bivariate scatterplot.

Figure 1.10 displays parallel coordinate plots of the *Iris* and earthquake data. The earthquake dataset represents the epicenters of 473 tremors beneath the Mount St. Helens volcano in the several months preceding its March 1982 eruption (Weaver et al., 1983). Clearly, the tremors are mostly small in magnitude, increasing in frequency over time, and clustered near the surface, although depth is clearly a bimodal variable. The longitude and latitude variables are least effective on this plot, because their natural spatial structure is lost.

1.3.4 Limitations

Tools such as Chernoff faces and scatter diagram glyphs tend to be most valuable with small datasets where individual points are "identifiable" or interesting. Such individualistic exploratory tools can easily generate "too much ink" (Tufte, 1983) and produce figures with black splotches, which convey little information. Parallel coordinates and Andrews' curves generate much ink. One obvious remedy is to plot

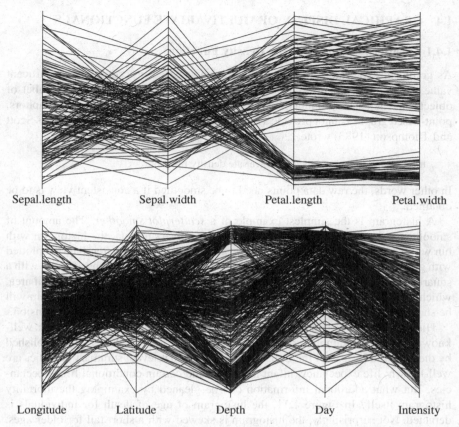

FIGURE 1.10 Parallel coordinate plot of the earthquake dataset.

only a subset of the data in a process known as "thinning." However, plotting random subsets no longer makes optimal use of all the data and does not result in precisely reproducible interpretations. Point-oriented methods typically have a range of sample sizes that is most appropriate: $n < 200$ for faces; $n < 2000$ for scatter diagrams.

Since none of these displays is truly d-dimensional, each has limitations. All pairwise scatterplots can detect distinct clusters and some two-dimensional structure (if perhaps in a rotated coordinate system). In the latter case, an interactive supplement such as brushing may be necessary to confirm the nature of the links among the scatterplots (not really providing any higher dimensional information). On the positive side, variables are treated symmetrically in the scatterplot matrix. But many different and highly dissimilar d-dimensional datasets can give rise to visually similar scatterplot matrix diagrams; hence the need for brushing. However, with increasing number of variables, individual scatterplots physically decrease in size and fill up with ink ever faster. Scatter diagrams provide a highly subjective view of data, with poor density perception and greatest emphasis on the tails of the data.

1.4 GRAPHICAL DISPLAY OF MULTIVARIATE FUNCTIONALS

1.4.1 Scatterplot Smoothing by Density Function

As graphical exploratory tools, each of the point-based procedures has significant value. However, each suffers from the problem of too much ink, as the number of objects (and hence the amount of ink) is linear in the sample size n. To mix metaphors, point-based graphs cannot provide a consistent picture of the data as $n \to \infty$. As Scott and Thompson (1983) wrote,

> the scatter diagram points to the bivariate density function.

In other words, the raw data points need to be smoothed if a consistent view is to be obtained.

A histogram is the simplest example of a *scatterplot smoother*. The amount of smoothness is controlled by the bin width. For univariate data, the histogram with bin width narrower than min $|x_i - x_j|$ is precisely a univariate scatter diagram plotted with glyphs that are tall, thin rectangles. For bivariate data, the glyph is a beam with a square base. Increasing the bin width, the histogram represents a count per unit area, which is precisely the unit of a probability density. In Chapter 3, the histogram will be shown to provide a consistent estimate of the density function in any dimension.

Histograms can provide a wealth of information for large datasets, even well-known ones. For example, consider the 1979–1981 decennial life table published by the U.S. and Bureau of the Census (1987). Certain relevant summary statistics are well-known: life expectancy, infant mortality, and certain conditional life expectancies. But what additional information can be gleaned by examining the mortality histogram itself? In Figure 1.11, the histogram of age of death for individuals is depicted. Not surprisingly, the histogram is skewed with a short tail for older ages. Not as well-known perhaps is the observation that the most common age of death is 85! The absolute and relative magnitude of mortality in the first year of life is made strikingly clear.

Careful examination reveals two other general features of interest. The first feature is the small but prominent bump in the curve between the ages of 13 and 27 years. This "excess mortality" is due to an increase in a variety of risky activities, the most notable being obtaining a driver's license. In the right frame of Figure 1.11, comparison of the 1959–1961 (Gross and Clark, 1975) and 1979–1981 histograms shows an impressive reduction of death in all preadolescent years. Particularly striking is the 60% decline in mortality in the first year and the 3-year difference in the locations of the modes.

These facts are remarkable when placed in the context of the *mortality histogram* constructed by John Graunt from the Bills of Mortality during the plague years. Graunt (1662) estimated that 36% of individuals died before attaining their sixth birthday! Graunt was a contemporary of the better-known William Petty, to whom some credit for these ideas is variously ascribed, probably without cause. The circumstantial evidence that Graunt actually invented the histogram while looking at these mortality data seems quite strong, although there is reason to infer that Galileo had used

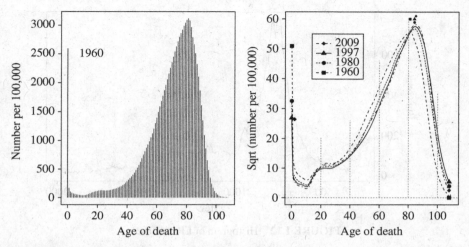

FIGURE 1.11 Histogram of the U.S. mortality data in 1960. Rootgrams (histograms plotted on a square-root scale) of the mortality data for 1960, 1980, and 1997.

histogram-like diagrams earlier. Hald (1990) recounts a portion of Galileo's *Dialogo*, published in 1632, in which Galileo summarized his observations on the star that appeared in 1572. According to Hald, Galileo noted the symmetry of the "observation errors" and the more frequent occurrence of small errors than large errors. Both points suggest Galileo had constructed a frequency diagram to draw those conclusions.

Many large datasets are in fact collected in binned or histogram form. For example, elementary particles in high-energy physics scattering experiments are manifested by small bumps in the frequency curve. Good and Gaskins (1980) considered such a large dataset ($n = 25,752$) from the Lawrence Radiation Laboratory (LRL) (see Figure 1.12). The authors devised an ingenious algorithm for estimating the odds that a bump observed in the frequency curve was real. This topic is covered in Chapter 9.

Multivariate scatterplot smoothing of time series data is also easily accomplished with histograms. Consider a univariate time series and smooth both the raw data $\{x_t\}$ as well as the lagged data $\{x_t, x_{t+1}\}$. Any strong elliptical structure present in the smoothed lagged-data diagram provides a graphical version of the first-order autocorrelation coefficient. Consider the Old Faithful geyser dataset listed in Table B.6. These data are the durations in minutes of 107 eruptions of the Old Faithful geyser (Weisberg, 1985). As there was a gap in the recording of data between midnight and 6 A.M., there are only 99 pairs $\{x_t, x_{t+1}\}$ available. The univariate histogram in Figure 1.13 reveals a simple bimodal structure—short and long eruption durations. The most notable feature in the bivariate (smoothed) histogram is the missing fourth bump corresponding to the short-short duration sequence. Clearly, graphs of $\hat{f}(x_{t+1}|x_t)$ would be useful for improved prediction compared to a regression estimate.

For more than two dimensions, only slices are available for viewing with histogram surfaces. Consider the Landsat data again. Divide the (jittered) data into four pieces using quartiles of x_1, which is the time of peak greenness. Examining a series of

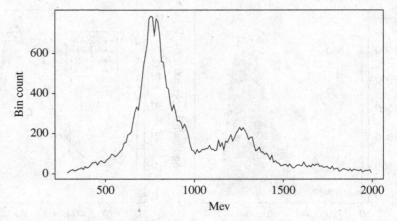

FIGURE 1.12   Histogram of LRL dataset.

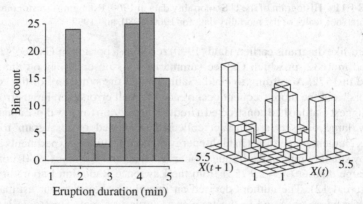

FIGURE 1.13   Histogram of $\{x_t\}$ for the Old Faithful geyser dataset, and a bivariate histogram of the lagged data (x_t, x_{t+1}).

bivariate pictures of (x_2, x_3) for each quartile slice provides a crude approximation of the four-dimensional surface $\hat{f}(x_1, x_2, x_3)$ (see Figure 1.14). The histograms are all constructed on the subinterval $[-5, 100] \times [-5, 100]$. Compare this representation of the Landsat data to that in Figure 1.3. From Figure 1.3, it is clear that most of the outliers are in the last quartile of x_1. How well can the relative density levels be determined from the scatter diagrams? Visualization of a smoothed histogram of these data will be considered in Section 1.4.3.

1.4.2   Scatterplot Smoothing by Regression Function

The term *scatterplot smoother* is most often applied to regression data. For bivariate data, either a nonparametric regression line can be superimposed upon the data, or the points themselves can be moved toward the regression line. Tukey (1977) presents

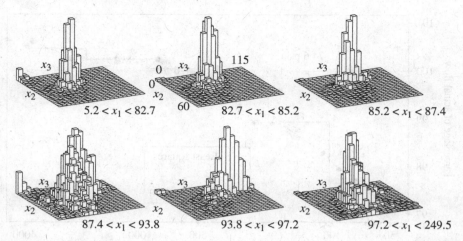

FIGURE 1.14 Bivariate histogram slices of the trivariate Landsat data. Slicing was performed at the quartiles of variable x_1.

the "3R" smoother as an example of the latter. Suppose that the n data points, $\{x_t\}$, are measured on a fixed time scale. The 3R smoothing algorithm replaces each point $\{x_t\}$ with the median of the three points $\{x_{t-1}, x_t, x_{t+1}\}$ recursively until no changes occur. This algorithm is a powerful filter that removes isolated outliers effectively. The 3R smoother may be applied to unequally spaced data or repeated data. Tukey also proposes applying a Hanning filter, by which $\tilde{x}_t \leftarrow 0.25 \times (x_{t-1} + 2x_t + x_{t+1})$. This filter may be applied several times as necessary. In Figure 1.15, the Tukey smoother (S function *smooth*) is applied to the gas flow dataset given in the Table B.5. Observe how the single potential outlier at $x = 187$ is totally ignored. The least-squares fit is shown for reference.

The simplest nonparametric regression estimator is the *regressogram*. The x-axis is binned and the sample averages of the responses are computed and plotted over the intervals. The regressogram for the gas flow dataset is also shown in Figure 1.15. The Hanning filter and regressogram are special cases of nonparametric kernel regression, which is discussed in Chapter 8.

The gas flow dataset is part of a larger collection taken at seven different pressures. A stick-pin plot of the complete dataset is shown in Figure 1.16 (the 74.6 psia data are second from the right). Clearly, the accuracy is affected by the flow rate, while the effect of psia seems small. These data will be revisited in Chapter 8.

1.4.3 Visualization of Multivariate Functions

Visualization of functions of more than two variables has not been common in statistics. The Landsat example in Figure 1.14 hints at the potential that visualization of 4-D surfaces would bring to the data analyst. In this section, effective visualization of surfaces in more than three dimensions is introduced.

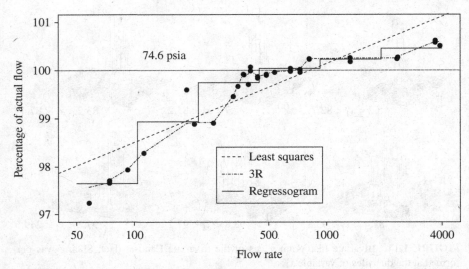

FIGURE 1.15 Accuracy of a natural gas meter as a function of the flow rate through the valve at 74.6 psia. The raw data ($n = 33$) are shown by the filled points. The three smooths (least squares, Tukey's 3R, and Tukey's regressogram) are superimposed.

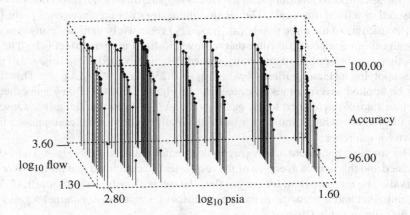

FIGURE 1.16 Complete 3-D view of the gas flow dataset.

Displaying a three-dimensional perspective plot of the surface $f(x, y)$ of a bivariate function requires one more dimension than the corresponding bivariate contour representation (see Figure 1.17). There are trade-offs. The contour representation lacks the exact detail and visual impact available in a perspective plot; however, perspective plots usually have portions obscured by peaks and present less precise height information. One way of expressing the difference is to say that a contour plot displays, loosely speaking, about 2.6–2.9 dimensions of the entire 3-D surface (more, as more contour lines are drawn). Some authors claim that one or the other representation is superior, but it seems clear that both can be useful for complicated surfaces.

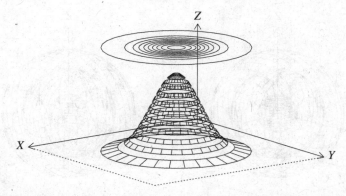

FIGURE 1.17 Perspective plot of bivariate normal density with a "floating" representation of the corresponding contours.

The visualization advantage afforded by a contour representation is that it lives in the *same dimension* as the data, whereas a perspective plot requires an additional dimension. Hence with trivariate data, the third dimension can be used to present a 3-D contour. In the case of a density function, the corresponding 3-D contour plot comprises one or more α-*level contour surfaces*, which are defined for $\mathbf{x} \in \Re^d$ by

$$\alpha\text{-Contour}: \qquad S_\alpha = \{\mathbf{x} : f(\mathbf{x}) = \alpha f_{\max}\}, \qquad 0 \le \alpha \le 1,$$

where $f_{\max}$ is the maximum or modal value of the density function.

For normal data, the general contour surfaces are hyper-ellipses defined by the easily verified equation (see Problem 1.14):

$$(\mathbf{x} - \mu)^T \Sigma^{-1} (\mathbf{x} - \mu) = -2 \log \alpha. \tag{1.1}$$

A trivariate contour plot of $f(x_1, x_2, x_3)$ would generally contain several "nested" surfaces, $\{S_{0.1}, S_{0.3}, S_{0.5}, S_{0.7}, S_{0.9}\}$, for example. For the independent standard normal density, the contours would be nested hyperspheres centered on the mode. In Figure 1.18, three contours of the trivariate standard normal density are shown in stereo. Many if not most readers, will have difficulty crossing their eyes to obtain the stereo effect. But even without the stereo effect, the three spherical contours are well-represented.

How effective is this in practice? Consider a smoothed histogram $\hat{f}(x, y, z)$ of 1000 trivariate normal points with $\Sigma = I_3$. Figure 1.19 shows surfaces of nine equally spaced bivariate slices of the trivariate estimate. Each slice is approximately bivariate normal but without rescaling. Of course, the surfaces are not precisely bivariate normal, due to the finite size of the sample.

A natural question to pose is: Why not plot the corresponding sequence of *conditional densities*, $\hat{f}(x, y | z = z_0)$, rather than the *slices*, $\hat{f}(x, y, z_0)$? If this were done, all the surfaces in Figure 1.19 would be nearly identical. (Theoretically, the condition

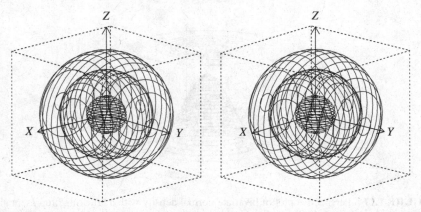

FIGURE 1.18   Stereo representation of three α-contours of a trivariate normal density. Gently crossing your eyes should allow the two frames to fuse in the middle.

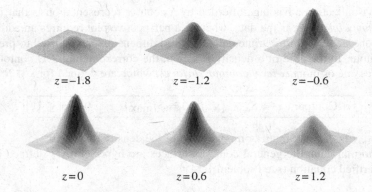

FIGURE 1.19   Sequence of bivariate slices of a trivariate smoothed histogram.

densities are all exactly $N(\mathbf{0}_2, I_2)$.) If the goal is to understand the 4-D density surface, then the sequence of conditional densities overemphasizes the (visual) importance of the tails and obscures information about the location of the "center" of the data. Furthermore, as nonparametric estimates in the tail will be relatively noisy, the estimates will be especially rough upon normalization (see Figure 1.20). For these reasons, it seems best to look at slices and to reserve normalization for looking at conditional densities that are particularly interesting.

Several trivariate contour surfaces of the same estimated density are displayed in Figure 1.21. Clearly, the trivariate contours give an improved "big picture"—just as a rotating trivariate scatter diagram improves on three static bivariate scatter diagrams. The complete density estimate is a 4-D surface, and the trivariate contour view in the final frame of Figure 1.21 may present only 3.5 dimensions, while the series of bivariate slices may yield a bit more, perhaps 3.75 dimensions, but without the visual impact. Examine the 3-D contour view for the Landsat data in the first frame of Figure 7.8 in comparison to Figures 1.3 and 1.14. The structure is quite complex.

$z = -3$ $z = -2.6$ $z = -2.2$

FIGURE 1.20 Normalized slices in the left tail of the smoothed histogram.

The presentation of clusters is stunning and shows multiple modes and multiple clusters. This detailed structure is not apparent in the scatterplot in Figure 1.3.

Depending on the nature of the variables, slicing can be attempted with four-, five-, or six-dimensional data. Of special importance is the 5-D surface generated by 4-D data, for example, space–time variables such as the Mount St. Helens data in Figure 1.10. These higher dimensional estimates can be animated in a fashion similar to Figure 1.19 (see Scott and Wilks (1990)).

In the 4-D case, the α-level contours of interest are based on the slices:

$$S_{\alpha,t} = \{(x,y,z) : f(x,y,z,t) = \alpha f_{\max}\},$$

where $f_{\max}$ is the global maximum over the 5-D surface. For a fixed choice of α, as the slice value t changes continuously, the contour shells will expand or contract smoothly, finally vanishing for extreme values of t. For example, a single theoretical contour of the $N(0, I_4)$ density would vanish outside a symmetric interval around the origin, but within that interval, the contour shell would be a sphere centered on the origin with greatest diameter when $t = 0$. With several α-shells displayed simultaneously, the contours would be nested spheres of different radii, appearing at different values of t, but of greatest diameter when $t = 0$.

One particularly interesting slice of the smoothed 5-D histogram estimate of the entire *Iris* dataset is shown in Figure 1.22. The $\alpha = 4\%$ contour surface reveals two well-separated clusters. However, the $\alpha = 10\%$ contour surface is trimodal, revealing the true structure in this dataset even with only 150 points. the *virginica* and *versicolor* data may not be separated in the point cloud but apparently can be separated in the density cloud.

The 3-D contour slices in Figure 1.22 were assembled from a 2-D contouring algorithm, then projected into the plane. The sequence of 2-D contour slices is shown in Figure 1.23. Study these two diagrams and think about the possibilities for exploring the entire five-dimensional surface.

To emphasize the potential value of additional variables, we conclude this vignette, we examine the *Iris* data excluding the sepal width variable. Figure 1.24 displays a 3-D scatterplot, as well as contours of the smoothed histogram at levels $\alpha = 0.17$ and $\alpha = 0.44$. A litle study supports the speculation that the data might contain a hybrid

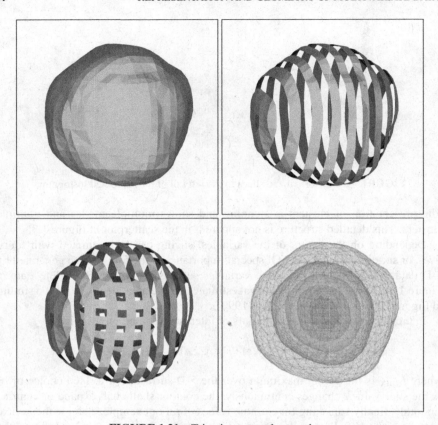

FIGURE 1.21   Trivariate normal examples.

species of the *versicolor* and *virginica* species. With such a small sample, that may be an embellishment.

With more than four variables, the most appropriate sequence of slicing is not clear. With five variables, bivariate contours of (x_4, x_5) may be drawn; then a sequence of trivariate slices may be examined tracing along one of these bivariate contours. With more than five or six variables, deciding where to slice at all is a difficult problem because the number of possibilities grows exponentially. That is why projection-based methods are so important (see Chapter 7).

1.4.3.1 Visualizing Multivariate Regression Functions   The same graphical representation can be applied to regression surfaces. However, the interpretation can be more difficult. For example, if the regression surface is monotone, the α-level contours of the surface will not be "closed" and will appear to "float" in space. If the regression surface is a simple linear function such as $ax + by + cz$, then a set of trivariate α-contours will simply be a set of parallel planes. Practical questions arise that do not appear for density surfaces. In particular, what is the natural extent of the regression surface; that is, for what region in the design space should the surface be

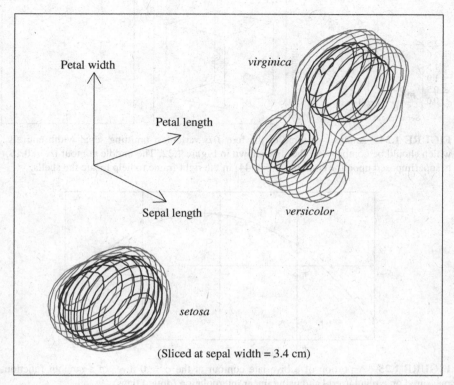

FIGURE 1.22 Two α-level contour surfaces from a slice of a five-dimensional averaged shifted histogram estimate, based on all 150 *Iris* data points. The displayed variables x, y, and z are sepal length, petal length and width, respectively, with the sepal width variable sliced at $t = 3.4$ cm. The (outer) darker $\alpha = 4\%$ contour reveals only two clusters, while the (inner) lighter $\alpha = 10\%$ contour reveals the three clusters.

1.4 Overview of Contouring and Surface Display

$x=4$	$x=4.15$	$x=4.3$	$x=4.45$	$x=4.6$	$x=4.75$	$x=4.9$	$x=5.05$
$x=5.2$	$x=5.35$	$x=5.5$	$x=5.65$	$x=5.8$			
					$x=5.95$	$x=6.1$	$x=6.25$
$x=6.4$	$x=6.55$	$x=6.7$	$x=6.85$	$x=7$	$x=7.15$	$x=7.3$	$x=7.45$

FIGURE 1.23 A detailed breakdown of the 3-D contours shown in Figure 1.22 taken from the ASH estimate $\hat{f}(x,y,z,t = 3.4)$ as the sepal length, x, ranges from 4.00 to 7.45 cm.

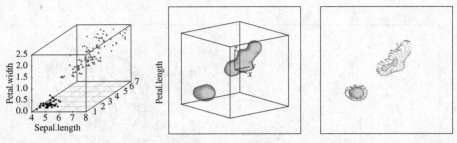

FIGURE 1.24 Analysis of three of the four *Iris* variables, omitting sepal width entirely, which should be compared to the slice shown in Figure 1.22. The middle contour ($\alpha = 0.17$) is superimposed upon the contour ($\alpha = 0.44$) in the right frame to help locate the shells.

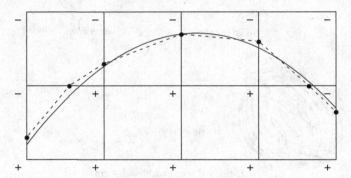

FIGURE 1.25 A portion of a bivariate contour at the $\alpha = 0$ level of a smooth function measured on a regular grid and using linear interpolation (dotted lines).

plotted? Perhaps one answer is to limit the plot to regions where there is sufficient data, that is, where the density of design points is above a certain threshold.

1.4.4 Overview of Contouring and Surface Display

Suppose that a general bivariate function $f(x, y)$ (taking on positive and negative values) is sampled on a regular grid, and the $\alpha = 0$ contour S_0 is desired; that is, $S_0 = \{(x, y) : f(x, y) = 0\}$. Label the values of the grid as $+$, 0, or $-$ depending on whether $f > 0$, $f = 0$, or $f < 0$, respectively. Then the desired contour is shown in Figure 1.25. The piecewise linear approximation and the true contour do not match along the bin boundaries since the interpolation is not exact.

However, bivariate contouring is not as simple a task as one might imagine. Usually, the function is sampled on a rectangular mesh, with no gradient information or possibility for further refinement of the mesh. If too coarse a mesh is chosen, then small local bumps or dips may be missed, or two distinct contours at the same level may be inadvertently joined. For speed and simplicity, one wants to avoid having to do any global analysis before drawing contours. A local contouring algorithm avoids multiple passes over the data. In any case, global analysis is based on certain

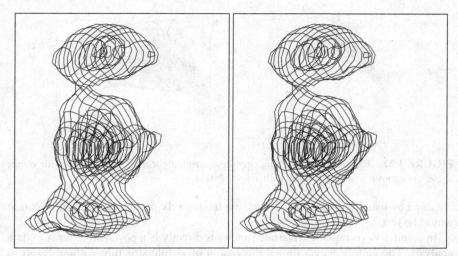

FIGURE 1.26 Simple stereo representation of four 3-D nested shells of the earthquake data.

smoothness assumptions and may fail. The difficulties and details of contouring are described more fully in Section A.1.

There are several varieties of 3-D contouring algorithms. It is assumed that the function has been sampled on a lattice, which can be taken to be cubical without loss of generality. One simple trick is to display a set of 2-D contour slices that result from intersecting the 3-D contour shell with a set of parallel planes along the lattice of the data, as was done in Figures 1.18 and 1.22. In this representation, a single spherical shell becomes a set of circular contours (Figure 1.26). This approach has the advantage of providing a shell representation that is "transparent" so that multiple α-level contour levels may be visualized. Different colors can be used for different contour levels (see Scott (1983, 1984, 1991a), Scott and Thompson (1983), Härdle and Scott (1988), and Scott and Hall (1989)).

More visually pleasing surfaces can be drawn using the *marching cubes* algorithm (Lorensen and Cline, 1987). The overall contour surface is represented by a large number of connected triangular planar sections, which are computed for each cubical bin and then displayed. Depending on the pattern of signs on the eight vertices of each cube in the data lattice, up to six triangular patches are drawn within each cube (see Figure 1.27). In general, there are 2^8 cases (each corner of the cube being either above or below the contour level). Taking into consideration certain symmetries reduces this number. By scanning through all the cubes in the data lattice, a collection of triangles is found that defines the contour shell. Each triangle has an inner and outer surface, depending on the gradient of the density function. The inner and outer surfaces may be distinguished by color shading. A convenient choice is various shades of red for surfaces pointing toward regions of higher (hotter) density, and shades of blue toward regions of lower (cooler) density; see the cover jacket of this book for an example. Each contour is a patchwork of several thousand triangles. Smoother surfaces may be

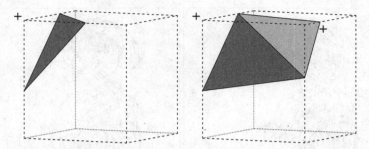

FIGURE 1.27 Examples of marching cube contouring algorithm. The corners with values above the contour level are labeled with a $+$ symbol.

obtained by using higher-order splines, but the underlying bin structure information would be lost.

In summary, visualizing trivariate functions directly is a powerful adjunct to data analysis. The gain of an additional dimension of visible structure without resort to slices greatly improves the ability of a data analyst to perceive structure. The same visualization applies to slices of density function with more than three variables. A demonstration tape that displays 4-D animation of $S_{\alpha,t}$ contours as α and t vary is available (Scott and Wilks, 1990).

1.5 GEOMETRY OF HIGHER DIMENSIONS

The geometry of higher dimensions provides a few surprises. In this section, a few standard figures are considered. This material is available in scattered references (see Kendall (1961), for example).

1.5.1 Polar Coordinates in d Dimensions

In d dimensions, a point $\mathbf{x}$ can be expressed in spherical polar coordinates by a radius r, a base angle θ_{d-1} ranging over $(0, 2\pi)$, and $d-2$ angles $\theta_1, \ldots, \theta_{d-2}$ each ranging over $(-\pi/2, \pi/2)$ (see Figure 1.28). Let $s_k = \sin\theta_k$ and $c_k = \cos\theta_k$. Then the transformation back to Euclidean coordinates is given by

$$x_1 = r\, c_1\, c_2 \cdots c_{d-3}\, c_{d-2}\, c_{d-1}$$
$$x_2 = r\, c_1\, c_2 \cdots c_{d-3}\, c_{d-2}\, s_{d-1}$$
$$x_3 = r\, c_1\, c_2 \cdots c_{d-3}\, s_{d-2}$$
$$\vdots$$
$$x_j = r\, c_1 \cdots c_{d-j}\, s_{d-j+1}$$
$$\vdots$$
$$x_d = r\, s_1\,.$$

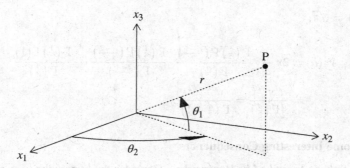

FIGURE 1.28   Polar coordinates (r, θ_1, θ_2) of a point P in $\Re^3$.

After some work (see Problem 1.11), the Jacobian of this transformation may be shown to be

$$J = r^{d-1} c_1^{d-2} c_2^{d-3} \cdots c_{d-2}. \tag{1.2}$$

1.5.2   Content of Hypersphere

The volume of the d-dimensional hypersphere $\{\mathbf{x} : \sum_{i=1}^{d} x_i^2 \le a^2\}$ is given by

$$V_d(a) = \int_{\sum_{i=1}^{d} x_i^2 \le a^2} 1 \, \mathbf{dx}$$

$$= \int_0^a dr \int_{-\pi/2}^{\pi/2} d\theta_1 \int_{-\pi/2}^{\pi/2} d\theta_2 \cdots \int_0^{2\pi} d\theta_{d-1} r^{d-1} c_1^{d-2} c_2^{d-3} \cdots c_{d-2}.$$

This can be simplified using the identity

$$\int_{-\pi/2}^{\pi/2} \cos^k \theta \, d\theta = 2 \int_0^{\pi/2} \cos^k \theta \, d\theta = 2 \int_0^{\pi/2} \cos^k \theta \, \frac{d(\cos^2 \theta)}{-2 \cos \theta \sin \theta},$$

which, using the change of variables $u = \cos^2 \theta$,

$$= \int_0^1 u^{k/2} \frac{du}{u^{1/2}(1-u)^{1/2}} = B\left(\tfrac{1}{2}, \tfrac{k+1}{2}\right) = \frac{\Gamma\left(\frac{1}{2}\right) \Gamma\left(\frac{k+1}{2}\right)}{\Gamma\left(\frac{k+2}{2}\right)}.$$

As $\Gamma\left(\frac{1}{2}\right) = \sqrt{\pi}$,

$$
\begin{aligned}
V_d(a) &= 2\pi \frac{a^d}{d} \cdot \frac{\Gamma\left(\frac{1}{2}\right)\Gamma\left(\frac{d-1}{2}\right)}{\Gamma\left(\frac{d}{2}\right)} \cdot \frac{\Gamma\left(\frac{1}{2}\right)\Gamma\left(\frac{d-2}{2}\right)}{\Gamma\left(\frac{d-1}{2}\right)} \cdots \frac{\Gamma\left(\frac{1}{2}\right)\Gamma(1)}{\Gamma\left(\frac{3}{2}\right)} \\
&= \frac{a^d \pi^{d/2}}{\frac{d}{2}\Gamma\left(\frac{d}{2}\right)} = \frac{a^d \pi^{d/2}}{\Gamma\left(\frac{d}{2}+1\right)}.
\end{aligned}
\tag{1.3}
$$

1.5.3 Some Interesting Consequences

1.5.3.1 Sphere Inscribed in Hypercube Consider the hypercube $[-a, a]^d$ and an inscribed hypersphere with radius $r = a$. Then using (1.3), the fraction of the volume of the cube contained in the hypersphere is given by

$$
f_d = \frac{\text{Volume sphere}}{\text{Volume cube}} = \frac{a^d \pi^{d/2}/\Gamma\left(\frac{d}{2}+1\right)}{(2a)^d} = \frac{\pi^{d/2}}{2^d \Gamma\left(\frac{d}{2}+1\right)}.
$$

For lower dimensions, the fraction f_d is as shown in Table 1.1. It is clear that the center of the cube becomes less important. As the dimension increases, the volume of the hypercube concentrates in its corners. This distortion of space (at least to our three-dimensional way of thinking) has many potential consequences for data analysis.

1.5.3.2 Hypervolume of a Thin Shell Wegman (1990) demonstrates the distortion of space in another setting. Consider two spheres centered on the origin, one with radius r and the other with slightly smaller radius $r - \epsilon$. Consider the fraction of the volume of the larger sphere in between the spheres. By Equation (1.3),

$$
\frac{V_d(r) - V_d(r - \epsilon)}{V_d(r)} = \frac{r^d - (r - \epsilon)^d}{r^d} = 1 - \left(1 - \frac{\epsilon}{r}\right)^d \xrightarrow[d\to\infty]{} 1.
$$

Hence, virtually all of the content of a hypersphere is concentrated close to its surface, which is only a $(d-1)$-dimensional manifold. Thus for data distributed uniformly over both the hypersphere and the hypercube, most of the data fall near the boundary and edges of the volume. Most statistical techniques exhibit peculiar behavior if the data fall in a lower dimensional subspace. This example illustrates one important aspect of the *curse of dimensionality*, which is discussed in Chapter 7.

TABLE 1.1 Fraction of the Volume of a Hypercube Lying in the Inscribed Hypersphere

Dimension (d)	1	2	3	4	5	6	7
Fraction volume (f_d)	1	0.785	0.524	0.308	0.164	0.081	0.037

1.5.3.3 Tail Probabilities of Multivariate Normal
The preceding examples make it clear that if we are trying to view uniform data over the hypercube in $\Re^{10}$, most (spherical) neighborhoods will be empty!

Let us examine what happens if the data follow the standard d-dimensional normal distribution:

$$f_d(\mathbf{x}) = (2\pi)^{-d/2} e^{-\mathbf{x}^T\mathbf{x}/2}.$$

Clearly, the origin (mode) is the most likely point and the equiprobable contours are spheres. Consider the spherical contour, $S_{0.01}(\mathbf{x})$, where the density value is only 1% of the value at the mode. Now

$$\frac{f(\mathbf{x})}{f(\mathbf{0})} = e^{-\mathbf{x}^T\mathbf{x}/2} \quad \text{and} \quad -2\log\frac{f(\mathbf{x})}{f(\mathbf{0})} = \sum_{i=1}^{d} x_i^2 \sim \chi^2(d);$$

therefore, the probability that a point is *within* the 1% spherical contour may be computed as

$$\Pr\left(\frac{f(\mathbf{x})}{f(\mathbf{0})} \geq \frac{1}{100}\right) = \Pr\left(\chi^2(d) \leq -2\log\frac{1}{100}\right). \tag{1.4}$$

Equation (1.4) gives the probability a random point will not fall in the "tails" or, in other words, will fall in the medium- to high-density region. In Table 1.2, these probabilities are tabulated for several dimensions. Around five or six dimensions, the probability mass of a multivariate normal begins a rapid migration into the extreme tails. In fact, more than half of the probability mass is in a very low-density region for 10-dimensional data. Silverman (1986) has dramatized this in 10 dimensions by noting that $\text{Prob}(\|\mathbf{x}\| \geq 1.6) = 0.99$. In very high dimensions, virtually the entire sample will be in the tails in a sense consistent with low-dimensional intuition. Table 1.2 is also applicable to normal data with a general full-rank covariance matrix, except that the contour is a hyper-ellipsoid.

1.5.3.4 Diagonals in Hyperspace
Pairwise scatter diagrams essentially project the multivariate data onto all the two-dimensional faces. Consider the hypercube $[-1,1]^d$ and let any of the diagonal vectors from the center to a corner be denoted by $\mathbf{v}$. Then $\mathbf{v}$ is any of the 2^d vectors of the form $(\pm 1, \pm 1, \ldots, \pm 1)^T$. The angle between

TABLE 1.2 Probability Mass *Not* in the "Tail" of a Multivariate Normal Density

d	1	2	3	4	5	6	7	8	9	10	15	20
$1000p$	998	990	973	944	899	834	762	675	582	488	134	20

a diagonal vector $\mathbf{v}$ and a Euclidean coordinate axis $\mathbf{e}_j = (0,\ldots,0,1,0,\ldots,0)^T$ is given by

$$\cos\theta_d = \frac{\langle \mathbf{v}, \mathbf{e}_j\rangle}{\sqrt{\langle \mathbf{v}, \mathbf{v}\rangle \langle \mathbf{e}_j, \mathbf{e}_j\rangle}} = \frac{\pm 1}{\sqrt{d}} \xrightarrow[d\to\infty]{} 0,$$

where $\langle \mathbf{u}, \mathbf{v}\rangle = \mathbf{u}^T\mathbf{v}$, so that $\theta_d \to \pi/2$ as $d \to \infty$. Thus the diagonals are nearly orthogonal to all coordinate axes for large d. Hence, any data cluster lying near a diagonal in hyperspace will be mapped into the origin in every paired scatterplot, while a cluster along a coordinate axis should be visible in some plot.

Thus the choice of coordinate system in high dimensions is critical in data analysis and intuition is highly dependent on a good choice. Real data structures may be missed due to overstriking. The general conclusion is that one- to two-dimensional intuition is valuable but not infallible when continuing on to higher dimensions.

1.5.3.5 Data Aggregate Around Shell

Returning to independent multivariate normal data, the mode is at the origin $\mathbf{x} = \mathbf{0}_d$. How far is a random point $\mathbf{X}$ from the origin for moderate to large dimensions d? Let $Z \sim N(0,1)$; then

$$\sqrt{\mathbf{X}^t\mathbf{X}} = \sqrt{\sum_{j=1}^d \mathbf{X}_j^2} = \sqrt{\chi^2(d)} \approx \sqrt{d + Z\sqrt{2d}} = \sqrt{d}\sqrt{1 + Z\sqrt{2/d}}$$

$$\approx \sqrt{d}\left(1 + \frac{1}{2}Z\sqrt{2/d}\right) = \sqrt{d} + \frac{1}{\sqrt{2}}Z \sim N\left(\sqrt{d}, \frac{1}{2}\right).$$

Thus while the highest density region is near the origin, virtually all (99.7%) of the data lie within a distance $\pm 3/\sqrt{2} = \pm 2.12$ of the hypersphere with radius $\sqrt{d}$. This is the data version of the volume result in Section 1.5.3.2.

1.5.3.6 Nearest Neighbor Distances

The derivation in Section 1.5.3.5 also addresses the question of the distribution of the closest pair of points in a random sample. As a conservative estimate, imagine one data point at the mode $\mathbf{x} = \mathbf{0}_d$ and compute the distribution of the distance from the origin to the closest of n sample points. Let D_i denote the distance the sample $\mathbf{X}_i$ is from the origin, and let D denote the minimum of $\{D_i\}$. Then

$$\Pr(D \le c) = 1 - \Pr(D > c) = 1 - \Pr(D_1 > c, D_2 > c, \ldots, D_n > c)$$
$$= 1 - \Pr(D_1 > c)^n = 1 - \Pr(D_1^2 > c^2)^n = 1 - \Pr(\chi_d^2 > c^2)^n$$
$$= 1 - \left(1 - \Pr(\chi_d^2 \le c^2)\right)^n.$$

Thus applying Leibniz's rule,

$$f_D(c) = \frac{d}{dc}\Pr(D \le c) = n\left(1 - \Pr(\chi_d^2 \le c^2)\right)^{n-1} \times 2c f_{\chi_d^2}(c^2). \qquad (1.5)$$

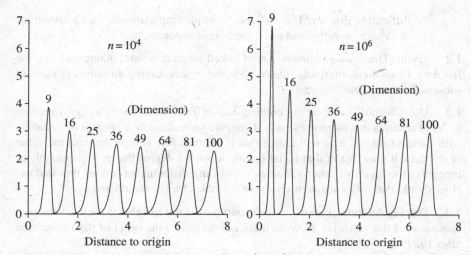

FIGURE 1.29 Densities of distance of closest point to the origin for sample sizes $n = 10^4$ and 10^6, for various dimensions $9 \leq d \leq 100$.

In Figure 1.29, this density is displayed for a sample of size $n = 10^4$ and several values of d. When $d > 25$, the closest pair of points is never closer than two units. Thus data in high dimensions are very sparse. A histogram bin will almost always be empty, or contain just one point. Increasing the sample size to 10^6 does not change the distribution very much. The sparseness of data in high dimensions is known as the *curse of dimensionality* (Bellman, 1961). This phenomenon will influence our thinking when analyzing data in more than five or six dimensions.

As a side note, there is a benefit to the curse of dimensionality in the field of biometrics. The uniqueness of fingerprints and other physical measurements (iris scans, for example) in a very large population is of much interest. What this analysis suggests is that if a feature space can be transformed into a reasonable number of independent measurements, and individuals may be viewed as independent samples from $N(\mathbf{0}_d, I_d)$, then unique identification is feasible with sufficiently accurate measurements of the features (see Kent and Millett (2002).)

PROBLEMS

1.1 A class of challenging data problems have a "hole" in them.

(a) Devise a simple way of creating radially symmetric trivariate data with a "hole" in it; that is, a region where the probability data points lie goes smoothly to zero at the center. *Hint*: Invent a rejection rule based on the distance a trivariate normal point is from the origin.

(b) Study a pairwise scatter diagram of 5000 trivariate data with either a "large" or a "small" hole in the middle. When does the hole become difficult to discern? Use "o" and "." as plotting symbols. Plot a histogram of $(x_1^2 + x_2^2 + x_3^2)^{1/2}$ and see if the hole is apparent.

1.2   Try the Diaconis–Friedman idea of linked bivariate scatter diagrams using the *Iris* data. Draw the scatterplots side-by-side and try connecting all points or random subsets. Evaluate your findings.

1.3   Use Chernoff faces on the college data in Table B.2. Try to assign variables to facial features in a memorable way. Compare your subjective choices of variables with those of others. You will notice that if one variable is near the extreme value of the data, it may distort that facial feature to such a degree that it is impossible to recognize the levels of other variables controlling different aspects of that feature. How should that influence your choice of variables for the mouth and eyes?

1.4   Display Andrews, curves for the economic dataset in Table B.1 for several permutations of the variables. How do these curves reflect the onset of the Depression after 1929?

1.5   Research problem: Generalize Andrews' representation so that the representation of a multidimensional point is a trajectory in the three-dimensional rectangle $[-\pi, \pi]^2 \times [0, 1]$.

1.6   Plot in parallel coordinates random samples of bivariate normal data with correlations ranging from -1 to 1. When the correlation $\rho = +1$, where does the point of intersection fall? Can you guess how trivariate correlated normal data will appear in parallel coordinates? Try it.

1.7   Investigate the appearance in parallel coordinates of data with clusters. For example, generate bivariate data with clusters centered at $(0, 0)$ and $(3, 3)$. Try centers at $(0, 0)$ and $(3, 0)$. Try centers of three clusters at $(0, 0)$, $(1, 0)$, and $(2, 0)$, where the data in each cluster have $\rho = -0.9$. The last example shows the duality between clusters and holes.

1.8   Prove that points falling on a straight line in Euclidean coordinates intersect in a point in parallel coordinates. What is the one exception? Superimposing Euclidean coordinates upon the parallel axes as shown in the right frame of Figure 1.9, find the (Euclidean) coordinates of the intersection point.

1.9   Investigate the literature for other ideas of data representation, including the star diagram, linear profile, weathervane, polygon star, and Kleiner–Hartigan faces.

1.10   What are the possible types of intersection of two planes (2-D) in four-space? *Hint*: Consider the two planes determined by pairs of coordinate axes (see Wegman (1990)).

1.11   Show that the Jacobian equals what is claimed in Equation (1.2). *Hint*: See Anderson (2003, p. 285). Interestingly, the signs of the determinant do not alternate as the dimension d increases, but go in pairs.

1.12   Verify Equation (1.3) for the well-known cases of a circle and a sphere.

1.13   Think of another way to represent high-dimensional data. Try using some other set of orthogonal functions for Andrews' curves (step functions, Legendre polynomials, or others). How sensitive is your method to permutations of the coordinate axes?

1.14   Show that the α-level contours of a multi-normal density are given by Equation (1.1). Use some of the techniques in Appendix A to display some contours when $d = 3$ with correlated and uncorrelated random variables.

1.15   (Problem 1.10 continued) What are the possible types of intersections of a k_1-dimensional subspace and a k_2-dimensional subspace in d dimensions? Think about the intersection of other types of hypersurfaces.

1.16   What fraction of a d-dimensional hypersphere lies in the inscribed d-dimensional hypercube? Find numerical values for dimensions up to 10.

1.17   Examine parallel coordinate plots of commonly observed bivariate and trivariate structure, including correlation and clustering. Summarize your findings.

1.18   Verify Equation (1.5) by simulation. What does Figure 1.29 look like for the world population of $n = 7.2 \times 10^9$? How many features would be required for a reliable biometrics system for the entire globe?

2

NONPARAMETRIC ESTIMATION CRITERIA

The focus of nonparametric estimation is different from that of parametric estimation. In the latter case, given a parametric density family $f(\cdot|\theta)$, such as the two-parameter normal family $N(\mu, \sigma^2)$ where $\theta = (\mu, \sigma^2)$, the emphasis is on obtaining the best estimator $\hat{\theta}$ of θ. In the nonparametric case, the emphasis is directly on obtaining a good estimate $\hat{f}(\cdot)$ of the *entire* density function $f(\cdot)$. In this chapter, an introduction to nonparametric estimation criteria is given, using only tools familiar from parametric analysis.

R. A. Fisher and Karl Pearson engaged in a lively debate on aspects of parametric estimation, with Pearson arguing in favor of nonparametric curves. Nonparametric curves are driven by structure in the data and are broadly applicable. Parametric curves rely on model building and prior knowledge of the equations underlying the data. Fisher (1922, 1932) called the two phases of parametric estimation the problems of *specification* and *estimation*. Fisher focused on achieving optimality in the estimation phase. His only remark on the problem of specification was that it was "entirely a matter for the practical statistician" to choose a form that "we know how to handle" based on experience. Fisher noted that misspecification could be detected by an *a posteriori* test, but offered no further instructions. If a parametric model is overparameterized in an effort to provide greater generality, then many choices of the vector θ will give nearly identical pointwise estimates $\hat{f}(x)$. An incorrectly specified parametric model has a bias that cannot be removed by large samples alone. Determining if the bias is too large to retain the parametric model can be tricky, since goodness-of-fit tests almost always reject quite reasonable models with large

Multivariate Density Estimation, First Edition. David W. Scott.
© 2015 John Wiley & Sons, Inc. Published 2015 by John Wiley & Sons, Inc.

samples. The "curse of optimality" is that incorrect application of "optimal" methods is preferred to more general but less-efficient methods. What Pearson failed to argue persuasively is that optimal estimators can become inefficient with only small perturbations in the assumptions underlying the parametric model. The modern emphasis on robust estimation correctly sacrifices a small percentage of parametric optimality in order to achieve greater insensitivity to model misspecification. However, in many multivariate situations, only vague prior information on an appropriate model is available. Nonparametric methods eliminate the need for model specification. The loss of efficiency need not be too large and is balanced by reducing the risk of misinterpreting data due to incorrect model specification.

2.1 ESTIMATION OF THE CUMULATIVE DISTRIBUTION FUNCTION

The simplest function to estimate nonparametrically is the cumulative distribution function (cdf) of a random variable X, defined by

$$F(x) = \Pr(X \leq x).$$

The obvious estimator from elementary probability theory is the *empirical cumulative distribution function* (ecdf), defined as

$$F_n(x) = \frac{\#\{x_i \leq x\}}{n} = \frac{\#x_i \in (-\infty, x]}{n} = \frac{1}{n} \sum_{i=1}^{n} I_{(-\infty, x]}(x_i), \qquad (2.1)$$

where $\{x_1, x_2, \ldots, x_n\}$ is a random sample from F and

$$I_A(x) = \begin{cases} 1 & \text{if } x \in A \\ 0 & \text{if } x \notin A. \end{cases}$$

This function has a staircase shape, as shown in Figure 2.1 of the geyser data given in Table B.6. It is easy to see that $F_n(x)$ has excellent mathematical properties for estimating the level of the function $F(x)$ for each fixed ordinate value x:

$$EF_n(x) = EI_{(-\infty, x]}(X) = 1 \times \Pr(X \in (-\infty, x]) = F(x).$$

In fact, $nF_n(x)$ is a binomial random variable, $B(n, p)$, with $p = F(x)$, so that $\text{Var}\{F_n(x)\} = p(1-p)/n$. There are no other unbiased estimators with smaller variance. This result follows since the order statistics form a complete sufficient statistic, and $F_n(x)$ is both unbiased and a function of the sufficient statistic. But notice that while the distribution function is often known to be continuous, the optimal estimator $F_n(x)$ is not.

A reasonable question to ask is whether the distribution function or its associated *probability density function* (pdf), $f(x) = F'(x)$, should be the focus for data analysis.

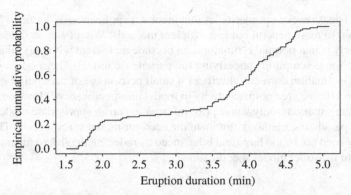

FIGURE 2.1   Empirical cumulative distribution function of the Old Faithful geyser dataset.

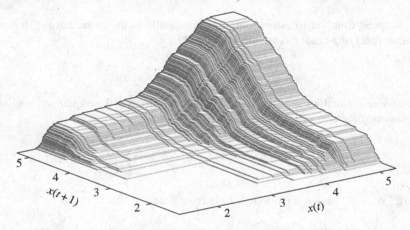

FIGURE 2.2   Empirical bivariate cdf of the lagged Old Faithful data $\{x_t, x_{t+1}\}$.

Certainly, the distribution function is quite useful. On the one hand, quite ordinary features such as skewness and multimodality are much more easily perceived in a graph of the density function than in a graph of the distribution function—compare Figures 1.13 and 2.1. On the other hand, it is difficult to ignore the fact that many social scientists are more focused on probabilities and therefore prefer to estimate the cdf rather than the pdf.

The distinction between the utility of the cdf and pdf becomes clearer in more than one dimension. The definition of a multivariate empirical cdf is simply

$$F_n(\mathbf{x}) = \frac{\#\{\mathbf{x}_i \leq \mathbf{x}\}}{n}, \quad \mathbf{x} \in \Re^d,$$

where the inequality $\{\mathbf{x}_i \leq \mathbf{x}\}$ is applied componentwise for each data point. The same optimality property holds as in the univariate case. However, few statisticians have even *seen* a bivariate empirical cdf. Consider the bivariate ecdf in Figure 2.2 of the geyser dataset corresponding to the bivariate pdf shown in Figure 1.13. Given the handful of gaps in the time series of 107 observations, there are 99 pairs $\{x_t, x_{t+1}\}$ included in this figure. The trimodal feature can be recognized with a little reflection, but not as easily as in Figure 1.13. The surface is perhaps unexpectedly complex given the small sample size. In particular, the number of jumps in the function is considered in Problem 2.1. Thus, the multivariate distribution function is of little interest for either graphical or data analytical purposes. Furthermore, ubiquitous multivariate statistical applications such as regression and classification rely on direct manipulation of the density function and not the distribution function.

2.2   DIRECT NONPARAMETRIC ESTIMATION OF THE DENSITY

Following the theoretical relationship $f(x) = F'(x)$, the *empirical probability density function* (epdf) is defined to be the derivative of the empirical cumulative distribution function:

$$f_n(x) = \frac{d}{dx} F_n(x) = \frac{1}{n} \sum_{i=1}^{n} \delta(x - x_i), \tag{2.2}$$

where $\delta(t)$ is the Dirac delta function. The epdf is always a discrete Uniform density over the data, that is, the probability mass is n^{-1} at each data point. The epdf is like a one-dimensional scatter diagram (dot plot) and is a useless estimate if the density is continuous (see Figure 2.3). The epdf is clearly inferior to the histogram from a graphical point of view. The primary modern use of the epdf has been as the sampling density for the bootstrap (Efron, 1982).

Does a uniformly minimum variance unbiased estimator of $f(x)$ exist? In the first theoretical treatment of the subject, Rosenblatt (1956) proved that no such estimator

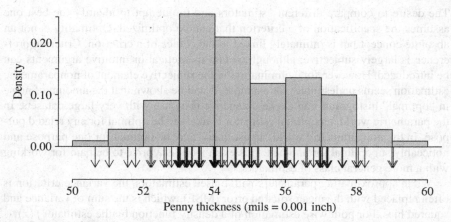

FIGURE 2.3   A histogram and empirical pdf (pointing down) of the US penny thickness dataset.

existed. Let $\mathbf{X}_n$ be a random sample of size n from f. Suppose that an estimator $T_n(x; \mathbf{X}_n)$ existed such that $\mathrm{E}[T_n(x; \mathbf{X}_n)] = f(x)$ for all continuous f and for all x and n. The data must appear symmetrically in $T_n(x; \mathbf{X}_n)$, as the nonparametric estimate cannot vary with the order in which the sample was collected. Now for all intervals (a, b),

$$\mathrm{E}\left[\int_a^b T_n(x)dx\right] = \int_a^b f(x)\,dx = F(b) - F(a) = \mathrm{E}[F_n(b) - F_n(a)],$$

using Fubini's theorem to justify switching the order of the expectation and integral operators. Both $T_n(x; \mathbf{X}_n)$ and $F_n(b) - F_n(a)$ are functions of the complete sufficient statistics, and since $F_n(b) - F_n(a)$ is the only symmetric unbiased estimator of $F(b) - F(a)$, it follows that

$$F_n(b) - F_n(a) = \int_a^b T_n(x)\,dx$$

for almost all samples $\{\mathbf{X}_n\}$. This result is a contradiction since the right-hand side is absolutely continuous while the left-hand side is not. At the time, this result was surprising and disappointing in a profession that had become accustomed to the pursuit of optimal unbiased estimators. Today, biased estimators are often preferred in situations where unbiased estimators are available, for example, with shrinkage estimators such as ridge regression (Hoerl and Kennard, 1970) and Stein (1956) estimators, or with penalty estimators such as the LASSO (Tibshirani, 1996).

2.3   ERROR CRITERIA FOR DENSITY ESTIMATES

The desire to compare different estimators and to attempt to identify the best one assumes the specification of a criterion that can be optimized. Optimality is not an absolute concept but is intimately linked to the choice of a criterion. Criterion preference is largely subjective, although certain theoretical or intuitive arguments can be introduced. However, total elimination of the subjective element of nonparametric estimation seems undesirable; for example, it will be shown that the amount of noise in "optimal" histograms can evoke a negative response with very large datasets. In the parametric world, an optimal estimator is likely to be optimal for any related purpose. In the nonparametric world, an estimator may be optimal for one purpose and noticeably suboptimal for another. This extra work is a price to be paid for working with a more general class of estimators.

When approximating parameters with biased estimators, the variance criterion is often replaced with the mean squared error (MSE), which is the sum of variance and squared bias. For pointwise estimation of a density function by the estimator $\hat{f}(x)$,

$$\mathrm{MSE}\{\hat{f}(x)\} = \mathrm{E}[\hat{f}(x) - f(x)]^2 = \mathrm{Var}\{\hat{f}(x)\} + \mathrm{Bias}^2\{\hat{f}(x)\},$$

where $\text{Bias}\{\hat{f}(x)\} = \mathrm{E}[\hat{f}(x)] - f(x)$. This equation treats the nonparametric density estimation problem as a standard point estimation problem with unknown parameter $\theta = f(x)$. While such pointwise analyses will prove interesting, the clear emphasis in this book will be on estimating and displaying the entire density surface. For some, the most intuitively appealing global criterion is the L_∞ norm:

$$\sup_x \left| \hat{f}(x) - f(x) \right| .$$

At the other end of the spectrum is the L_1 norm:

$$\int \left| \hat{f}(x) - f(x) \right| dx .$$

Neither of these criteria is as easily manipulated as the L_2 norm, which in this context is referred to as the *integrated squared error* (ISE):

$$\text{ISE} = \int \left[\hat{f}(x) - f(x) \right]^2 dx . \tag{2.3}$$

Even so, the integrated squared error is a complicated random variable that depends on the true unknown density function, the particular estimator, and the sample size; furthermore, even with these three quantities fixed, the ISE is a function of the particular realization of n points. For most purposes, it will be sufficient to examine the average of the ISE over these realizations; that is, the mean of the random variable ISE or *mean integrated squared error* (MISE):

$$\text{MISE} \equiv \mathrm{E}\left[\text{ISE}\right] = \mathrm{E}\left\{ \int [\hat{f}(x) - f(x)]^2 dx \right\}$$

$$= \int \mathrm{E}[\hat{f}(x) - f(x)]^2 dx = \int \text{MSE}\{\hat{f}(x)\} dx \equiv \text{IMSE},$$

where the interchange of the integral and expectation operators is justified by an application of Fubini's theorem. The last quantity is the IMSE, which is an abbreviation for the *integrated mean squared error*. Thus the MISE error criterion has two different though equivalent interpretations: it is a measure of both the average global error and the accumulated pointwise error. This criterion could be modified by including a weight function that would emphasize the tails, for example, or perhaps a local interval. Many students may be tempted to compute the definition of the MISE by expanding

$$\mathrm{E}\left\{ \int [\hat{f}(x) - f(x)]^2 dx \right\} \stackrel{?}{=} \int \mathrm{E}[\hat{f}(x) - f(x)]^2 f(x) \, dx,$$

which is a weighted average MSE criterion. But this is *wrong!* This common mistake might be avoided if the density estimator was written in the more complete form $\hat{f}(x | x_1, x_2, \ldots, x_n)$ that emphasizes the conditioning on all the data.

Other possible candidates for measuring error include information numbers such as the Kullback–Leibler criterion, which is defined as $\int f \log(f/\hat{f})$, Hellinger distance, which is defined as $[\int (\hat{f}^{1/p} - f^{1/p})^p]^{1/p}$ usually with $p = 2$, Akaike's information criterion, or even other L_p distances. These alternatives are not pursued further here.

2.3.1 MISE for Parametric Estimators

Few statisticians have much intuition about the MISE criterion, since it is almost never discussed in the parametric framework. However, it is entirely possible to evaluate the quality of *parametric* density estimators by the MISE of the entire parametric density estimate rather than the MSE of the parameter alone. In most cases, both the MSE of the parameter and the MISE of the density estimate decrease at the rate $O(n^{-1})$ as the sample size increases.

2.3.1.1 *Uniform Density Example*   Consider the Uniform density $f = U(0, \theta)$, where θ is estimated by the maximum likelihood estimator $\hat{\theta} = x_{(n)}$, the nth-order statistic. Thus $\hat{f} = U(0, x_{(n)})$. Without loss of generality, choose $\theta = 1$. Following Equation (2.3),

$$\text{ISE} = \left(\frac{1}{x_{(n)}} - 1 \right)^2 \cdot x_{(n)} + (0 - 1)^2 \cdot \left(1 - x_{(n)} \right) = \frac{1}{x_{(n)}} - 1.$$

As $f(x_{(n)}) = n x_{(n)}^{n-1}$, for $0 < x_{(n)} < 1$, it follows that

$$\text{MISE} = \int_0^1 \left(\frac{1}{x_{(n)}} - 1 \right) \cdot n x_{(n)}^{n-1} \, dx_{(n)} = \frac{1}{n-1} = O(n^{-1}). \tag{2.4}$$

Consider the one-dimensional family of estimators $\hat{f} = U(0, c x_{(n)})$, where c is a positive constant. There are two cases depending on whether $c x_{(n)} < 1$ or not, but the ISE is easily computed in each instance (see Problem 2.2). Taking expectations yields

$$\text{MISE}(c) = \begin{cases} n/[(n-1)c] - 1 & c < 1 \\ \left[2 - n c^{n-1} + (n-1)c^n \right] / [(n-1)c^n] & c > 1 \end{cases} \tag{2.5}$$

As both $\text{MISE}(c)$ and its derivative are continuous at $c = 1$ with $\text{MISE}'(c = 1) = -n/(n-1)$, the minimum MISE is realized for some $c > 1$, namely,

$$c^* = 2^{1/(n-1)} \approx 1 + n^{-1} \log(2).$$

This result is similar to the well-known fact that the MSE of the parametric estimator is minimized when $c = [1 + (n+1)^{-1}]$. However, in this case, the parametric MSE converges at the unusually fast rate $O(n^{-2})$ (see Romano and Siegel (1986, p. 212)).

The $O(n^{-1})$ MISE rate reflects more accurately the average error for the entire density curve and not just its endpoint.

2.3.1.2 General Parametric MISE Method with Gaussian Application

Consider an unbiased estimator $\hat{\theta}$ of a parameter θ. Writing the parametric ISE as $I(\hat{\theta})$ and taking a Taylor's series gives us

$$I(\hat{\theta}) = \int \left[f(t|\hat{\theta}) - f(t|\theta) \right]^2 dt = \sum_k \frac{1}{k!} (\hat{\theta} - \theta)^k I^{(k)}(\theta).$$

Now $I(\theta) = 0$ and $I'(\theta) = 2 \int \left[f(t|\theta) - f(t|\theta) \right] f'(t|\theta) dt = 0$ as well; hence,

$$\text{MISE}(\theta) = \text{E}[I(\hat{\theta})] = \frac{1}{2} \text{Var}(\hat{\theta}) \, I''(\theta) + \cdots.$$

Omitting higher order terms in the MISE expansion leaves the *asymptotic mean integrated squared error* (AMISE), which is given by

$$\text{AMISE}(\theta) = \frac{1}{2} \, I''(\theta) \text{Var}(\hat{\theta}).$$

This result can easily be extended to a vector of parameters by a multivariate Taylor's expansion. Consider estimation of the two-parameter normal density

$$\phi(x|\mu, \sigma^2) = \frac{1}{\sqrt{2\pi\sigma^2}} \exp\left[-\frac{(x-\mu)^2}{2\sigma^2} \right],$$

and an estimator of the form

$$\hat{\phi}(x) \equiv \phi(x|u, v);$$

for example, $\hat{\phi}(x) = \phi(x|\bar{x}, \hat{s}^2)$, using unbiased estimates of μ and σ^2. If the true density is a standard normal, then the bivariate ISE may be expanded as

$$I(u,v) \sim I(0,1) + u I_u + (v-1) I_v + \frac{u^2}{2} I_{uu} + u(v-1) I_{uv} + \frac{(v-1)^2}{2} I_{vv}, \tag{2.6}$$

where the partial derivatives $(I_u, I_v, I_{uu}, I_{uv}, I_{vv})$ are evaluated at $(u,v) = (0,1)$. Then (see Problem 2.4)

$$I_u(u,v) = 2 \int (\hat{\phi} - \phi) \left(\frac{x-u}{v} \hat{\phi} \right) dx$$

$$I_{uu}(u,v) = 2 \int (\hat{\phi} - \phi) \left(\frac{x-u}{v} \hat{\phi} \right)' dx + 2 \int \left(\frac{x-u}{v} \hat{\phi} \right)^2 dx. \tag{2.7}$$

Now the factor $(\hat{\phi} - \phi) = 0$ when evaluated at $(u, v) = (0, 1)$. The final integral becomes $2 \int \left[x\phi(x|0, 1) \right]^2 dx = 1/(2\sqrt{\pi})$. Similarly, $I_{vv}(0, 1) = 3/(16\sqrt{\pi})$ and $I_{uv} = 0$. Hence, taking the expectation of Equation (2.6) yields

$$\text{AMISE} = \frac{1}{2} I_{uu} \text{Var}(u) + \frac{1}{2} I_{vv} \text{Var}(v) \approx \frac{1}{4n\sqrt{\pi}} + \frac{3}{16n\sqrt{\pi}} = \frac{7}{16n\sqrt{\pi}}, \qquad (2.8)$$

since the variances of $\bar{x}$ and s^2 are $1/n$ and $2/(n-1) \approx 2/n$, respectively. It is of interest to note that the AMISE in Equation (2.8) is the sum of

$$\text{AMISE}\{\phi(x|\bar{x}, 1)\} + \text{AMISE}\{\phi(x|0, s^2)\} = \frac{1}{4n\sqrt{\pi}} + \frac{3}{16n\sqrt{\pi}} \approx \frac{0.2468}{n}. \qquad (2.9)$$

Therefore, the parametric MISE is larger if the mean is unknown than if the variance is unknown, because the shifted normal curves are further apart than when the normal curves cross as when only their variances differ. This simple calculation has been verified by a Monte Carlo simulation estimate of $1/(4n)$, which is only 1.2% larger.

2.3.2 The L_1 Criterion

2.3.2.1 L_1 versus L_2
Intuitively, one appeal of the L_1 criterion $\int |\hat{f} - f| dx$ is that it pays more attention to the tails of a density than the L_2 criterion, which de-emphasizes the relatively small-density values there by squaring. This de-emphasis can be checked in specific cases and by simulation. Theoretically, the L_1 criterion enjoys several other important advantages. First, consider a dimensional analysis. Since a density function has *inverse length* as its unit, L_1 is a dimensionless quantity after integration. The L_2 criterion, on the other hand, retains the units of inverse length after integration of the squared density error. It is possible to try to make L_2 dimensionless in a manner similar to the construction of such quantities for the skewness and kurtosis, but no totally satisfactory method is available (see Section 7.2.1). Furthermore, L_1 is invariant to monotone continuous changes of scale, as shown below. It is also easy to see that $0 \le L_1 \le 2$ while $0 \le L_2 \le \infty$ (see Problem 2.7). On the other hand, in practical situations the estimators that optimize these criteria are similar. The point to remember is that there is not a canonical L_1–*method* or L_2–*method* estimator. Devroye and Györfi (1985) have compiled an impressive theoretical treatment based on L_1 error, but the analytical simplicity of squared error and its adequacy in practical applications makes it the criterion of choice here. Some asymptotic results for L_1 estimates by Hall and Wand (1988a) and Scott and Wand (1991) support the notion that the practical differences between L_1 and L_2 criteria are reasonably small except in extreme situations.

2.3.2.2 Three Useful Properties of the L_1 Criterion
Several theoretically appealing aspects of absolute error are discussed in detail later. Each holds in the multivariate setting as well (Devroye, 1987).

The first and most appealing property is its interpretability, since L_1 is dimensionless and invariant under any smooth monotone transformation. Hence it is possible to compare the relative difficulty of estimating different densities. To see this, suppose $X \sim f$ and $Y \sim g$, and define $X^* = h(X)$ and $Y^* = h(Y)$; then $f^* = f[h^{-1}(x^*)]|J|$ with a similar expression for g^*, where J is the Jacobian. A simple change of variables gives

$$\int_u |f^*(u) - g^*(u)| \, du = \int_u |f[h^{-1}(u)] - g[h^{-1}(u)]| \, |J| \, du$$

$$= \int_v |f(v) - g(v)| \, dv.$$

An incorrect interpretation of this result is to conclude that all measurement scales are equally difficult to estimate or that variable transformation does not affect the quality of a density estimate. The second result is called Scheffé's Lemma:

$$2 \sup_A \left| \int_A f - \int_A g \right| = \int |f - g| = 2 \int_{f>g} (f - g). \tag{2.10}$$

To prove this, consider the set $B = \{x : f(x) > g(x)\}$; then for any set A,

$$2 \int_A (f - g) = 2 \int_{AB} (f - g) - 2 \int_{AB^*} (g - f) \le 2 \int_B (f - g) - 0$$

$$= \int_B (f - g) + \left(1 - \int_{B^*} f \right) - \left(1 - \int_{B^*} g \right)$$

$$= \int_B (f - g) + \int_{B^*} (g - f) = \int |f - g|,$$

which establishes the second equality in the lemma; taking the supremum over A establishes the "$\le$" for the first equality. The "$\ge$" follows directly from

$$2 \sup_A \left| \int_A (f - g) \right| \ge 2 \int_{A=B} (f - g).$$

The equality (2.10) provides a connection with statistical classification. Suppose that data come randomly from two densities, f and g, and that a new point x is to be classified. Using a Bayesian rule of the form: assign x to f if $x \in A$, for some set A, and to g otherwise, the probability of misclassification is

$$\Pr(\text{error}) = \frac{1}{2} \int_{A^*} f + \frac{1}{2} \int_A g = \frac{1}{2} \left(1 - \int_A f \right) + \frac{1}{2} \int_A g$$

$$= \frac{1}{2} - \frac{1}{2} \int_A (f - g).$$

Choosing A to minimize this error leads to $A = B$ using the lemma and gives us the third result

$$\Pr(\text{error}) = \frac{1}{2} - \frac{1}{4} \int |f - g|. \qquad (2.11)$$

Thus minimizing the L_1 distance between $g = \hat{f}$ and f is equivalent to maximizing the probability in (2.11); that is, optimizing L_1 amounts to maximizing the *confusion* between $\hat{f}$ and f. This optimization is precisely what is desired. The probability interpretation holds in any dimension. In Problem 2.8, this expression is verified in the two extreme cases.

2.3.3   Data-Based Parametric Estimation Criteria

In this section, the parametric estimator will be denoted by $f_\theta(x)$, but the true density will be denoted by $g(x)$. This will emphasize that the model is only an approximation to the true density. What is the theoretical basis of maximum likelihood? The answer is the KL divergence (Kullback and Leibler, 1951) between f_θ and g:

$$\begin{aligned} d_{KL}(f_\theta, g) &= \int g(x) \log \frac{g(x)}{f_\theta(x)} dx \\ &= \int g(x) \log g(x)\, dx - \int g(x) \log f_\theta(x)\, dx, \end{aligned} \qquad (2.12)$$

which is nonnegative by an application of Jensen's Inequality (Example 7.7, Lehmann and Casella, 1998), and is zero if and only if $f_\theta = g$. The first integral in (2.12) is constant, and we try replacing the unknown true density $g(x)$ with the empirical probability density function $f_n(x) = \frac{1}{n} \sum_{i=1}^{n} \delta(x - x_i)$ (see Equation (2.2)). Now $\int \delta(x - x_i) \log f_\theta(x)\, dx = \log f_\theta(x_i)$; thus

$$\hat{d}_{KL}(f_\theta, g) = \text{constant} - \frac{1}{n} \sum_{i=1}^{n} \log f_\theta(x_i).$$

Thus minimizing the KL divergence is equivalent to maximizing the log likelihood.

Next, consider one of the dimensionless criteria, namely, Hellinger Distance:

$$d_H(f_\theta, g)^2 = \int \left[\sqrt{f_\theta(x)} - \sqrt{g(x)} \right]^2 dx \geq 0.$$

There is no obvious quantity to substitute for the unknown $g(x)$, save a nonparametric estimate such as a histogram. Aside from the computational complexity of performing the numerical minimization that would be required, different choices of the histogram would result in different estimates of θ. By way of contrast, the use of the epdf rather than a histogram in the KL divergence gives a well-defined fully data-based criterion. These same observations apply to the dimensionless L_1 criterion.

Finally, we consider a criterion that is not dimensionless, namely, the integrated squared error (cf. Eq. 2.3):

$$
\begin{aligned}
\mathrm{ISE}(\theta) &= \int [f_\theta(x) - g(x)]^2 \, dx \\
&= \int f_\theta(x)^2 \, dx - 2 \int f_\theta(x) g(x) \, dx + \int g(x)^2 \, dx \\
&= \int f_\theta(x)^2 \, dx - 2\, \mathrm{E} f_\theta(X) + \text{constant} .
\end{aligned} \tag{2.13}
$$

Assuming that the model is square integrable in a convenient form, and choosing the obvious unbiased estimator for the expectation term, we arrive at the fully data-based criterion

$$
\hat{\theta} = \arg\min_\theta \widehat{\mathrm{ISE}}(\theta) = \arg\min_\theta \left[\int f_\theta(x)^2 \, dx - \frac{2}{n} \sum_{i=1}^n f_\theta(x_i) \right] . \tag{2.14}
$$

This minimum L_2 distance criterion was denoted as $L_2 E(\theta)$ by Scott (2001) and is a special case of a divergence family investigated by Basu et al. (1998).

As an example, consider fitting a normal density, $\phi(x|\mu, \sigma^2)$, to the Rayleigh data ($n = 15$) that measured the weight of a standard volume of nitrogen (see Tukey (1977)). The criterion (2.14) to be minimized over $\theta = (\mu, \sigma)$ is

$$
\hat{\theta} = \arg\min_\theta L_2 E(\theta) = \arg\min_\theta \left[\frac{1}{2\sqrt{\pi}\sigma} - \frac{2}{n} \sum_{i=1}^n \phi(x_i|\mu, \sigma^2) \right] . \tag{2.15}
$$

The left frame of Figure 2.4 displays a histogram of this small dataset and the two normal fits. A careful examination of the data along the axis shows two clusters. The four-parameter mixture model,

$$
f_\theta(x) = w\, \phi(x|\mu_1, \sigma^2) + (1 - w)\, \phi(x|\mu_2, \sigma^2), \tag{2.16}
$$

was fit by $L_2 E$ and is shown in the right frame of Figure 4.2, together with a different bimodal histogram. The data were blurred as $L_2 E$ is sensitive to ties in the data. Notice that the two blurred values around 2.3013 are not included in the left mixture (treated as outliers). Lord Rayleigh's recognition of the second cluster resulted in his winning the 1904 Nobel Prize in Physics for the discovery of the noble gas argon.

Note that the ISE / $L_2 E$ criterion is intuitively satisfying in that the goal is to make the two curves as close together as possible. We will focus primarily on this criterion for our nonparametric curve estimators as a result. Donoho and Liu (1988) showed that the ISE criterion is by nature robust, a property shared by all minimum distance criteria. The two points ignored by $L_2 E$ in the mixture fit is evidence of this property.

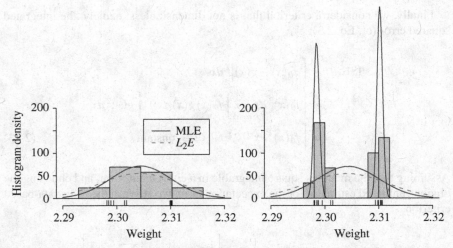

FIGURE 2.4 (Left frame) Histogram with MLE and L_2E normal fits to the Rayleigh data. (Right frame) L_2E normal mixture fit to blurred Rayleigh with common variance.

2.4 NONPARAMETRIC FAMILIES OF DISTRIBUTIONS

2.4.1 Pearson Family of Distributions

Karl Pearson provided impetus for nonparametric estimators in two ways. He coined the word *histogram*, and he studied the density functions that are solutions to the differential equation

$$\frac{d \log f(x)}{dx} = \frac{x - a}{b + cx + dx^2}. \tag{2.17}$$

Pearson (1902a, b) identified seven types of solutions to this equation, depending on the roots of the denominator and which parameters were 0. Interestingly, this class contains most of the important classical distributions: Normal, Student's t, Beta, and Snedecor's F (see Problem 2.9). Pearson proposed using the first four sample moments to estimate the unknown constants (a, b, c, d) in Equation (2.17). Today, maximum likelihood might be used.

Pearson motivated the differential Equation (2.17) by appealing to the discrete hypergeometric distribution, which describes a large urn with N balls, pN of which are black. A sample of n balls is drawn, and the number of black balls is recorded as X; then

$$f(x) = \Pr(X = x) = \binom{Np}{x} \binom{N(1-p)}{n-x} \div \binom{N}{n}, \quad x = 0, \ldots, n.$$

Pearson's differential Equation (2.17) emerges after a bit of computation (see Problem 2.11), keeping p fixed as the urn grows:

$$\frac{d\log f(x)}{dx} \approx \frac{\Delta f(x)}{f(x)} = \frac{f(x)-f(x-1)}{f(x)} = \frac{x-a}{b+cx+dx^2}, \qquad (2.18)$$

where

$$a = b = \frac{-(n+1)(Np+1)}{N+2}; \quad c = \frac{(Np+n+2)}{N+2}; \quad d = \frac{-1}{N+2}. \qquad (2.19)$$

Pearson's contribution and insight was to use the data not to fit a particular parametric form of a density but rather to compute the coefficients of the differential equation from the data via the sample moments. Mathematically, the Pearson family could be considered parametric; but philosophically, it is nonparametric. Pearson devoted considerable resources to computing percentiles for his distributions. Given the lack of computing power at that time, it is remarkable that the Pearson family gained wide acceptance. It still appears regularly in practical applications today.

Many other families of distributions, some multivariate, have been proposed. The Johnson (1949) family is notable together with the multivariate generalizations of Marshall and Olkin (1985).

2.4.2 When Is an Estimator Nonparametric?

A parametric estimator is defined by the model $\hat{f}(x|\theta)$, where $\theta \in \Theta$. It has proven surprisingly difficult to formulate a working definition for what constitutes a nonparametric density estimator. A heuristic definition may be proposed based on the necessary condition that the estimator "work" for a "large" class of true densities. One useful notion is that a nonparametric estimator should have many parameters, in fact, perhaps an infinite number, or a number that diverges as a function of the sample size. The Pearson family does not fully qualify as nonparametric under any of these definitions, although the dimension of the Pearson family is larger than for most parametric families. Tapia and Thompson (1978) tend to favor the notion that the nonparametric estimator should be infinite dimensional. Silverman (1986) simply indicates that a nonparametric approach makes "less rigid assumptions ... about the distribution of the observed data." But how many parameters does a histogram have? What about the naive orthogonal series estimator with an infinite number of terms that is equal (in distribution) to the empirical density function for any sample size? (See Section 6.1.3.)

A surprisingly elegant definition is implicit in the work of Terrell (Terrell and Scott, 1992), who shows that all estimators, parametric or nonparametric, are *generalized kernel estimators*, at least asymptotically (see Section 6.4). Terrell introduces the idea of the influence of a data point x_i on the point density estimate at x. If $\hat{f}(x)$ is a nonparametric estimator, the influence of a point should vanish asymptotically if $|x - x_i| > \epsilon$ for any $\epsilon > 0$, while the influence of distant points does not vanish for a parametric estimator. Roughly speaking, nonparametric estimators are asymptotically local, while parametric estimators are not. However, nonparametric estimators must not be *too* local in order to be consistent.

PROBLEMS

2.1 Devise an algorithm for plotting the empirical bivariate cdf. Give a range for the possible number of jumps there can be in this function. Give simple examples for the two extreme cases. What is the order of your algorithm (i.e., the number of arithmetic operations as a function of the sample size)?

2.2 Verify Equation (2.5). Plot it for several sample sizes and compare the actual minimizer to the asymptotic formula.

2.3 Show that the expected Kullback–Leibler distance for the parametric estimator $\hat{f} = U\left(0, x_{(n)}\right)$ of $f = U(0, 1)$ is $1/(n-1)$.

2.4 Verify the equations in (2.7) and that $I_{vv}(0, 1) = 3/(16\sqrt{\pi})$.

2.5 Complete the calculations for the normal parametric AMISE. Find the optimal AMISE estimators of the form $c\bar{x}$ and $c\hat{s}^2$ for standard normal data.

2.6 Assuming standard normal data, compute the *exact* MISE of the estimator $N(\bar{x}, 1)$, and check that the series approximations match. Research problem: Can you find any similar closed-form MISE expressions for the estimators $N(0, \hat{s}^2)$ and $N(\bar{x}, \hat{s}^2)$?

2.7 Verify that the ranges of L_1 and L_2 errors are $(0, 2)$ and $(0, \infty)$, respectively. Give examples where the extreme values are realized.

2.8 Verify that the probability of error given in (2.11) is correct in the two extreme cases.

2.9 Verify directly that the parametric densities mentioned in the text actually satisfy the differential equation for Pearson's system.

2.10 Use a symbolic program to verify that

$$f(x) \propto (1+x^2)^{-k} \exp(-\alpha \arctan x)$$

is a possible solution to Pearson's differential equation. Plot $f(x)$.

2.11 Verify the calculations in Equations (2.18) and (2.19).

2.12 Verify the L_2E criterion in Equation (2.15). Derive the four-parameter mixture L_2E criterion used in Figure 2.4. Use R function *nlminb* to find $\hat{\theta}$. Hint: Use the identity $\int \phi(x|\mu_1, \sigma_1^2) \phi(x|\mu_2, \sigma_2^2)\, dx = \phi(0|\mu_1 - \mu_2, \sigma_1^2 + \sigma_2^2)$.

2.13 Consider a five-parameter normal mixture like Equation (2.16) with σ^2 replaced by σ_1^2 and σ_2^2. Suppose $\mu_2 = x_1$. Show that the likelihood diverges to $+\infty$ as $\sigma_2 \to 0$. Derive the L_2E criterion in Equation (2.14) for the five-parameter mixture model. What is the limit of the L_2E criterion for this scenario?

3

HISTOGRAMS: THEORY AND PRACTICE

The framework of the classic histogram is useful for conveying the general flavor of nonparametric theory and practice. From squared-error theory to state-of-the-art cross-validation algorithms, these topics should be studied carefully. Discussing these issues in the context of the simplest and best-known nonparametric density estimator is of practical interest, and will be of value when developing the theory of more complicated estimators.

The histogram is most often displayed on a nondensity scale: either as bin counts or as a stem-and-leaf plot (Tukey, 1977). Some authors have drawn a distinction between the purposes of a histogram as a density estimator and as a data presentation device (Emerson and Hoaglin, 1983, p. 22). Such a distinction seems artificial, as there is no reason to disregard a histogram constructed in a less-than-optimal fashion. The examination of both undersmoothed and oversmoothed histograms should be routine. A histogram conveys visual information of both the frequency and relative frequencies of observations; that is, the essence of a density function.

3.1 STURGES' RULE FOR HISTOGRAM BIN-WIDTH SELECTION

The classical frequency histogram is formed by constructing a complete set of nonoverlapping intervals, called *bins*, and counting the number of points in each bin. In order for the bin counts to be comparable, the bins should all have the same width. If so, then the histogram is completely determined by two parameters, the *bin width, h,*

Multivariate Density Estimation, First Edition. David W. Scott.
© 2015 John Wiley & Sons, Inc. Published 2015 by John Wiley & Sons, Inc.

and the *bin origin*, t_0, which is any conveniently chosen bin interval endpoint. Often, the bin origin is chosen to be $t_0 = 0$.

Although the idea of grouping data in the form of a histogram is at least as old as Graunt's work in 1662, no systematic guidelines for designing histograms were given until Herbert Sturges' short note in 1926. His work made use of a device that has been advocated more generally by Tukey (1977); namely, taking the normal density as a point of reference when thinking about data. Sturges simply observed that the binomial distribution, $B(n, p = 0.5)$, could be used as a model of an optimally constructed histogram with appropriately scaled normal data (see Figure 3.1).

Construct a frequency histogram with k bins, each of width 1 and centered on the points $i = 0, 1, \ldots, k - 1$. Choose the bin count of the ith bin to be the binomial coefficient $\binom{k-1}{i}$. As k increases, this ideal frequency histogram assumes the shape of a normal density with mean $(k - 1)/2$ and variance $(k - 1)/4$. The total sample size is

$$n = \sum_{i=0}^{k-1} \binom{k-1}{i} = (1+1)^{k-1} = 2^{k-1}$$

by the binomial expansion. Sturges' rule follows immediately:

> Sturges' number-of-bins rule: $\qquad k = 1 + \log_2 n.$ (3.1)

Of course, any constant multiple of the binomial coefficients would also look normal, but note that the boundary condition $n = 1$ gives $k = 1$ in Equation (3.1).

In practice, Sturges' rule is applied by dividing the sample range of the data into the prescribed number of equal-width bins. Technically, Sturges' rule is a number-of-bins rule rather than a bin-width rule. Much simpler is to adopt the convention that all histograms have an infinite number of bins, only a finite number of which are nonempty. Further, adaptive histograms, which are considered in Section 3.2.8, do not use equal-width bins. Thus the focus on bin width rather than number of bins seems appropriate.

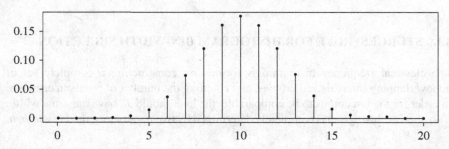

FIGURE 3.1 Binomial pdf with $p = 0.5$ used by Sturges to determine the number of histogram bins.

Sturges' rule is widely recommended in introductory statistics texts and is often used in statistical packages as a default. If the data are not normal, but are skewed or leptokurtotic, additional bins may be required. For example Doane (1976) proposed increasing the number of bins in (3.1) by $\log_2(1 + \hat{\gamma}\sqrt{n/6})$, where $\hat{\gamma}$ is an estimate of the standardized skewness coefficient (see Problem 3.1).

3.2 THE L_2 THEORY OF UNIVARIATE HISTOGRAMS

3.2.1 Pointwise Mean Squared Error and Consistency

In this section are presented the mean squared error (MSE) properties of a density histogram. The difference between a frequency histogram and a density histogram is that the latter is normalized to integrate to 1. As noted earlier, the histogram is completely determined by the sample $\{x_1, \ldots, x_n\}$ from $f(x)$ and a choice of mesh $\{t_k, -\infty < k < \infty\}$. Let $B_k = [t_k, t_{k+1})$ denote the kth bin. Suppose that $t_{k+1} - t_k = h$ for all k; then the histogram is said to have fixed bin width h. A frequency histogram is built using blocks of height 1 and width h stacked in the appropriate bins. The integral of such a figure is clearly equal to nh. Thus a density histogram uses building blocks of height $1/(nh)$, so that each block has area equal to $1/n$. Let ν_k denote the bin count of the kth bin, that is, the number of sample points falling in bin B_k (see Figure 3.2). Then the histogram is defined as follows:

$$\hat{f}(x) = \frac{\nu_k}{nh} = \frac{1}{nh}\sum_{i=1}^{n} I_{[t_k, t_{k+1})}(x_i) \quad \text{for } x \in B_k. \tag{3.2}$$

The analysis of the histogram random variable, $\hat{f}(x)$, is quite simple, once it is recognized that the bin counts, $\{\nu_k\}$, are binomial random variables:

$$\nu_k \sim B(n, p_k), \quad \text{where} \quad p_k = \int_{B_k} f(t)\,dt.$$

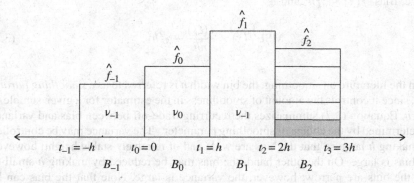

FIGURE 3.2 Notation for construction of an equally spaced histogram.

Consider the MSE of $\hat{f}(x)$ for $x \in B_k$. Now $E[v_k] = np_k$ and Var $[v_k] = np_k(1-p_k)$. Hence,

$$\text{Var}\,\hat{f}(x) = \frac{\text{Var } v_k}{(nh)^2} = \frac{p_k(1-p_k)}{nh^2} \tag{3.3}$$

and

$$\text{Bias}\,\hat{f}(x) = E\hat{f}(x) - f(x) = \frac{1}{nh}Ev_k - f(x) = \frac{p_k}{h} - f(x). \tag{3.4}$$

To proceed with the fewest assumptions, suppose that $f(x)$ is Lipschitz continuous over the bin B_k.

> **Definition:**   *A function is said to be* Lipschitz continuous *over an interval* B_k *if there exists a positive constant* γ_k *such that* $|f(x) - f(y)| < \gamma_k|x - y|$ *for all* $x, y \in B_k$.

Then by the mean value theorem (MVT),

$$p_k = \int_{B_k} f(t)dt = hf(\xi_k) \qquad \text{for some} \quad \xi_k \in B_k. \tag{3.5}$$

It follows that

$$\text{Var}\,\hat{f}(x) \leq \frac{p_k}{nh^2} = \frac{f(\xi_k)}{nh} \tag{3.6}$$

and

$$|\text{Bias}\,\hat{f}(x)| = |f(\xi_k) - f(x)| \leq \gamma_k|\xi_k - x| \leq \gamma_k h;$$

hence, $\text{Bias}^2\,\hat{f}(x) \leq \gamma_k^2 h^2$ and

$$\text{MSE}\,\hat{f}(x) \leq \frac{f(\xi_k)}{nh} + \gamma_k^2 h^2. \tag{3.7}$$

In the literature on smoothing, the bin width h is referred to as a *smoothing parameter*, since it controls the amount of smoothness in the estimator for a given sample of size n. Equation (3.7) summarizes the recurring trade-off between bias and variance as determined by the choice of smoothing parameter. The variance may be controlled by making h large so that the bins are wide and of relatively stable height; however, the bias is large. On the other hand, the bias may be reduced by making h small so that the bins are narrow; however, the variance is large. Note that the bias can be eliminated by choosing $h = 0$, but this very rough histogram is exactly the empirical probability density function $f_n(x)$, which has infinite (vertical) variance. The bias and variance may be controlled simultaneously by choosing an intermediate value of the bin width, and allowing the bin width to slowly decrease as the sample size increases.

Definition: *A density estimator is said to be* consistent in the mean square *if the* MSE$\{\hat{f}(x)\} \to 0$ *as* $n \to \infty$.

An *optimal smoothing parameter* h^* is defined to be that choice that minimizes the (asymptotic) MSE. The following results are consequences of Equation (3.7).

Theorem 3.1:   *Assume that $x \in B_k$ is a fixed point and that f is Lipschitz continuous in this bin with constant γ_k. Then the histogram estimate $\hat{f}(x)$ is mean square consistent if, as $n \to \infty$, then $h \to 0$ and $nh \to \infty$.*

The first condition ensures that the bias vanishes asymptotically, while the second condition ensures that the variance goes to zero. Duda and Hart (1973) suggest choosing $h = n^{-1/2}$, for example.

Corollary 3.2:   *The MSE(x) bound* (3.7)   *is minimized when*

$$h^*(x) = \left[\frac{f(\xi_k)}{2\gamma_k^2 n} \right]^{1/3} ; \tag{3.8}$$

the resulting MSE$^*(x)$ *is* $O(n^{-2/3})$.

These results deserve careful examination. The optimal bin width decreases at a rate proportional to $n^{-1/3}$. This rate is much faster than Sturges' rule, which suggests the rate $\log_2^{-1}(n)$ (see Table 3.1). The optimal rate of decrease of the MSE does not attain the Cramer–Rao lower bound rate of $O(n^{-1})$ for parametric estimators.

The noise inherent in the histogram varies directly with the square root of its height, since Var$\{\hat{f}(x)\} \approx f(x)/(nh)$ from Equation (3.6). This heteroscedasticity (unequal variance) across the histogram estimate may be eliminated by using a variance-stabilizing transformation. Each bin count is approximately a Poisson random variable. It is well-known that the square root is the variance-stabilizing transformation for Poisson data. Suppose that Y_n has moments (μ_n, σ_n^2) with $\mu_n > 0$ and $\sigma_n \to 0$. Then Var $\{g(Y_n)\} \approx g'(\mu_n)^2 \sigma_n^2$. Choosing $Y_n = \hat{f}(x)$ so that $\mu_n \approx f(x)$ and $\sigma_n^2 \approx f(x)/(nh)$, then $g(y) = \sqrt{y}$ and $g'(y) = 1/(2\sqrt{y})$ or $1/(2\sqrt{f(x)})$ at $y = \mu_n$. Therefore,

$$\sqrt{\operatorname{Var}\sqrt{\hat{f}(x)}} \approx \frac{1}{2\sqrt{f(x)}}\sqrt{\frac{f(x)}{nh}} = \frac{1}{2\sqrt{nh}}, \tag{3.9}$$

which is independent of the unknown $f(x)$.

Thus plotting the histogram on a square root scale allows for easy comparison of noise in the histogram in regions of high and low density. Tukey (1977) called the resulting estimate the *rootgram*. Of course, the rootgram no longer accurately portrays the relative frequencies of the observations. In more than one dimension, the rootgram is still variance-stabilizing. However, since the contours of the bivariate histogram and bivariate rootgram are identical, the practical applications are limited.

One consequence of Corollary 3.2 is that the use of a fixed bandwidth over the entire range of the data is not generally optimal. For example, the bin width should be relatively wider in regions of higher density to reduce the variance in Equation (3.7). Now if the width of bin B_k is sufficiently narrow, then the Lipschitz constant γ_k is essentially the magnitude of the slope of $f(x)$ in that bin. Therefore, from Equation (3.8), the bin width should be narrower in regions where the density is changing rapidly, and vice versa. These notions are confirmed in Section 3.2.8. However, in practice, there are no reliable algorithms for constructing adaptive histogram meshes. Therefore, the study of fixed-width histograms remains important.

3.2.2   Global L_2 Histogram Error

Consistency results based on upper bounds are not useful in practice, since the upper bounds may be quite far from truth. More useful approximations can be made by assuming the existence of derivatives of f. These results can be useful even in the practical situation where f is unknown, by employing a variety of techniques called cross-validation algorithms (see Section 3.3).

Computing the mear integrated square error (MISE) is accomplished by aggregating the MSE over each bin and summing over all bins. Consider first the integrated variance (IV):

$$\mathrm{IV} = \int_{-\infty}^{\infty} \operatorname{Var}\hat{f}(x)\,dx = \sum_{k=-\infty}^{\infty} \int_{B_k} \operatorname{Var}\hat{f}(x)\,dx. \tag{3.10}$$

From Equation (3.3), the last integral over B_k is simply $p_k(1-p_k)/(nh)$. Now $\sum p_k = \int f(x)\,dx = 1$. Recall that $\sum \phi(\xi_k)\cdot h = \int \phi(x)dx + o(1)$ by standard Riemannian integral approximation. Therefore, using the approximation (3.5) for p_k, $\sum p_k^2 = \sum f(\xi_k)^2 h^2 = h\sum f(\xi_k)^2 \cdot h = h[\int f(x)^2 dx + o(1)]$. Combining, we have

$$\mathrm{IV} = \frac{1}{nh} - \frac{R(f)}{n} + o(n^{-1}), \tag{3.11}$$

where the following notation is adopted for the squared L_2-norm of ϕ:

$$R(\phi) \equiv \int \phi(x)^2\, dx.$$

The squared L_2-norm is only one possible measure of the *roughness* (R) of the function ϕ. Alternatives include $R(\phi')$ and $R(\phi'')$. $R(\phi)$ in this context refers more to the *statistical roughness* than to the *mathematical roughness* of the function ϕ. The latter usually refers to the number of continuous derivatives in the function. The former refers loosely to the number of wiggles in the function. Statistical roughness does, however, take account of the first few derivatives of the function. In summary, neither definition would describe a normal density as rough, but the lognormal density is a very rough function from the statistical point of view, even though it is infinitely differentiable. A low-order polynomial density, such as one from the Beta family, is statistically smooth even though it possesses only a few derivatives.

In order to compute the bias, consider a typical bin $B_0 = [0, h)$. The bin probability, p_0, may be approximated by

$$p_0 = \int_0^h f(t)\, dt = \int_0^h \left[f(x) + (t-x)f'(x) + \frac{1}{2}(t-x)^2 f''(x) + \cdots \right] dt$$

$$= hf(x) + h\left(\frac{h}{2} - x\right)f'(x) + O(h^3)$$

so that

$$\text{Bias}\, \hat{f}(x) = \frac{p_0}{h} - f(x) = \left(\frac{h}{2} - x\right)f'(x) + O(h^2). \tag{3.12}$$

See Figure 3.3. For future reference, note that (3.12) implies that the bias is of higher-order $O(h^2)$ at the center of a bin, when $x = h/2$.

Using the generalized mean value theorem (GMVT), the leading term of the integrated squared bias (ISB) for this bin is

$$\int_{B_0} \left(\frac{h}{2} - x\right)^2 f'(x)^2\, dx = f'(\eta_0)^2 \int_0^h \left(\frac{h}{2} - x\right)^2 dx = \frac{h^3}{12} f'(\eta_0)^2, \tag{3.13}$$

for some $\eta_0 \in B_0$. This result generalizes to other bins for some collection of points $\eta_k \in B_k$. Hence the total ISB is

$$\text{ISB} = \frac{h^2}{12} \sum_{k=-\infty}^{\infty} f'(\eta_k)^2 \times h = \frac{h^2}{12} \int_{-\infty}^{\infty} f'(x)^2\, dx + o(h^2), \tag{3.14}$$

which follows from standard Riemannian convergence of sums to integrals. A note on assumptions: if the total variation of $f'(\cdot)^2$ is finite, then the remainder term in (3.14)

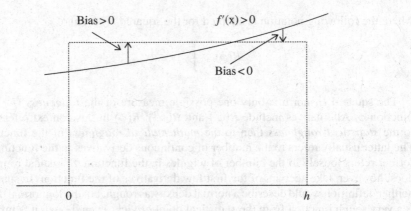

FIGURE 3.3 Bias of histogram estimator in a typical bin.

becomes $O(h^3)$ (see Problem 3.3). Assuming the existence of an absolutely continuous second derivative, the error term is $O(h^4)$. There is little practical difference among these observations.

Rather than persist in keeping track of the order of the remainder term explicitly, the following notation will be adopted. The main terms in the ISB will be referred to as the *asymptotic* ISB (AISB). Thus, ISB = AISB + $o(h^2)$ and

$$\text{AISB} = \frac{1}{12}h^2 \int_{-\infty}^{\infty} f'(x)^2 \, dx = \frac{1}{12}h^2 R(f').$$

Similarly, the *asymptotic integrated variance* (AIV) and *asymptotic mean integrated squared error* (AMISE) refer to the main terms in the approximations to the IV and MISE, respectively.

The following theorem, which summarizes Equations (3.11) and (3.14), is due to Scott (1979) and Freedman and Diaconis (1981).

Theorem 3.3:   *Suppose that f has an absolutely continuous derivative and a square-integrable first derivative. Then the asymptotic MISE is*

$$\text{AMISE}(h) = \frac{1}{nh} + \frac{1}{12}h^2 R(f'); \qquad \text{hence,}$$
$$h^* = [6/R(f')]^{1/3} n^{-1/3}$$
$$\text{AMISE}^* = (3/4)^{2/3} R(f')^{1/3} n^{-2/3}. \qquad (3.15)$$

Thus the asymptotically optimal bin width depends on the unknown density only through the roughness of its first derivative. This result holds irrespective of the choice of bin origin, which must have a secondary role in MISE compared to the bin width. The optimal bin width decreases at a relatively slow rate. The corresponding optimal error,

$$\text{AMISE}^* \equiv \text{AMISE}(h^*),$$

decreases at the same rate as the bound in Corollary 3.2, far from the desirable $O(n^{-1})$ rate.

3.2.3 Normal Density Reference Rule

It will be convenient to fix some of these results for the special case of normal data. If $f = N(\mu, \sigma^2)$, then $R(f') = 1/(4\sqrt{\pi}\sigma^3)$ (see Problem 3.2). Hence, from Theorem 3.3,

$$h^* = (24\sqrt{\pi}\sigma^3/n)^{1/3} \approx 3.5\sigma n^{-1/3}. \tag{3.16}$$

Scott (1979) proposed using the normal density as a reference density for constructing histograms from sampled data by using the sample standard deviation $\hat{\sigma}$ in (3.16) to obtain the

Normal bin width reference rule: $\quad \hat{h} = 3.5\hat{\sigma}n^{-1/3}.$	(3.17)

Since $\hat{\sigma} \to \sigma$ faster than $O(n^{-1/3})$, this rule is very stable. Freedman and Diaconis (1981) proposed a more robust rule, replacing the unknown scale parameter σ by a multiple of the interquartile range (IQR):

$$\hat{h} = 2\,(\text{IQR})\,n^{-1/3}. \tag{3.18}$$

If the data are in fact normal, the Freedman–Diaconis rule is about 77% of the Scott's rule as the IQR $= 1.348\sigma$ in this case.

3.2.3.1 Comparison of Bandwidth Rules
How does the optimal rule compare to Sturges' rule in the case where the situation is most favorable to the latter? In Table 3.1, the number of bins suggested by the three rules is given, assuming that the data are normal and that the sample range is $(-3, 3)$. Perhaps the most remarkable observation obtainable from this table is how closely the rules agree for samples between 50 and 500 points. For larger samples, Sturges' rule gives too few bins, corresponding to a much oversmoothed histogram estimate that wastes much of the information in the data. As shown in Section 3.3.1, almost any other non-normal sampling density will require even more bins. This result is a consequence of the fact that the normal case is very close to a theoretical lower bound on the number of bins, or equivalently, an upper bound on the bin width. The Freedman–Diaconis rule has 35% more bins than Scott's rule and the histogram will be rougher.

TABLE 3.1 Comparison of number of bins from three normal reference rules

n	Sturges' rule	Scott's rule	F–D rule
50	5.6	6.3	8.5
100	7.6	8.0	10.8
500	10.0	13.6	18.3
1,000	11.0	17.2	23.2
5,000	13.3	29.4	39.6
10,000	14.3	37.0	49.9
100,000	17.6	79.8	107.6

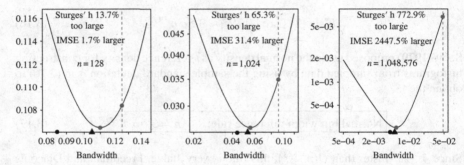

FIGURE 3.4 AMISE versus bandwidth for a $B(5,5)$ density. The best and Sturges' bandwidths are indicated by points on the curves. The Freedman–Diaconis and Scott reference bandwidths are shown as semicircular and triangular points, respectively, along the x-axis.

As another example, consider the beta density, $B(5,5) = 630x^4(1-x)^4$, which has finite support $(0,1)$ and is similar to a normal density. Asymptotic MISE versus bandwidth curves are shown in Figure 3.4 for several sample sizes. Again, Sturges' rule leads to severe oversmoothing as the sample size increases. Also shown along the axis are the reference rules of Scott and Freedman–Diaconis. The Freedman–Diaconis (robust) rule is calibrated to be more aggressive than a purely normal reference rule, being 18% narrower than the Scott reference rule for this density. The IQR and standard deviation are 0.216 and $1/\sqrt{44} = 0.151$, so that bandwidth constants in Equations (3.17) and (3.18) are $2\,\mathrm{IQR} = 0.432$ and $3.5\sigma = 0.528$, respectively, a difference of 18%.

3.2.3.2 Adjustments for Skewness and Kurtosis

Scott (1979) also proposed using the lognormal and t distributions as reference densities to modify the normal rule when the data are skewed or heavy-tailed. Suppose Y is a non-normal random variable with density $g(y)$ and moments σ_y^2, β_1, and β_2 (the standardized skewness and kurtosis, respectively). Consider using the normal rule with $\sigma = \sigma_y$ compared to the optimal bin width in Theorem 3.3. The ratio of these two bin widths is given by

$$\frac{h_y^*}{h_N} = \left[\frac{R\left(\phi'\left(y|0,\sigma_y^2\right)\right)}{R(g'(y))} \right]^{1/3} ; \qquad (3.19)$$

if this ratio is much different from 1, then the normal reference rule should be modified by this ratio.

For example, let $g(y)$ be the lognormal density of the random variable $Y = \exp(X)$, where $X \sim N(0,\sigma^2)$. Then treating σ^2 as a parameter, we have

$$\sigma_y^2 = e^{\sigma^2}(e^{\sigma^2} - 1); \quad \beta_1 = (e^{\sigma^2} - 1)^{1/2}(e^{\sigma^2} + 2); \quad R(g') = \frac{(\sigma^2 + 2)e^{9\sigma^2/4}}{8\sqrt{\pi}\sigma^3}.$$

Since $R(\phi'(y)) = 1/(4\sqrt{\pi}\sigma_y^3)$, the normal rule should be multiplied by the factor in (3.19) given by

$$\text{Skewness factor}\{\beta_1(\sigma)\} = \frac{2^{1/3}\sigma}{e^{5\sigma^2/4}(\sigma^2 + 2)^{1/3}(e^{\sigma^2} - 1)^{1/2}}. \qquad (3.20)$$

This factor is plotted in Figure 3.5. Clearly, any skewness requires smaller bin widths than given by the normal reference rule.

A similar calculation can be performed for kurtosis assuming that $Y \sim t_\nu$, which is the t distribution with ν degrees of freedom. Let the excess kurtosis be denoted by $\tilde{\beta}_2 = \beta_2 - 3$. Treating ν as a parameter yields

$$\sigma_y^2 = \frac{\nu}{\nu - 2}; \quad \tilde{\beta}_2 = \frac{6}{\nu - 4}; \quad R(g') = \frac{2\Gamma(\nu + \frac{3}{2})\Gamma(\frac{\nu+3}{2})^2}{\sqrt{\pi}\nu^{3/2}\Gamma(\nu + 3)\Gamma(\frac{\nu}{2})^2}.$$

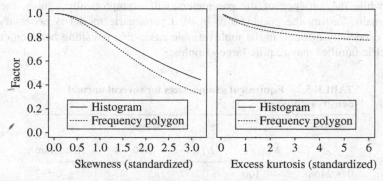

FIGURE 3.5 Factors that modify the normal reference bin width rules for the histogram and frequency polygon as a function of the standardized skewness and excess kurtosis.

Substituting into (3.19) gives us

$$
\text{Kurtosis factor}\left\{\tilde{\beta}_2(\nu)\right\} = \frac{\sqrt{\nu-2}}{2}\left[\frac{\Gamma(\nu+3)\Gamma\left(\frac{\nu}{2}\right)^2}{\Gamma\left(\nu+\frac{3}{2}\right)\Gamma\left(\frac{\nu+3}{2}\right)^2}\right]^{1/3}. \tag{3.21}
$$

See Figure 3.5. The modification to the bin width is not as great as for large skewness.

3.2.4 Equivalent Sample Sizes

In comparison with parametric estimators, histograms enjoy the advantage of quite general consistency without what Fisher called the "specification problem." How great a penalty is incurred if an optimal histogram is used rather than the corresponding optimal parametric estimator?

Consider again the example of Section 3.2.3. From Theorem 3.3, it follows that if $f = N(0,1)$, then the optimal AMISE of the histogram is

$$
\text{AMISE}^* = [9/(64\sqrt{\pi})]^{1/3}n^{-2/3} \approx 0.4297 n^{-2/3}.
$$

Suppose a sample of size 100 is available for the parametric estimator $N(\bar{x},\sigma^2)$, for which AMISE $= 0.002468$ from Equation (2.9). Table 3.2 gives the equivalent sample sizes for the histogram and several parametric estimators. The parametric estimators are the clear winners, more so for smaller errors.

On the other hand, if the true density is only approximately normal, then the parametric MISE will never be less than the true integrated squared bias level

$$
\text{ISB} = \int \left[\phi\left(x\,|\,\mu_f,\sigma_f^2\right) - f(x)\right]^2 dx,
$$

where $\left(\mu_f,\sigma_f^2\right)$ are the actual moments of the unknown density $f(x)$. The maximum likelihood estimator assuming normality will asymptotically match those moments. Thus, while the variance of the parameters will asymptotically vanish, the ISB will remain. Testing the goodness-of-fit of a parametric model is necessary and can be especially difficult in the multivariable case; distinguishing between certain parametric families may require large samples.

TABLE 3.2 Equivalent sample sizes for several normal density estimators

AMISE	Estimator			
	$N(\bar{x},s^2)$	$N(\bar{x},1)$	$N(0,s^2)$	Histogram
0.002468	100	57	43	2,297
0.000247	1,000	571	429	72,634

3.2.5 Sensitivity of MISE to Bin Width

How important is it that the smoothing parameter be optimally chosen? If it happens that virtually any bin width "works," then less effort would need to be applied to the calibration problem.

3.2.5.1 Asymptotic Case

In practice, given real data from an unknown density, the smoothing parameter chosen will not be h^*, but instead be of the form $h = ch^*$. On average, if $c \ll 1$, then the histogram will have high variance and will be "too rough"; if $c \gg 1$, then the estimate will have high bias or systematic error and will be "too smooth." How sensitive is the MISE to local deviations of c from 1?

The asymptotic MISE of the histogram in Theorem 3.3 is of the form

$$\text{AMISE}(h) = \frac{a}{nh} + \frac{b}{2}h^2, \tag{3.22}$$

where a and b are positive constants. Rather than considering only this special case, a more general form of the AMISE will be considered:

$$\text{AMISE}(h) = \frac{a}{(d+2r)nh^{d+2r}} + \frac{b}{2p}h^{2p}, \tag{3.23}$$

where (d, p, r) are positive integers, and a and b are positive constants that depend on the density estimator and unknown density function. In general, the triple (d, p, r) refers to, respectively: (1) the dimension of the data, (2) the "order" of the estimator's bias, and (3) the order of the derivative being estimated. Comparing Equations (3.22) and (3.23), $(d, p, r) = (1, 1, 0)$ for the case of the histogram. It follows from minimizing Equation (3.23) that

$$h^* = (a/bn)^{1/(d+2r+2p)}$$

$$\text{AMISE}^* = \text{AMISE}(h^*) = \left(\frac{d+2p+2r}{2p(d+2r)} \right) \left(\frac{a^{2p}b^{d+2r}}{n^{2p}} \right)^{1/(d+2p+2r)}. \tag{3.24}$$

In Problem 3.5, it is shown that the variance portion of the histogram's AMISE* is twice that of the squared bias. Finally, it is easy to check (see Problem 3.6) that

$$\frac{\text{AMISE}(ch^*)}{\text{AMISE}(h^*)} = \frac{2p + (d+2r)c^{d+2p+2r}}{(d+2p+2r)c^{d+2r}} \quad \left(= \frac{2+c^2}{3c} \text{ for a histogram} \right). \tag{3.25}$$

A consequence of this expression is that departures of h from h^* should be measured in a *multiplicative* rather than an *additive* fashion. This may also be clear from a dimensional analysis of h^* in Theorem 3.3. In the author's experience, a 10–15% change in h produces a small but noticeable change in a histogram. Nonetheless, in Table 3.3, it is clear that the histogram, which corresponds to the case $p = 1$, is fairly insensitive to a choice of bin width within 33% of optimal. Notice that the L_2 criterion is less affected by high variance than high bias errors (for example, $c = 1/2$ vs. $c = 2$).

TABLE 3.3   Sensitivity of AMISE to error in bin width choice
$h = ch^*$

$(d = 1, r = 0)$ c	$p = 1$ $(c^3 + 2)/(3c)$	$p = 2$ $(c^5 + 4)/(5c)$	$p = 4$ $(c^9 + 8)/(9c)$
1/2	1.42	1.61	1.78
3/4	1.08	1.13	1.20
1	1	1	1
4/3	1.09	1.23	1.78
2	1.67	3.60	28.89

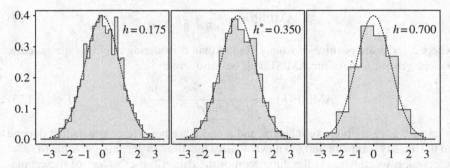

FIGURE 3.6   Three histograms of 1000 normal observations with bin widths $h = (\frac{1}{2}h^*, h^*, 2h^*)$.

Looking forward, larger values of p correspond to "higher order" methods that have faster MISE rates. However, the sensitivity to choice of smoothing parameters is much greater. Increased sensitivity is also apparent with increasing dimension d and derivative order r (see Problem 3.7).

3.2.5.2   Large-Sample and Small-Sample Simulations
The previous analysis focused only on average behavior. The actual ISE for an individual sample is considered in this section.

Figure 3.6 displays three histograms of 1000 normal observations. When $h = 2h^*$, the estimate incurs substantial bias because of the broad bins. In informal questioning, most statisticians prefer the histogram with $h = \frac{1}{2}h^*$, even though it contains several spurious small bumps. However, it is easy to visually smooth the local noise on that estimate. The converse is not true, since it is not possible to visualize the lost detail in the histogram with bin width $h = 2h^*$.

Generalizing too much from small-sample experience is a temptation that should be avoided. Figure 3.7 displays four histograms of a sample of a million normal points. The exact ISE is given in each frame. Examine the changes in ISE. Locally, the histogram remains very sensitive to changes in h even when n is large. This may seem

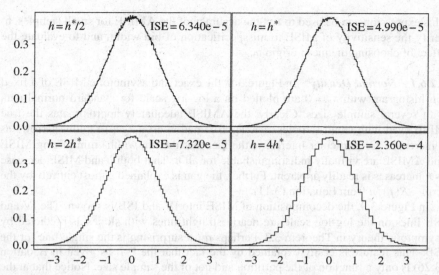

FIGURE 3.7 Exact ISE of four histograms of a million normal points with bin widths $h = (h^*/2, h^*, 2h^*, 4h^*)$, where $h^* = 0.035$.

counterintuitive initially, especially given the very general conditions required for consistency. But large samples do not automatically guarantee good estimates. The histogram with $h = \frac{1}{2}h^*$ contains many local modes. The behavior of sample modes in a histogram will be reexamined in Section 3.5.2. Even the "optimal" histogram with $h = h^*$ appears noisy (locally) compared to the $h = 2h^*$ histogram. The observation that optimally smoothed estimates appear on the verge of a form of instability seems a common occurrence.

3.2.6 Exact MISE versus Asymptotic MISE

For a general unequally spaced mesh, where the width of bin B_k is h_k, it is straightforward to show from Equations (3.3) and (3.4) that

$$IV = \frac{1}{n}\sum_k \frac{p_k(1-p_k)}{h_k} \quad \text{and} \quad ISB = R(f) - \sum_k \frac{p_k^2}{h_k} \qquad (3.26)$$

exactly (see Problem 3.8). Now, $\sum_k p_k = 1$; hence, for the special case of an equally spaced mesh, $h_k = h$,

$$MISE(h, t_0, n) = \frac{1}{nh} - \frac{n+1}{nh}\sum_k p_k^2 + R(f).$$

This expression can be used to test the accuracy of the AMISE for small samples, to verify the sensitivity of MISE to misspecification of bin width, and to evaluate the effect of choosing the mesh origin t_0.

3.2.6.1 Normal Density

3.2.6.1 Normal Density   In Figure 3.8, the exact and asymptotic MISE of a fixed bin histogram with $t_0 = 0$ are plotted on a log-log scale for standard normal data with several sample sizes. Clearly, the AMISE adequately approximates the true MISE even with small samples. The approximation error (gap) rapidly narrows as n increases. Of particular interest is the fact that the bin widths minimizing MISE and AMISE are virtually indistinguishable for all n. That both h and MISE decrease as n increases is readily apparent. Further, the gap is explained almost entirely by the term $-R(f)/n$ from Equation (3.11).

In Figure 3.9, the decomposition of MISE into IV and ISB is shown. The IV and ISB lines on the log-log scale are nearly straight lines with slopes as predicted by asymptotic theory in Theorem 3.3. Perhaps most surprising is the single line for the ISB. This feature is easily explained by the fact that the ISB as given in Equation (3.26) is only a function of the partition and not of the sample size. Notice that at the optimal choice of bin width, the contribution of IV exceeds ISB, by a ratio approximately of 2:1, as shown in Problem 3.5. The L_2 criterion is more sensitive when $h > h^*$.

Figure 3.10 displays a picture of the MISE curve over a much wider range of bin widths when $n = 100$. The MISE(h) is unbounded as $h \to 0$, but

$$\lim_{h \to \infty} \text{MISE}(h) = \lim_{h \to \infty} \text{ISB}(h) = R(f),$$

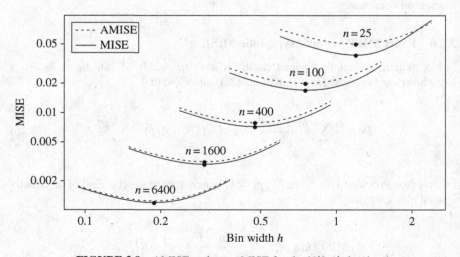

FIGURE 3.8   AMISE and exact MISE for the $N(0,1)$ density.

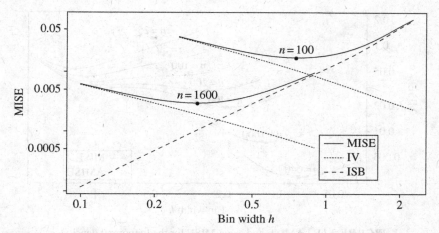

FIGURE 3.9 Integrated squared-bias/variance decomposition of MISE for the $N(0,1)$ density.

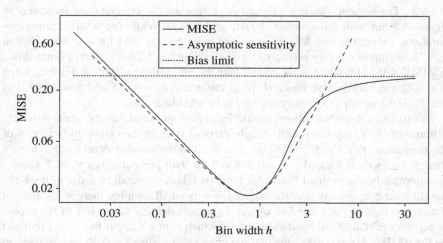

FIGURE 3.10 Complete MISE curve for the $N(0,1)$ density when $n = 100$. The asymptotic sensitivity relationship holds over a wide range of bin widths.

as is clearly evident. Asymptotic theory predicts the manner in which MISE varies with h (see Equation (3.25)). The dashed line in Figure 3.10 is a plot of

$$\left(h, \frac{c^3 + 2}{3c} \text{MISE}^* \right),$$

where $c = h/h^*$ and $\text{MISE}^* = \text{MISE}(h^*)$. This approximation is accurate far beyond the immediate neighborhood of h^*.

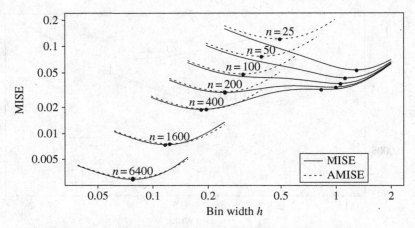

FIGURE 3.11   AMISE and exact MISE for the lognormal density.

3.2.6.2 Lognormal Density   Figure 3.11 repeats the computation indicated in Figure 3.8 but with a lognormal density and $t_0 = 0$. While the AMISE retains its parabolic shape, the true MISE curves are quite complicated for $n < 1000$. When $n = 25$, asymptotic theory predicts $h^* = 0.49$ while $h = 1.28$ is in fact optimal. Also, AMISE$^* = 0.1218$, which is far greater than the true MISE$^* = 0.0534$. Further, when $n = 200$, the MISE curve has 2 (!) local minima at $h = 0.245$ and $h = 0.945$. For $n > 1000$ the asymptotic theory appears to be adequate.

What can explain these observations? First, the lognormal density, while infinitely differentiable, is very statistically rough, particularly near the origin. In fact, 90% of the roughness, $R(f') = 3e^{9/4}/(8\sqrt{\pi})$, comes from the small interval $[0, 0.27]$, even though the mode is located at $x = 0.368$ and the 99th percentile is $x = 10.2$. Hence, asymptotically, the optimal bin width must be relatively small in order to track the rapidly changing density near the origin. For very small samples, there is insufficient data to accurately track the rise of the lognormal density to the left of the mode; hence the optimal MISE bandwidth $h > 1$, which is much larger than that predicted by the AMISE. For $n > 1000$, the optimal bin width satisfies $h < 0.15$ and the rise can be tracked. For $n \approx 200$, the two locally optimal bin widths indicate a situation where the rise can be equally well approximated by either a narrow or wide bin, or almost any bin width in between since the MISE curve is relatively flat.

Figure 3.11 illustrates an important practical and generalizable observation about nonparametric methods. Sample sizes may be grouped into three categories: inadequate, transitional, and sufficient. For lognormal data, the inadequate sample sizes are for $100 < n < 400$ with equally spaced histograms, and transitional for $400 < n < 1000$. For other densities, the transitory region may extend to even larger samples, although the lognormal density is reasonably extreme. Unfortunately, in practice with a particular dataset, it is not possible to be certain into which category n falls. However, as can readily be seen, features narrower than h cannot be detected reliably below certain sample sizes. Donoho (1988) neatly summarizes the situation in nonparametric estimation as one-sided inference—only larger sample sizes can definitively answer whether smaller structure exists in the unknown density.

3.2.7 Influence of Bin Edge Location on MISE

3.2.7.1 General Case The asymptotic MISE analysis in Theorem 3.3 indicates that the choice of the bin origin t_0 is a lower order effect, as long as the conditions of the theorem are satisfied. For $N(0,1)$ data with optimal smoothing, the exact MISE is minimized if the point $x = 0$ is in the *middle* of a bin rather than at the boundary. The difference in MISE when using these two bin edge locations is minuscule, being 1.09% when $n = 25$ and less than 10^{-5} when $n > 100$.

With lognormal data, the edge effect is also less than 10^{-5} when $n > 400$. But when $n = 25$ and $h = 1.28$, the MISE ranges over $(0.042, 0.149)$—best for the mesh $(-1.20, 0.08, 1.36, 2.64, \ldots)$ and worst for the mesh $(-0.65, 0.63, 1.91, 3.19, \ldots)$. Clearly, the choice of bin edge is negligible for sufficiently large sample sizes but can be significant for inadequate sample sizes. Graphically, the choice of t_0 for fixed h provides a range of possible histograms with quite different subjective appearances (see Chapter 5 and Figure 5.1).

3.2.7.2 Boundary Discontinuities in the Density The approximation properties of a histogram are not affected by a simple jump in the density, *if* the jump occurs at the boundary of a histogram bin. This follows from a careful examination of the derivation of the MISE in Section 3.2.2. Discontinuities can adversely affect all density estimators. The adverse effect can be demonstrated for the histogram by example.

Consider $f(x) = e^{-x}, x \geq 0$ and mesh $t_k = kh$ for integer $k \geq 0$. Then Theorem 3.3 holds on the interval $(0, \infty)$, on which $R(f') = 1/2$. Therefore,

$$h^* = (12/n)^{1/3} \quad \text{and} \quad \text{AMISE}^* = 0.6552 n^{-2/3}.$$

Suppose the discontinuity at zero was not known *a priori* and the mesh $t_k = (k - \frac{1}{2})h$ was chosen. Then attention focuses on the bias in bin $B_0 = [-h/2, h/2)$ where the density is discontinuous (see Figure 3.12).

Note that the probability mass in bin $B_0 = (-h/2, h/2]$ is

$$p_0 = \int_0^{h/2} e^{-x} dx = 1 - e^{-h/2}.$$

Now $E\hat{f}(x) = p_0/h$; therefore,

$$\int_{-h/2}^{h/2} \text{Bias}(x)^2 \, dx = \int_{-h/2}^{0} \left(\frac{p_0}{h} - 0\right)^2 dx + \int_{0}^{h/2} \left(\frac{p_0}{h} - e^{-x}\right)^2 dx, \quad (3.27)$$

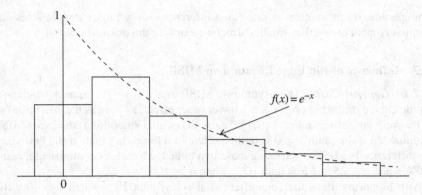

FIGURE 3.12   Illustration of discontinuity boundary bin problem.

which equals $(1 - e^{-h})/2 + (2e^{-h/2} - e^{-h} - 1)/h \approx h/4 - h^2/8 + \cdots$. Thus over the interval $(-h/2, \infty)$,

$$\text{ISB} = \frac{h}{4} - \frac{h^2}{8} + O(h^3) + \frac{1}{12}h^2 \int\limits_{h/2}^{\infty} f'(x)^2 dx + o(h^2). \qquad (3.28)$$

The worst outcome has been realized—the ISB is entirely dominated by the contribution from the bin containing the discontinuity at $x = 0$. The asymptotic integrated variance is unchanged so that

$$\text{AMISE}(h) = \frac{1}{nh} + \frac{h}{4} \quad \Rightarrow \quad h^* = \frac{2}{\sqrt{n}} \quad \text{and} \quad \text{AMISE}^* = \frac{1}{\sqrt{n}}, \qquad (3.29)$$

which is significantly worse than $O(n^{-2/3})$. In fact, the rate is as slow as for bivariate data, as will be shown in Section 3.4. Table 3.4 illustrates the costs. The histogram tries to accommodate the discontinuity by choosing a narrower bin width. Compare this situation to that of the very rough behavior of the lognormal density near the origin with small samples.

3.2.8   Optimally Adaptive Histogram Meshes

An optimally calibrated fixed-bin-width histogram with real data often *appears* rough in the tails due to the paucity of data. This phenomenon is one of several reasons for considering histograms with adaptive meshes. Evaluating the reduction in MISE with adaptive meshes is the task of this section. Creating adaptive meshes is a familiar exercise for anyone who has ever performed a χ^2 goodness-of-fit test: to satisfy the recommendation that the expected count in every cell exceeds 5, cells in the tails are usually combined or, alternatively, cells are formed so that each contains exactly the same number of points (see Section 3.2.8.4).

TABLE 3.4   Potential impact on AMISE of lack of knowledge of boundary discontinuities

n	$X > 0$ known		$X > 0$ unknown		Error ratio
	h^*	AMISE*	h^*	AMISE*	
10	1.063	0.14116	0.3162	0.31623	2.24
100	0.493	0.03041	0.1	0.1	3.29
1,000	0.229	0.00655	0.0316	0.03162	4.83
10,000	0.106	0.00141	0.01	0.01	7.09
100,000	0.049	0.00030	0.0032	0.00316	10.54

3.2.8.1   Bounds on MISE Improvement for Adaptive Histograms   A lower bound, which is asymptotically correct, for the reduction in MISE of an optimally adaptive histogram follows from Equations (3.6) and (3.14). Construct an approximation to the asymptotically adaptive pointwise histogram MSE (AAMSE):

$$\text{AAMSE}(x) \approx \frac{f(x)}{nh} + \frac{1}{12}h^2 f'(x)^2. \qquad (3.30)$$

Minimize AAMSE(x) for each point x. Therefore,

$$h^*(x) = \left[\frac{6f(x)}{nf'(x)^2} \right]^{1/3} \quad \Rightarrow \quad \text{AAMSE}^*(x) = \left[\frac{3f(x)f'(x)}{4n} \right]^{2/3}.$$

Integrating AAMSE$^*(x)$ over x gives the following result.

Theorem 3.4:   *Asymptotically, for an optimally adaptive histogram,*

$$\text{AAMISE}^* = (3/4)^{2/3} \left(\int_{-\infty}^{\infty} [f'(x)f(x)]^{2/3} dx \right) n^{-2/3}. \qquad (3.31)$$

This result has been discussed by Terrell and Scott (1983, 1992) and Kogure (1987). Comparing Equations (3.15) and (3.31), the improvement of the adaptive mesh is guaranteed if

$$\int [f'(x)f(x)]^{2/3} \leq \left[\int f'(x)^2 \right]^{1/3}$$

TABLE 3.5   Reduced AMISE using an optimally adaptive histogram mesh

Density	$\int (f^2 f'^2)^{1/3} \div [\int \int f'^2]^{1/3}$
$N(0,1)$	$0.4648/0.5205 = 89.3\%$
$3/4\,(1-x^2)_+$	$0.8292/1.1447 = 72.4\%$
$15/16\,(1-x^2)_+^2$	$2.1105/2.8231 = 74.8\%$
$315/256\,(1-x^2)_+^4$	$1.4197/1.6393 = 86.6\%$
Cauchy	$0.3612/0.4303 = 84.0\%$
Lognormal	$0.6948/1.2615 = 55.1\%$

or equivalently, if

$$\mathrm{E}\left[\frac{f'(\mathbf{X})^2}{f(\mathbf{X})}\right]^{1/3} \leq \left[\mathrm{E}\frac{f'(\mathbf{X})^2}{f(\mathbf{X})}\right]^{1/3}; \tag{3.32}$$

but this last inequality follows from Jensen's inequality (the concave function version) (see Problem 3.11). Table 3.5 shows numerical computations of this ratio for some simple densities. At first glance the gains with adaptive meshes are surprisingly small.

3.2.8.2   Some Optimal Meshes   Since the expression for the exact MISE for arbitrary meshes is available in Equation (3.26), the optimal adaptive mesh may be found numerically. Certain features in these optimal meshes may be anticipated [review the discussion following Equation (3.8)]. Compared to the optimal fixed bandwidth, bins will be relatively wider not only in the tails but also near modes where $f'(x)$ is small. In regions where the density is rapidly changing, the bins will be narrower.

This "accordion" pattern is apparent in the following example with a Beta(5,5) density transformed to the interval $(-1, 1)$, that is, $f(x) = \frac{315}{256}(1 - x^2)_+^4$ (see Figure 3.13). Since the cdf of this density is a polynomial, it is more amenable than the normal cdf to numerical optimization over the exact adaptive MISE given in Equation (3.26). Clearly, the optimal mesh is symmetric about 0 in this example. But the optimal mesh does not include 0 as a *knot* (mesh node). Forcing 0 to be a knot increases the MISE, for example, by 4.8% when $n = 10{,}000$; the best MISE for an equally spaced mesh ($h = 0.0515$) is 8.4% greater. These results are very similar to those for the normal density.

3.2.8.3   Null Space of Adaptive Densities   There exists a class of density functions for which there is no improvement asymptotically with an adaptive histogram procedure. This class may be called the *null space of adaptive densities*. Examining

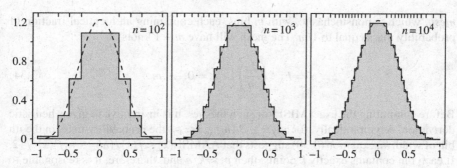

FIGURE 3.13 Representation of optimal adaptive meshes for the transformed Beta(5,5) density, which equals $315/256\,(1-x^2)^4_+$. The optimal adaptive mesh is also indicated by tick marks above each graph.

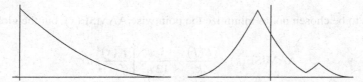

FIGURE 3.14 Examples of piecewise-quadratic densities for which the asymptotically optimal adaptive mesh is in fact equally spaced.

Equation (3.32), any density that results in equality in Jensen's inequality is in the null space. This occurs when the argument in brackets is constant; that is,

$$\frac{f'(x)^2}{f(x)} = c \quad \Rightarrow \quad f(x) = \frac{1}{4}c(x-a)^2, \qquad (3.33)$$

where "a" is an arbitrary constant. The densities $f_1(x) = 3(1-x)^2 I_{[0,1]}(x)$ and $f_2(x) = \frac{3}{2}(1-|x|)^2 I_{[-1,1]}(x)$ also satisfy the differential equation piecewise. In fact, any density that is obtained by piecing together densities of the form (3.33) in a continuous manner is in the null space (see Figure 3.14). As before, the best adaptive mesh may be found by numerical optimization. However, with $n = 100$ and the density $f_1(x)$ defined earlier, the computed optimal mesh nodes differ by less than 2% from the equally spaced mesh with $h = 1/6$. And the MISE of the computed optimal mesh is only 0.06% better than the equally spaced mesh.

3.2.8.4 Percentile Meshes or Adaptive Histograms with Equal Bin Counts In practice, finding the adaptive mesh described in the preceding section is difficult. Thus some authors have proposed (nonoptimal) adaptive meshes that are easier to implement. One of the more intuitively appealing meshes has an equal number (or equal fraction) of points in each bin. This idea can be modeled by a *percentile* *mesh*, which is nonstochastic, with m bins, each containing an identical fraction of probability mass equal to $1/m$. The mesh will have $m+1$ knots at

$$t_k = F_X^{-1}\left(\frac{k}{m}\right), \quad k = 0,\dots,m. \tag{3.34}$$

Before computing the exact MISE for such meshes, it is instructive to give a heuristic derivation. Asymptotically, for any $x \in B_i, h_i f(x) \approx p_i$, the probability mass in the ith bin of width h_i. Equation (3.30) suggests that $\text{AAMSE}(x) \approx f(x)/(nh_i) + h_i^2 f'(x)^2/12$. If each bin contains exactly k points, then $p_i \approx k/n$ and, therefore, it is reasonable to assign

$$h_i = \frac{k}{nf(x)} \quad \Rightarrow \quad \text{AAMSE}(x) = \frac{f(x)^2}{k} + \frac{1}{12}\frac{k^2}{n^2}\left[\frac{f'(x)}{f(x)}\right]^2. \tag{3.35}$$

Now k is to be chosen not to minimize the pointwise $\text{AAMSE}(x)$, but the global

$$\text{AAMISE}(k) = \frac{R(f)}{k} + \frac{1}{12}\frac{k^2}{n^2}\int\frac{f'(x)^2}{f(x)^2}\,dx. \tag{3.36}$$

Now $k^* = O(n^{2/3})$ with $\text{AAMISE}^* = O(n^{-2/3})$. But computing the integral in the bias term when $f = \phi$, the standard normal density, gives

$$\int\frac{f'(x)^2}{f(x)^2}\,dx = \int\frac{[-x\phi(x)]^2}{\phi(x)^2}\,dx = \int_{-\infty}^{\infty} x^2\,dx = \infty \quad (!?). \tag{3.37}$$

This puzzling result may only be the result usin of the approximation (3.35) combined with the infinite support of ϕ, or it may indicate that these meshes suffer unexpectedly large biases. The results depicted in Figure 3.15 clearly demonstrate the poor performance of the percentile meshes when compared with fixed-width bins, with the gap increasing with sample size. The consequences, if any, of this negative result for chi-squared tests are unknown (see Problem 3.13).

3.2.8.5 *Using Adaptive Meshes versus Transformation*   In the univariate setting, adaptive meshes are most useful when the data are skewed or when the data are clustered with different scaling within each cluster. For heavy-tailed data alone, adaptive meshes may provide some visual improvement, but the MISE criterion is largely insensitive to those features. The best recommendation is to use (Tukey)'s (1977) transformation ladder or Box and Cox (1964) methods to reduce skewness before applying a fixed mesh. The Tukey ladder is a subset of the power family of transformations, x^λ, where the transformation is defined to be $\log(x)$ when $\lambda = 0$:

$$\dots, x^{-2}, x^{-1}, x^{-1/2}, x^{-1/4}, \log(x), x^{1/4}, x^{1/2}, x, x^2, x^3, \dots \tag{3.38}$$

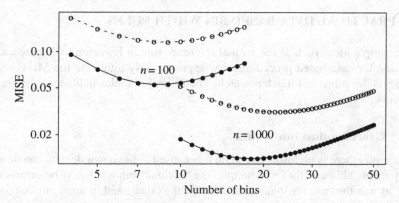

FIGURE 3.15 Exact MISE for fixed (•••••) and percentile (ooooo) meshes with k bins over $(-1, 1)$ for the transformed Beta(5,5) density with $n = 100$ and 1000.

The Box–Cox family is similar, but continuous in λ for $x > 0$:

$$x^{(\lambda)} = \begin{cases} \frac{x^\lambda - 1}{\lambda} & \lambda \neq 0 \\ \log x & \lambda = 0. \end{cases} \tag{3.39}$$

Simple transformations such as $\sqrt{x}$ or $\log(x + c)$ are often sufficient. In the multivariate setting, working with strongly skewed marginal variables results in data clustered near the faces of a surrounding hypercube. Even simple marginal variable transformations can move the data cloud towards the middle of the hypercube, greatly reducing one of the effects of the *curse of dimensionality* (see Chapter 7). In the univariate setting, Wand et al. (1991) have proposed finding an optimal transformation (within the Box–Cox transformation family, for example) with respect to AMISE in the transformed coordinates, and then transforming the density estimate back to the original scale. In many applications, the choice is made to work directly in the transformed coordinates. Economists, for example, usually work with log (income) rather than income to cope with the strong skewness present in income data.

3.2.8.6 Remarks Adaptive histograms that are optimally constructed *must* be better than non-adaptive fixed-width histograms. But adaptive histograms constructed in an *ad hoc* fashion need not be better and in fact can be much worse. In particular, an adaptive histogram algorithm may be inferior if it does not include fixed bin widths as a special case. Even so, in practice, an adaptive procedure may be worthwhile. Of particular practical concern is the introduction of additional smoothing parameters defining the adaptivity that must be estimated *from the data* by some method such as *cross-validation* (CV) (which is introduced later). Given the performance of these algorithms in the simplest case of only one smoothing parameter in the fixed-bandwidth setting, caution seems in order. Other adaptive histogram algorithms have been proposed by Van Ryzin (1973) and Wegman (1970).

3.3   PRACTICAL DATA-BASED BIN WIDTH RULES

While simple ideas such as the normal reference rule in Equation (3.17) are useful, the quest for data-based procedures that approximately minimize the MISE and/or the ISE is the subject of much research. The following topics indicate the variety of that work.

3.3.1   Oversmoothed Bin Widths

In principle, there is no lower bound on h because the unknown density can be arbitrarily rough, although for finite samples the "optimal" bin width may be surprisingly large, as was the case for lognormal data. On the other hand, it turns out that useful *upper* bounds exist for the bin width, depending on some data-based knowledge of the scale of the unknown density (Terrell and Scott, 1985).

3.3.1.1   Lower Bounds on the Number of Bins   Reexamining the expression for the minimizer of AMISE in Equation (3.15),

$$h^* = \left(\frac{6}{R(f')} \right)^{1/3} n^{-1/3}, \tag{3.40}$$

it follows that any *lower bound* on $R(f')$ leads to an *upper bound* on the bin width. The simplest prior knowledge of scale is that the density is zero outside an interval (a, b). his suggests the following optimization problem:

$$\min_f \int_{-\infty}^{\infty} f'(x)^2 dx \quad \text{s/t} \quad \text{support of } f = [-0.5, 0.5]. \tag{3.41}$$

Hodges and Lehmann (1956) solved this optimization problem in another setting and showed that the solution is

$$f_1(x) = \frac{3}{2}(1 - 4x^2)I_{[-.5,.5]}(x) = \frac{3}{2}(1 - 4x^2)_+ \, .$$

It is instructive to show that f_1 indeed minimizes $R(f')$. Define $g(x) = f_1(x) + e(x)$, where $e(x)$ is a perturbation function satisfying the following conditions: (i) $e(x) = 0$ outside the interval $[-0.5, 0.5]$; (ii) $e(\pm 0.5) = 0$ and is continuous, as otherwise $g(x)$ would be discontinuous and $R(g') = \infty$; (iii) $\int e(x)\,dx = 0$ so that g integrates to 1.

The solution must have support exactly on $(-0.5, 0.5)$ and not on a strict subinterval. If not, rescaling the density to the interval $(-0.5, 0.5)$ would reduce the roughness, leading to a contradiction. Likewise, the solution must be symmetric around 0. If g is asymmetric, consider the symmetric density $[g(x) + g(-x)]/2$, whose derivative is $[g'(x) - g'(-x)]/2$. Using Minkowski's inequality, which states that $[\int (f+g)^2]^{1/2} \leq [\int f^2]^{1/2} + [\int g^2]^{1/2}$, the roughness of the first derivative of this symmetric density equals

$$\sqrt{\int \frac{1}{4}[g'(x) - g'(-x)]^2\, dx} \leq \frac{1}{2}\left[\sqrt{R(g')} + \sqrt{R(g'(-x))}\right] = \sqrt{R(g')},$$

which contradicts the optimality of g.

To verify the optimality of f_1, directly compute

$$\int g'(x)^2 dx = \int f_1'(x)^2 dx + \int e'(x)^2 dx + 2\int f_1'(x)\, e'(x)\, dx. \qquad (3.42)$$

The last integral vanishes as

$$\int\limits_{-1/2}^{1/2} f_1'(x)\, e'(x)\, dx = f_1'(x)\, e(x)\Big|_{x=-1/2}^{x=1/2} - \int\limits_{-1/2}^{1/2} f_1''(x)\, e(x)\, dx = 0,$$

because $f_1''(x) = -12$, a constant. Therefore, from Equation (3.42), $R(g') = R(f_1') + R(e')$, so that $R(g') \geq R(f_1')$, which proves the result. Notice that f_1 is the unique quadratic density that satisfies all the side conditions.

To apply this result, note that $R(f') = 12$. If f_1 is linearly mapped to the interval (a, b), then the lower bound is $R(f') \geq 12/(b-a)^3$. Therefore,

$$h^* = \left(\frac{6}{nR(f')}\right)^{1/3} \leq \left(\frac{6(b-a)^3}{n \cdot 12}\right)^{1/3} = \frac{b-a}{\sqrt[3]{2n}} \equiv h_{OS},$$

which is the so-called *oversmoothed bandwidth*. Rearranging gives

$$\boxed{\text{Number of bins} = \frac{b-a}{h^*} \geq \frac{b-a}{h_{OS}} = \sqrt[3]{2n}\,a.} \qquad (3.43)$$

Terrell and Scott (1985) showed that the sample range may be used if the interval (a, b) is unknown or even if $b - a = \infty$ and the tail is not too heavy.

EXAMPLE: For the Buffalo snowfall dataset ($n = 63$; see Table B.9) and the LRL data ($n = 25{,}752$; see Table B.8), rule (3.43) suggests 5 and 37 bins, respectively. For the snowfall data, the sample range is $126.4 - 25.0 \approx 100$; therefore, the oversmoothed bandwidth $h_{OS} = 20$ inches of snow is suggested (see Figure 3.16). There is no hint of the trimodal behavior that appears to be in the histogram with more bins. The LRL data were recorded in 172 consecutive bins, so that $h_{OS} = (2000 - 280)/37 = 46.5$. Since the original data were rounded to bins of width of 10 MeV, the only choices available for h_{OS} are 40 or 50 MeV. Strictly speaking, combining only four adjacent bins ($h = 40$) is not oversmoothing, so the slightly

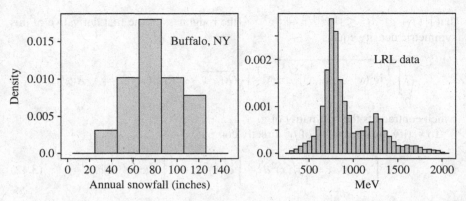

FIGURE 3.16 Oversmoothed histograms of the Buffalo snowfall and LRL data.

conservative choice $h = 50$ is adopted (see Figure 3.16). At least three groups appear in this histogram, the third around $x = 1700$.

3.3.1.2 Upper Bounds on Bin Widths

The previous result is most notable for its simplicity and mnemonic value. However, the focus is on choosing the smoothing parameter rather than the number of bins. Oversmoothing provides a solution in that setting as well.

Consider the optimization Problem (3.41) except with the fixed-range constraint replaced by the constraint that the variance of f equal σ^2. Terrell (1990) showed that the solution to this problem is

$$f_2(x) = \frac{15}{16\sqrt{7}\sigma} \left(1 - \frac{x^2}{7\sigma^2} \right)^2 I_{[-\sqrt{7}\sigma, \sqrt{7}\sigma]}(x).$$

It can be checked that $R(f_2') = 15\sqrt{7}/(343\sigma^3)$; hence,

$$h^* \leq \left(\frac{6}{nR(f_2')} \right)^{1/3} = \left(\frac{686\sigma^3}{5\sqrt{7}n} \right)^{1/3} \approx 3.729\,\sigma\,n^{-1/3} \equiv h_{\mathrm{OS}}. \tag{3.44}$$

Terrell (1990) considered other less common scale constraints. The version based on the interquartile range (IQR) is particularly robust:

$$h^* \leq 2.603(\mathrm{IQR})n^{-1/3} \equiv h_{\mathrm{OS}}. \tag{3.45}$$

EXAMPLE: For $N(\mu, \sigma^2)$ data,

$$h^* = 3.5\,\sigma\,n^{-1/3} < 3.729\,\sigma\,n^{-1/3} = h_{\mathrm{OS}} \quad (= \text{upper bound}), \tag{3.46}$$

but h^* is only 93.6% of h_{OS}. Apparently, the normal density is very smooth and using the simple rule (3.17) is further justified. In the case of lognormal data, for which $\sigma^2 = e(e-1)$,

$$h^* = 1.44n^{-1/3} < 3.729 \times 2.161 \times n^{-1/3} = 8.059n^{-1/3}, \quad (3.47)$$

which is 18% of h_{OS}. On the other hand, for $n = 25$, the best $h \approx 1.28$ from Figure 3.11, which is 46% of $h_{OS} = 2.756$. For Cauchy data, the bound based on variance is asymptotically meaningless, whereas the oversmoothed rule based on IQR still applies.

3.3.2 Biased and Unbiased CV

The CV approach to the automatic data-based calibration of histograms is introduced in this section. CV algorithms *reuse the data*. The goal is not simply to produce a consistent sequence of bin widths—even Sturges' rule is consistent. Rather, the goal is to reliably produce bin widths $\hat{h}_{CV}$ that are close to h^* for *finite samples* or perhaps even more optimistically, bin widths that minimize ISE errors for *individual* samples.

3.3.2.1 Biased CV

The only unknown quantity in the AMISE is $R(f')$, which may be estimated using the data at hand. An approximation of $f'(t_k)$ is available based on a finite difference of the histogram at the midpoints of bins B_k and B_{k+1}, namely, $\hat{f}'(t_k) = [\nu_{k+1}/(nh) - \nu_k/(nh)]/h$. A potential estimate of $R(f')$ is

$$\hat{R}_1 = \sum_k [\hat{f}'(t_k)]^2 \cdot h = \frac{1}{n^2 h^3} \sum_k (\nu_{k+1} - \nu_k)^2. \quad (3.48)$$

It can be shown (see Problem 3.16) that

$$E[\hat{R}_1] = R(f') + 2/(nh^3) + O(h). \quad (3.49)$$

With optimal smoothing, $2/(nh^3)$ converges to $R(f')/3$ by Theorem 3.3. It follows that $\hat{R}_1$ is a biased estimator of $R(f')$, too large by a factor of a third, so that $\frac{3}{4}\hat{R}_1$ is an asymptotically unbiased estimator of $R(f')$. Alternatively,

$$\hat{R}_h(f') = \frac{1}{n^2 h^3} \sum_k (\nu_{k+1} - \nu_k)^2 - \frac{2}{nh^3}. \quad (3.50)$$

Substituting (3.50) into the AMISE expression (3.15) gives a *biased CV* (BCV) estimate of the MISE(h):

$$\boxed{BCV(h) = \frac{5}{6nh} + \frac{1}{12n^2 h} \sum_k (\nu_{k+1} - \nu_k)^2,} \quad (3.51)$$

where ν_k is recomputed as h and the mesh vary (for definiteness, the bin origin, t_0, remains fixed). The bias moniker simply refers to the fact that the AMISE(h) is a

biased approximation to the true MISE(h). The BCV bin width, $\hat{h}_{BCV}$, is defined
to be the minimizer of BCV(h), subject to the constraint $h \leq h_{OS}$: The theoretical
properties of these random variables have been studied by Scott and Terrell (1987),
who showed that $\hat{h}_{BCV}$ converges to h^* with a relative error of $O(n^{-1/6})$. This rate
of (relative) convergence is especially slow. However, the algorithm seems useful in
practice, especially for large samples. Examples are given in Section 3.3.2.4.

3.3.2.2 Unbiased CV

Rudemo (1982) undertook the more ambitious task of
attempting to minimize the actual L_2 ISE error of a histogram for an individual
sample. Expanding yields

$$\text{ISE}(h) = \int \left[\hat{f}(x) - f(x) \right]^2 dx = R(\hat{f}) - 2 \int \hat{f}(x)f(x)\,dx + R(f).$$

Rudemo noted that the minimizer of ISE(h) did not depend on the unknown quantity $R(f)$ and that $R(\hat{f})$ could be computed easily in closed form. Furthermore, he
observed that the second integral could be written as $E[\hat{f}(X)]$, where the expectation
is with respect to the point of evaluation and not over the random sample $x_1, \ldots, x_n$.
CV methodology suggests removing one data point and using the remaining $n - 1$
points to construct an estimator of $E[\hat{f}(X)]$. The nth data point is then evaluated in
the estimate for the purpose of determining the quality of fit. This step is repeated n
times, once for each data point, and the results averaged.

Rudemo considered the histogram $\hat{f}_{-i}(x)$, which is the histogram based on the
$n - 1$ points in the sample excluding x_i. It is easy to check that the observable random variable $\hat{f}_{-i}(X_i)$ has the same mean as the unobservable random variable $E\hat{f}(X)$,
although based on a sample of $n - 1$ rather than n points. By leaving out each of the n
data points one at a time, Rudemo obtained a stable estimate of $E\hat{f}(X)$ by averaging
over the n cases. With this estimate, he proposed minimizing the least-squares CV or
unbiased CV (UCV) function

$$\text{UCV}(h) = R(\hat{f}) - \frac{2}{n} \sum_{i=1}^{n} \hat{f}_{-i}(x_i). \tag{3.52}$$

For the histogram, the terms in the UCV are easily evaluated. For example, $\hat{f}_{-i}(x_i) = (\nu_k - 1)/[(n - 1)h]$ if $x_i \in B_k$. The final computational formula for the UCV is

$$\boxed{\text{UCV}(h) = \frac{2}{(n-1)h} - \frac{n+1}{n^2(n-1)h} \sum_k \nu_k^2.} \tag{3.53}$$

See Problem 3.17. The similarity of the UCV and BCV formulas is striking and par-
tially justifies the BCV label, since BCV is based on AMISE rather than MISE.
Methods relying on AMISE formula are often referred to as *plug-in* (PI) methods. Scott and Terrell (1987) also studied the asymptotic distributional properties of
the UCV and $\hat{h}_{UCV}$ random variables. They showed that the UCV and BCV bandwidths were consistent, but that the convergence was slow. Specifically, $\sigma_{h_{CV}}/h_{CV} =$

$O(n^{-1/6})$. However, Hall and Marron (1987a, b) showed that this is the same rate of convergence of the smoothing parameter that actually minimizes ISE. Hall and Marron (1987c) also studied the estimation of functionals like $R(f')$.

3.3.2.3 End Problems with BCV and UCV

For a fixed sample, BCV(h) and UCV(h) exhibit behavior not found in MISE(h) for large and small h, respectively. For example, as $h \to \infty$, all bins except one or two will be empty. It is easy to see that

$$\lim_{h \to \infty} \text{BCV}(h) = \lim_{h \to \infty} \text{UCV}(h) = 0. \qquad (3.54)$$

Clearly, ISE(h) $\to R(f)$ as $h \to \infty$. Hence, for UCV(h), the limit in (3.54) is correct since the term $R(f)$ is omitted in its definition. However, for BCV(h), which is asymptotically nonnegative near h^*, the limiting value of 0 means that the global minimizer is actually at $h = \infty$. In practice, $\hat{h}_{\text{BCV}}$ is chosen to be a local minimizer constrained to be less than the oversmoothed bandwidth. If there is no local minimizer within that region, the oversmoothed bandwidth itself is the *constrained* minimizer.

On the other hand, as $h \to 0$, the bin counts should all be zero or one, assuming the data are continuous (no ties). In that case, both BCV(h) and UCV(h) approximately equal $1/(nh)$, the correct expression for the integrated variance (IV). Observe that IV(h) $\to +\infty$ as $h \to 0$. However, if the data contain many ties, perhaps the result of few significant digits or rounding, UCV(h) can diverge to $-\infty$. For example, suppose the n data points consist of n/m distinct values, each replicated m times, then UCV(h) is approximately $(2 - m)/(nh)$, which diverges to $-\infty$ (as opposed to $+\infty$) if $m > 2$ as $h \to 0$. Thus the global minimizer of UCV(h) occurs at $h = 0$ in that case. In such situations, the divergence may occur only in a small neighborhood around $h = 0$, and $\hat{h}_{\text{UCV}}$ is again defined to be the appropriate local minimum, if it exists, closest to the oversmoothed bandwidth; see, however, Figure 3.19.

3.3.2.4 Applications

Figure 3.17 displays the BCV and UCV curves for a $N(0, 1)$ sample of 1000 points. The vertical scales are comparable, since UCV is shifted by a fixed constant, $R(f)$. Thus the UCV function turns out to be much noisier than the BCV function; however, for other more difficult data, the minimizer of the BCV function may be quite poor—typically, it is biased towards larger bin widths. The UCV function may be noisy, but its minimizer is correct on average. For these data, the two minimizers are $h_{\text{BCV}} = 0.32$ and $h_{\text{UCV}} = 0.36$, while the oversmoothed bin width is $h_{\text{OS}} = 0.37$, using the variance rule in Equation (3.44). Choosing h_{UCV} from the many strong local minimizers in the UCV function is a practical problem.

Taken together, oversmoothing, BCV, and UCV are a powerful set of tools for choosing a bin width. Specifically, the UCV and BCV curves should be plotted on comparable scales, marking the location of the upper bound given by the oversmoothed bandwidth. A log-log plot is recommended, but UCV is negative, so only $\log(h)$ is plotted. The minimizer is located and the *quality* of that CV is evaluated subjectively by examining how well-articulated the minimizer appears. Watch for obvious failures: no local minimizer in BCV(h) or the degenerate $h = 0$ UCV solution. If there is no agreement among the proposed solutions, examine plots of the

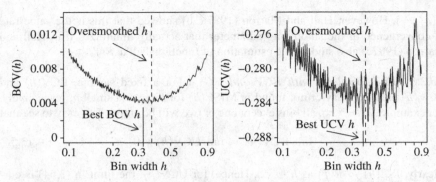

FIGURE 3.17   BCV and UCV functions for a $N(0,1)$ sample of 1000 points.

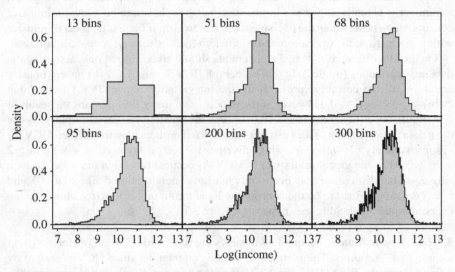

FIGURE 3.18   Six histograms of the 1983 German income sample of 5625 households.

corresponding histograms and choose the one that exhibits a small amount of local noise, particularly near the peaks.

The second example focuses on 1983 income sampled from 5626 German households (DIW, 1983). The data were transformed to a log scale, as is often done by economists, in order to reduce the MISE penalty due to extreme skewness. Four histograms of these transformed data points are plotted in Figure 3.18. For more details, see Scott and Schmidt (1988, 1989).

Which histogram is best? Sturges' rule suggests only 13 bins, which is clearly oversmoothed. The oversmoothed rule (maximum bin width) gives $h = 0.160$, corresponding to 68 bins over the sample range. The BCV and UCV functions are plotted in Figure 3.19. The minimizer of the BCV criterion occurs at $h = 0.115$, suggesting 95 bins. The UCV function appears to have its global minimum at $h = 0$, but has no

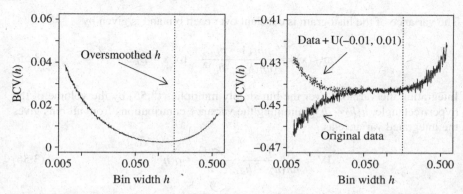

FIGURE 3.19 BCV and UCV functions of the 1983 German income sample of 5625 households. A second UCV curve is shown for the blurred logarithmic data.

strong local minimum in the region of interest. Indeed, examination of the original data revealed only 3322 unique incomes, so the duplicated data problem is clearly in effect. It was determined empirically that this effect could be eliminated by adding some small uniform noise to the logarithmic values, as shown in Figure 3.19. However, a wide range of possible minima still exists. Even the small bin width $h = 0.055$ seems feasible, which corresponds to 200 bins. When $U(-0.005, 0.005)$ noise was added, the UCV curve was horizontal, splitting the two curves shown in the figure. For large amounts of additive noise, the UCV curve was visually affected for bin widths near h_{OS}. Thus there is a bit of subjectiveness in the process of removing the duplicate data.

The lesson is that all three algorithms should be examined simultaneously even with very large samples. This topic will be continued in Chapter 6.

3.4 L_2 THEORY FOR MULTIVARIATE HISTOGRAMS

The derivation of the MISE for the multivariate histogram is only slightly more complicated than in the univariate case. Given a sample from $f(\mathbf{x})$, where $\mathbf{x} \in \Re^d$, the histogram is determined by a partition of the space. Consider a regular partition by hyper-rectangles of size $h_1 \times h_2 \times \cdots \times h_d$. Choosing hypercubes as bins would be sufficient if the data were properly scaled, but in general that will not be the case. Further improvements may be obtained by considering nonregular or rotated bins (see Scott (1988a) and Hüsemann and Terrell (1991)).

Consider a generic hyper-rectangular bin labeled B_k containing ν_k points. As usual, $\sum_k \nu_k = n$. Then

$$\hat{f}(\mathbf{x}) = \frac{\nu_k}{nh_1h_2\cdots h_d} \quad \text{for } \mathbf{x} \in B_k.$$

The variance of the histogram is constant over each bin and is given by

$$\text{Var}\,\hat{f}(\mathbf{x}) = \frac{np_k(1-p_k)}{(nh_1h_2\cdots h_d)^2} \quad \text{for } \mathbf{x} \in B_k. \tag{3.55}$$

Integrating the variance over the bin simply multiplies (3.55) by the volume of the hyper-rectangle, $h_1h_2\cdots h_d$. Summing the variance contributions from all bins gives the integrated variance as

$$\text{IV} = \frac{1}{nh_1h_2\cdots h_d} - \frac{R(f)}{n} + o(n^{-1}), \tag{3.56}$$

where $R(f) = \int_{\mathfrak{R}^d} f(\mathbf{x})^2 d\mathbf{x}$. The first term in (3.56) is exact, since $\sum_k p_k = 1$; the remainder is obtained by noting that $p_k^2 \approx [h_1h_2\cdots h_d f(\xi_k)]^2$, where $\xi_k \in B_k$, and using standard multivariate Riemannian integration approximations.

The following outline for the bias calculation can be made rigorous. Consider the bin B_0 centered on the origin $\mathbf{x} = \mathbf{0}_d$. Now

$$f(\mathbf{x}) = f(\mathbf{0}) + \sum_{i=1}^{d} x_i f_i(\mathbf{0}) + \frac{1}{2}\sum_{i=1}^{d}\sum_{j=1}^{d} x_i x_j f_{ij}(\mathbf{0}) + O(h^3),$$

where $f_i(\mathbf{x}) = \partial f(\mathbf{x})/\partial x_i$ and $f_{ij}(\mathbf{x}) = \partial^2 f(\mathbf{x})/\partial x_i \partial x_j$. Hence,

$$p_0 = \int_{-h_d/2}^{h_d/2} \cdots \int_{-h_1/2}^{h_1/2} f(\mathbf{x})d\mathbf{x} = h_1h_2\cdots h_d f(\mathbf{0}) + O(h^{d+2}), \tag{3.57}$$

where $h = \min_i(h_i)$. Computing the Bias$\{\hat{f}(x)\}$ gives

$$\text{E}\hat{f}(\mathbf{x}) - f(\mathbf{x}) = \frac{p_0}{h_1h_2\cdots h_d} - f(\mathbf{x}) = -\sum_{i=1}^{d} x_i f_i(\mathbf{0}) + O(h^2). \tag{3.58}$$

Squaring and integrating over B_0, the integrated squared bias for the bin is

$$h_1h_2\cdots h_d \left[\sum_{i=1}^{d} \frac{1}{12} h_i^2 f_i(\mathbf{0})^2 + O(h^4) \right]. \tag{3.59}$$

A similar expression holds for all other bins, with the origin $\mathbf{x} = \mathbf{0}$ replaced by the respective multivariate bin centers. Summing over all bins yields

$$\text{AISB}(\mathbf{h}) = \frac{1}{12}\sum_{i=1}^{d} h_i^2 R(f_i). \tag{3.60}$$

These approximations are collected in a theorem.

Theorem 3.5: *For a sufficiently smooth density function $f(\mathbf{x})$, the multivariate MISE is asymptotically*

$$\text{AMISE}(\mathbf{h}) = \text{AIV} + \text{AISB} = \frac{1}{nh_1 h_2 \cdots h_d} + \frac{1}{12} \sum_{i=1}^{d} h_i^2 R(f_i). \tag{3.61}$$

The asymptotically optimal bin widths, h_k^*, and resulting AMISE* are

$$h_k^* = R(f_k)^{-1/2} \left(6 \prod_{i=1}^{d} R(f_i)^{1/2} \right)^{1/(2+d)} n^{-1/(2+d)}, \tag{3.62}$$

$$\text{AMISE}^* = \frac{d+2}{2} 6^{-d/(2+d)} \left(\prod_{i=1}^{d} R(f_i) \right)^{1/(2+d)} n^{-2/(2+d)}. \tag{3.63}$$

EXAMPLE: Suppose that $X \sim N(\mu, \Sigma), \Sigma = \text{Diag}(\sigma_1^2, \sigma_2^2, \ldots, \sigma_d^2)$. Then, letting $c_d = \sigma_1 \sigma_2 \cdots \sigma_d$, we have

$$R(f_i) = (2^{d+1} \pi^{d/2} \sigma_i^2 c_d)^{-1} \tag{3.64}$$

and

$$h_k^* = 2 \cdot 3^{1/(2+d)} \pi^{d/(4+2d)} \sigma_k n^{-1/(2+d)}$$
$$\text{AMISE}^* = (2+d) 2^{-(1+d)} 3^{-d/(2+d)} \pi^{-d^2/(4+2d)} c_d^{-1} n^{-2/(2+d)}. \tag{3.65}$$

Note that the constant in the bandwidth increases quickly from 3.4908 in one dimension to the limiting value of $2\sqrt{\pi} = 3.5449$ as $d \to \infty$. Hence, a very useful formula to memorize is

Normal reference rule: $\qquad h_k^* \approx 3.5 \, \sigma_k n^{-1/(2+d)}$. $\qquad (3.66)$

3.4.1 Curse of Dimensionality

Bellman (1961) first coined the phrase "curse of dimensionality" to describe the exponential growth in combinatorial optimization as the dimension increases. Here, it is the number of bins that grows exponentially as the dimension increases. It is important to try to see how histograms are affected by this phenomenon.

A relatively simple density to estimate is the multivariate normal with $\Sigma = I_d$. Following up on the results in the earlier example, we have Table 3.6. Clearly, the rate of

TABLE 3.6   Example of asymptotically optimal bin widths and errors for $f = N(0_d, I_d)$

Dimension d	h_d^*	AMISE_d^*
1	$3.491n^{-1/3}$	$0.430n^{-2/3}$
2	$3.504n^{-1/4}$	$0.163n^{-2/4}$
3	$3.512n^{-1/5}$	$0.058n^{-2/5}$
4	$3.518n^{-1/6}$	$0.020n^{-2/6}$

TABLE 3.7   Equivalent sample sizes across dimensions for the multivariate normal density, based on epanechnikov's criterion

d	Equivalent sample sizes (read down each column)		
1	10	100	1,000
2	39	838	18,053
3	172	7,967	369,806
4	838	83,776	8,377,580
5	4,446	957,834	206,359,075

decrease of the MISE with respect to the sample size degrades rapidly as the dimension increases compared to the ideal parametric rate $O(n^{-1})$. A strong advantage of parametric modeling is that the rate of decrease of MISE is independent of dimension. However, several data analysts have used histograms when classifying quadrivariate data from remote sensing with satisfactory results (Wharton, 1983). One possible optimistic observation is that the constants in the AMISE in Table 3.6 are also decreasing as the dimension increases. Unfortunately, MISE is not a dimensionless quantity and, hence, these coefficients are not directly comparable.

Epanechnikov (1969) described a procedure for comparing histogram errors and performance across dimensions. He considered one possible dimensionless rescaling of MISE:

$$\epsilon_d \equiv \frac{\text{MISE}}{R(f)} \quad \left\{ \approx \frac{2+d}{2} 3^{\frac{-d}{2+d}} \pi^{\frac{d}{2+d}} n^{-\frac{2}{2+d}} \quad \text{when } f = N(\mathbf{0}_d, I_d) \right\}. \tag{3.67}$$

Again, for the normal case, a table of equivalent sample sizes may be computed (see Table 3.7). This table graphically illustrates the curse of dimensionality and the intuition that density estimation in more than two or three dimensions will not work. This conclusion is, however, much too pessimistic, and other evidence will be presented in Section 7.2.

Another approach towards understanding this problem is to count bins in a region of interest. For this same normal example, consider the region of interest to be a

TABLE 3.8 Approximate number of bins in the region of
interest for a multivariate histogram of 1000 normal points

d	h_d^*	r_d	Number of bins
1	0.35	2.57	15
2	0.62	3.03	75
3	0.88	3.37	235
4	1.11	3.64	573
5	1.30	3.88	1254

sphere with radius r_d, containing 99% of the probability mass. The radius r_d is the solution to

$$\text{Prob}\left(\sum_{i=1}^{d} Z_i^2 \leq r_d^2\right) = 0.99 \quad \Rightarrow \quad r_d = \sqrt{\chi_{.99}^2(d)}, \tag{3.68}$$

that is, the square root of the appropriate chi-squared quantile. The volume of the sphere is given by Equation (1.3). The number of hypercube bins over the data in this sphere is approximately equal to the volume of the sphere divided by the bin volume, h^d, plus the bins covering the 1% of points outside the sphere. Table 3.8 illustrates these calculations for a sample of size 1000. In five dimensions, the optimally smoothed histogram has approximately 1250 bins in the region of interest. Since there are only 1000 points, it is clear that most of these bins will be empty and that the histogram will be rather rough. Scott and Thompson (1983) have called this the "empty space phenomenon." As the dimension grows, the histogram provides reasonable estimates only near the mode and away from the tails.

3.4.2 A Special Case: $d = 2$ with Nonzero Correlation

The effects, if any, of correlation have been ignored up to this point. Consider the simple bivariate normal case, $f(x_1, x_2) = N(\mu_1, \mu_2, \sigma_1^2, \sigma_2^2, \rho)$, for which $R(f_1) = \left[8\pi(1-\rho^2)^{3/2}\sigma_1^3\sigma_2\right]^{-1}$ and $R(f_2) = \left[8\pi(1-\rho^2)^{3/2}\sigma_1\sigma_2^3\right]^{-1}$. From Theorem 3.5,

$$h_i^* = 3.504\,\sigma_i\,(1-\rho^2)^{3/8}\,n^{-1/4}$$

$$\text{AMISE}^* = \frac{0.122}{\sigma_1\sigma_2}(1-\rho^2)^{-3/4}n^{-1/2}. \tag{3.69}$$

The effect of the correlation ρ is to introduce powers of the quantity $(1-\rho^2)$ into the equations. Thus, if the data are not independent but are clustering along a line, smaller bin widths are required to "track" this feature. If the density is degenerate (i.e., $\rho = \pm 1$), then the MISE blows up. This result also indicates that if the data fall onto any lower dimensional manifold, a histogram will never be consistent! This inconsistency is a second and perhaps more important aspect of the "curse of dimensionality" as it applies to statistics. Therefore, a significant portion of the effort in good data analysis should be to check the (local) rank of the data and identify nonlinear structure falling in lower dimensions (see Chapter 7).

3.4.3   Optimal Regular Bivariate Meshes

Consider tiling the plane with regular polygons as bins for a bivariate histogram. Other than squares, there are only two other regular polygon meshes: equilateral triangles and hexagons (see Figure 3.20). Nonregular tiling of the plane is of course feasible, but is of secondary interest. Scott (1988a) compared the bivariate AMISE for these three regular tile shapes. If the individual bins are parameterized so that each has the same area h^2, then Scott showed that

$$\text{AMISE}(h) = \frac{1}{nh^2} + ch^2 \left[R(f_x) + R(f_y) \right], \tag{3.70}$$

where $f_x = \partial f(x,y)/\partial x$ and $f_y = \partial f(x,y)/\partial y$, and

$$c = \left[\frac{1}{12}, \frac{1}{6\sqrt{3}}, \frac{5}{36\sqrt{3}} \right] = \left[\frac{1}{12}, \frac{1}{10.39}, \frac{1}{12.47} \right], \tag{3.71}$$

for the square, triangle, and hexagonal tiles, respectively (see Problem 3.25). Therefore, hexagonal bins are in fact the best, but only marginally. The triangular meshes are quite inferior. Scott also considered meshes composed of right triangles. He noted that these were inferior to equilateral triangles and that the ISB sometimes included cross-product terms such as $\int \int f_x(x,y)f_y(x,y)\,dx\,dy$.

Carr et al. (1987) suggested using hexagonal bins for a different reason. They were evaluating plotting bivariate glyphs based on counts in bivariate bins. Using square bins resulted in a visually distracting alignment of the glyphs in the vertical and horizontal directions. This distraction was virtually eliminated when the glyphs were located according to a hexagonal mesh. The elimination is readily apparent in the examples given in Figure 3.21. Note that the size of each hexagon glyph is drawn so that its area is proportional to the number of observations in that bin. The authors note the method is especially useful for very large datasets. Observe that while the MISE

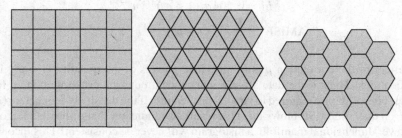

FIGURE 3.20   The three possible regular polygon meshes for bivariate histograms.

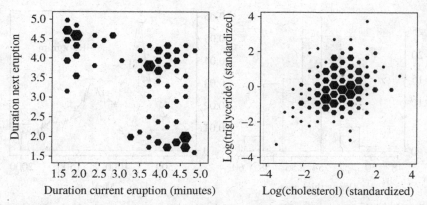

FIGURE 3.21 Hexagon glyphs for the lagged Old Faithful duration data and for the lipid dataset for 320 males with heart disease.

criterion does not strongly differentiate between square or hexagonal bins, subjective factors may strongly influence the equation one way or the other.

3.5 MODES AND BUMPS IN A HISTOGRAM

Many of the examples to this point have shown multimodal data. Good and Gaskins (1980) discussed advanced methods for finding modes and bumps in data. In 1-D, a *mode* is a set (a collection of points) where $f'(x) = 0$ and $f''(x) < 0$. A *bump* is a set (a collection of disjoint intervals) where $f''(x) < 0$. Thus bump-hunting is more general than estimating the mode. Figure 3.22 shows a histogram which contains one mode and one bump. A bump does not necessarily contain a mode, although a mode is always located in a bump. In practical bump-hunting situations, the density is often thought to be a mixture of several component densities (see Izenman and Sommer (1988), for example). The mixture problem is even more general than bump-hunting, since a normal mixture density such as

$$f(x) = \sum_{i=1}^{q} w_i \phi(x|\mu_i, \sigma_i^2) \quad \text{where} \sum_{i=1}^{q} w_i = 1, \tag{3.72}$$

need not exhibit modes or bumps (see Problem 3.26). However, the estimation of the parameters $\{q, w_i, \mu_i, \sigma_i^2\}$ is often ill-conditioned, especially if q is larger than the true number of densities (see Day (1969), Everitt and Hand (1981), Hathaway (1985), and Redner and Walker (1984)).

With histogram data, it is natural to examine plots of (standardized) first and second differences for evidence of modes and bumps, respectively:

$$\frac{\nu_{k+1} - \nu_k}{\sqrt{\nu_{k+1} + \nu_k}} \quad \text{and} \quad \frac{\nu_{k+1} - 2\nu_k + \nu_{k-1}}{\sqrt{\nu_{k+1} + 4\nu_k + \nu_{k-1}}},$$

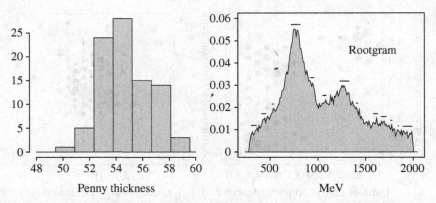

FIGURE 3.22 Histogram of US penny thickness with one mode and a bump. In the right frame, the histogram of the LRL dataset with $h = 10$ MeV is plotted on a square root scale; the 13 bumps found by Good and Gaskins are indicated by the line segments above the histogram.

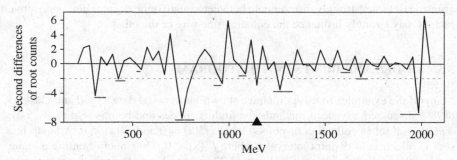

FIGURE 3.23 Second differences of square root of LRL histogram dataset with bin width of 30 MeV. The 13 bumps found by Good and Gaskins are indicated as before. The dashed line indicates the approximate 5% cutoff level for a bump to be significant.

since the bin counts can be approximately modeled as independent Poisson random variables. For example, $\text{Var}\,(\nu_{k+1} - \nu_k) \approx \nu_{k+1} + \nu_k$. Alternatively, we may focus on finite differences of the rootgram, using the fact that the square root is the variance stabilizing transformation for Poisson random variables, with variance of $1/4$:

$$\sqrt{\nu_{k+1}} - \sqrt{\nu_k} \quad \text{and} \quad \sqrt{\nu_{k+1}} - 2\sqrt{\nu_k} + \sqrt{\nu_{k-1}}; \qquad (3.73)$$

which have approximate variances $1/2 = \frac{1}{4} + \frac{1}{4}$ and $3/2 = \frac{1}{4} + (-2)^2 \frac{1}{4} + \frac{1}{4}$, respectively, assuming independence. Good and Gaskins found 13 bumps in their LRL dataset (see Figure 3.22). A plot of the second differences of the root bin counts with $h = 30$ MeV is shown in Figure 3.23. Under the null hypothesis that there is no bump, that is, the second difference is nonnegative, the one-sided 5% test level will be at $-1.645\sqrt{3/2} = -2.015$, which is indicated by the dashed line in Figure 3.23. Recall that these data were collected in bins of width 10 MeV. For narrower bin widths, the figure was too noisy. Both CV criteria give $h = 10$ MeV as the best bin width. Most of the bumps found by Good and Gaskins can be seen in the plot. However, a few bumps are not readily apparent. One large negative portion of the curve near $x = 1145$ MeV (indicated by the solid triangle) seems to have been missed.

3.5.1 Properties of Histogram "Modes"

Suppose, without loss of generality, that the true density f has a mode at 0. Consider an equally spaced histogram with bin B_0 centered on $x = 0$. The bin count $\nu_0 \sim B(n, p_0) \approx P(\lambda_0)$, which is the Poisson density with $\lambda_0 = np_0$. Asymptotically, the adjacent bin counts, $(\nu_{-k}, \ldots, \nu_0, \ldots, \nu_k)$, are independent and normally distributed with $\nu_i \approx N(\lambda_i, \lambda_i)$. Consider the following question: What is the probability that the histogram will have a sample mode in bin B_0?

Conditioning on the observed bin count in B_0, we obtain

$$\Pr\left(\nu_0 = \arg\max_{|j| \leq k} \nu_j\right) = \sum_{x_0} \Pr\left(\nu_0 = \arg\max_{|j| \leq k} \nu_j \,\Big|\, \nu_0 = x_0\right) f_{\nu_0}(x_0)$$

$$= \sum_{x_0} \Pr(\nu_j < x_0; |j| \leq k, j \neq 0) f_{\nu_0}(x_0)$$

$$\approx \int_{x_0} \prod_{\substack{j=-k \\ j \neq 0}}^{k} \Phi\left(\frac{x_0 - \lambda_j}{\sqrt{\lambda_j}}\right) \phi\left(\frac{x_0 - \lambda_0}{\sqrt{\lambda_0}}\right) \frac{1}{\sqrt{\lambda_0}} \, dx_0, \tag{3.74}$$

using the normal approximation for $\Pr(\nu_j < x_0)$, a normal approximation for the discrete Poisson density $f_{\nu_0}(x_0)$, and replacing the sum by an integral. Now $\lambda_j = np_j$. Noting that $f_0' \equiv f'(0) = 0$, we have as an approximation for p_j:

$$p_j = \int_{(j-1/2)h}^{(j+1/2)h} \left[f_0 + \frac{x^2}{2} f_0'' + \cdots\right] dx = hf_0 + \frac{h^3}{6}\left(3j^2 + \frac{1}{4}\right) f_0'' + \cdots. \tag{3.75}$$

Making the change of variables $y = (x_0 - \lambda_0)/\sqrt{\lambda_0}$, it follows that (3.74) equals

$$\Pr\left(\nu_0 = \arg\max_{|j| \leq k} \nu_j\right) \approx \int_y \prod_{\substack{j=-k \\ j \neq 0}}^{k} \Phi\left(\frac{\lambda_0 - \lambda_j + y\sqrt{\lambda_0}}{\sqrt{\lambda_j}}\right) \phi(y) \, dy$$

$$\approx \int_y \prod_{\substack{j=-k \\ j \neq 0}}^{k} \Phi\left(y - \frac{j^2 h^{5/2}\sqrt{n}}{2} \frac{f''(0)}{\sqrt{f(0)}} + \cdots\right) \phi(y) \, dy. \tag{3.76}$$

In the case of an optimal histogram, $h = cn^{-1/3}$, so that

$$\lim_{n \to \infty} [h^{5/2}\sqrt{n}] = O(n^{-1/3}) \to 0;$$

hence,

$$\lim_{n \to \infty} \Pr\left(\nu_0 = \arg\max_{|j| \leq k} \nu_j\right) = \int_y \Phi(y)^{2k} \phi(y) dy = \frac{1}{2k+1} \quad (!!) \qquad (3.77)$$

from Equation (3.76). The probability $1/(2k+1)$ is far from a more desirable value close to 1. The correct interpretation of this result is that optimal MISE-smoothing results in bins with widths too narrow to estimate modes. In fact, in a neighborhood of the mode, the density looks flat (and the bins have essentially the same expected height to first order), so that each of the $2k+1$ bins is equally likely to be the *sample mode*. In a simulation of optimally smoothed normal data with $n = 256,000$, the average number of sample modes in the histogram was 20! Now admittedly, most of these "modes" were just small aberrations, but the result is most unexpected (see Figure 3.24).

Next, suppose that the origin is not a mode; then $f'(0) \neq 0$. A similar analysis shows the probability that bin B_0 is a mode (which it is not!) converges to a fixed nonzero probability as $n \to \infty$. The probability is smaller the larger the magnitude of $f'(0)$. If wider bins are used, then the probabilities can be made to converge to the desired values of 1 and 0, respectively. An interesting special case for the limit in Equation (3.77) follows from the use of a much larger bin width $h = cn^{-1/5}$. This choice will be examined in Chapter 4, which deals with the closely related frequency polygon density estimator.

3.5.2   Noise in Optimal Histograms

Consider again the histograms of a million normal points shown in Figure 3.7. Plotting a rootgram, which is the variance-stabilized histogram plotted on a square root scale, for the cases $h = h^*$ and $h = h^*/2$ clearly displays the many local false modes

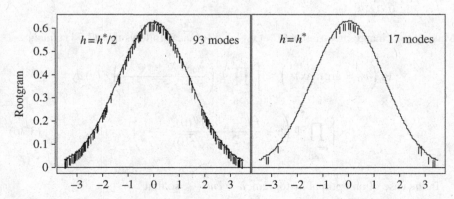

FIGURE 3.24   Rootgram of two histograms of a million normal points $(h^*/2, h^*)$.

(see Figure 3.24). The dominant mode at zero is not, however, lost among all the noisy modes.

The bottom line is that histograms which are "optimal" with respect to MISE may not be "optimal" for other purposes. This specific insight concerning modes should not be surprising given the fixed asymptotic noise in $R(\hat{f}')$ encountered in the biased CV derivation. But this general feature of nonparametric procedures is quite different than in the parametric setting, in which a maximum likelihood density estimate will be optimal for virtually any specific application. In the nonparametric setting, the "optimal" calibration will depend on the purpose. For small samples, the noise in the histogram is seldom attributed to the correct source and, hence, is not well understood. In any case, optimally smooth histograms do not provide a powerful tool for bump-hunting (look again at the LRL example). In fact, the second differences of an optimally smoothed histogram diverge from the true second derivative. This result will be easier to see later with estimators that are themselves differentiable.

3.5.3 Optimal Histogram Bandwidths for Modes

In this section, we search for a bandwidth that provides better estimates of the derivative of the unknown density. Using the bin notation as in Figure 3.2, a natural estimate of the first derivative $f'(x)$ is the first finite difference

$$\hat{f}'(x) = \frac{\hat{f}_0 - \hat{f}_{-1}}{h} = \frac{\nu_0 - \nu_{-1}}{nh^2} \qquad -h/2 < x < h/2.$$

A Taylor's series of $f(x)$ around $x = 0$ leads to the approximations $hf(0) \pm h^2 f'(0)/2 + h^3 f''(0)/6 \pm h^4 f'''(0)/24 + O(h^5)$ for p_0 and p_{-1}. Since $E\nu_i = np_i$,

$$E\hat{f}'(x) = \frac{np_0 - np_{-1}}{nh^2} = f'(0) + \frac{1}{12}h^2 f'''(0) + \cdots, \qquad -h/2 < x < h/2.$$

Now $f'(x) = f'(0) + xf''(0) + x^2 f'''(0)/2 + \cdots$, so the bias is $-xf''(0) + O(h^2)$. Then the integrated squared bias over the bin is $f''(0)^2 h^2/12 \times h$. Replacing 0 with kh in the kth bin and summing, we obtain AISB $= h^2 R(f'')/12$. Similarly,

$$\begin{aligned}
\text{Var}\hat{f}'(x) &= \text{Var}\left(\frac{\nu_0 - \nu_{-1}}{nh^2}\right) = \frac{\text{Var}\,\nu_0 - 2\,\text{Cov}\,(\nu_0, \nu_{-1}) + \text{Var}\,\nu_{-1}}{n^2 h^4} \\
&= \frac{p_0(1 - p_0) - 2(-p_0 p_{-1}) + p_{-1}(1 - p_{-1})}{nh^4} = \frac{2f(0)}{nh^3} + \cdots.
\end{aligned}$$

Integrating over the bin multiplies by h and summing over all the bins gives AIV $= 2/(nh^3)$, since $\int f(x)dx = 1$. Thus we have shown

Theorem 3.6: *Suppose that f has an absolutely continuous second derivative that is square-integrable. Then, the asymptotic MISE of $\hat{f}'(x)$ is*

$$\text{AMISE}_{\hat{f}'}(h) = \frac{2}{nh^3} + \frac{1}{12}h^2 R(f'''); \quad therefore,$$

$$h^* = 6^{2/5} R(f''')^{-1/5} n^{-1/5}$$

$$\text{AMISE}^* = 5 \cdot 6^{-6/5} R(f''')^{3/5} n^{-2/5}.$$

If $f \sim N(\mu, \sigma^2)$, then $h^* = 2 \cdot 3^{1/5} \pi^{1/10} \sigma n^{-1/5} \approx 2.8 \sigma n^{-1/5}$.

In Figures 3.25 and 3.26, portions of ten histograms computed from very large samples from $N(0,1)$ are displayed using the optimal bin width for the density versus

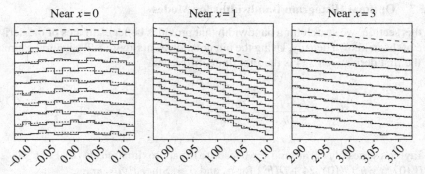

FIGURE 3.25 For 10 $N(0,1)$ samples of size $n = 10^7$, (vertically shifted) blowups of portions of the 10 histograms using $h^* = 0.01625$ in the vicinity of $x = 0, 1,$ and 3. Each histogram snippet shows the bin of interest and seven bins on either side.

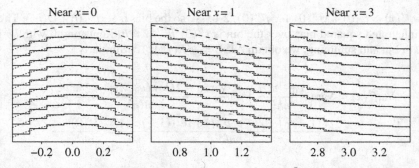

FIGURE 3.26 For the same 10 $N(0,1)$ samples of size $n = 10^7$, (vertically shifted) blowups of portions of the 10 histograms using $h^* = 0.1115$ in the vicinity of $x = 0, 1,$ and 3. Each histogram snippet shows the bin of interest and three bins on either side.

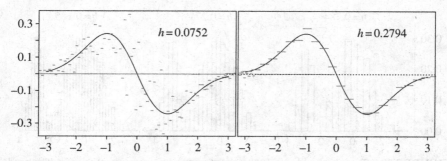

FIGURE 3.27 For a standard normal sample $n = 10^5$ points, comparison of the histogram "derivative" estimates using the optimal density and derivative bandwidths.

the optimal bin width for the derivative of the density. Clearly, the latter bandwidth gives much more stable estimates of the slopes in the true density (although at the cost of higher bias). Where the density is nearly flat, many false sample modes are apparent in Figure 3.26, while all the histograms are monotone decreasing in a neighborhood of $x = 1$. Even with sample sizes of this magnitude, the bandwidth problem should always be regarded as potentially important.

In Figure 3.27, the histogram "derivative" estimates are shown for the entire $N(0, 1)$ density for a sample of size $n = 10^5$ using the optimal bandwidths for the density and derivative. Clearly, the noise is greatly reduced using the wider bandwidth. Which bandwidth should be chosen depends on the purpose of the analysis.

3.5.4 A Useful Bimodal Mixture Density

The study of the performance of nonparametric estimators has been evaluated using a range of true densities, for example, the 15 normal mixtures proposed by Marron and Wand (1992). However, one may simplify this task by focusing on only two densities: a standard normal density and a particular mixture of two normals given by

$$f_M(x) = \frac{3}{4}\phi(x|0, 1) + \frac{1}{4}\phi(x|0, 1/3^2), \tag{3.78}$$

which is a simplified version of MW#14 (see Figure 3.28). This density embodies both skewness and different amounts of curvature. However, the value of the density at the two modes is virtually identical. Later on, this test density will prove a difficult challenge in many instances.

In Figure 3.28, three histograms for a sample of $n = 1000$ points from the mixture are shown. The first uses the optimal bandwidth for the left component (and hence the right component is oversmoothed). The second uses the optimal bandwidth for the mixture (and hence does slightly worse for both). Finally, the third histogram uses the optimal bandwidth for the right component (and hence severely undersmooths the left component). We shall return to the this density from time to time.

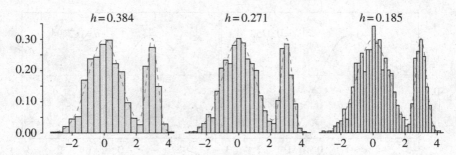

FIGURE 3.28   Three histograms of 1000 points from the two-component mixture. The bandwidths (from left to right) are optimal for the left component, the mixture, and the right component, respectively.

3.6   OTHER ERROR CRITERIA: $L_1, L_4, L_6, L_8,$ AND L_∞

3.6.1   Optimal L_1 Histograms

Recall the several theoretically appealing aspects of absolute rather than squared error discussed in Section 2.3.2.2. The primary difficulty with absolute error is that it lacks an exact variance–bias decomposition [although Hall and Wand (1988a) have shown how to directly estimate the total error asymptotically]. Hjort (1986) has shown that

$$\mathrm{E} \int |\hat{f} - f| \leq \mathrm{E} \int |\hat{f} - \mathrm{E}\hat{f}| + \int |\mathrm{E}\hat{f} - f|$$
$$\approx \sqrt{\frac{2}{\pi n h}} \int \sqrt{f} + \frac{1}{4} h \int |f'|.$$

Therefore, minimizing this asymptotic upper bound for MIAE leads to

$$h^* = 2\pi^{-1/3} \left[\int f^{1/2} \div \int |f'| \right]^{2/3} n^{-1/3}$$
$$= 2.717 \sigma n^{-1/3} \quad \text{for } N(\mu, \sigma^2) \text{ data.} \tag{3.79}$$

However, numerical simulations with normal data show that in fact $h^* \approx 3.37 \sigma n^{-1/3}$, which is 3.5% *narrower* than the optimal L_2 bandwidth. Minimizing upper bounds to MIAE leads to bin widths that can be either larger or smaller than h^*. Therefore, contrary to intuition, absolute error does not always provide wider bins in the tails. What would happen with optimal adaptive L_1 meshes is unknown? Recently, Hall and Wand (1988a) have examined the optimal bandwidths for many densities, and found that with some heavier tailed densities, the L_1 bandwidth can be wider than the optimal L_2 bandwidth.

TABLE 3.9 Optimal bandwidths for $N(0,1)$ data with different L_p criteria

Error criterion	Optimal bin width	Expected error
L_1 (upper bound)	$2.72n^{-1/3}$	$1.6258n^{-1/3}$
L_1 (numerical)	$3.37n^{-1/3}$	$1.1896n^{-1/3}$
L_2	$3.49n^{-1/3}$	$(0.6555n^{-1/3})^2$
L_4	$3.78n^{-1/3}$	$(0.6031n^{-1/3})^4$
L_6	$4.00n^{-1/3}$	$(0.6432n^{-1/3})^6$
L_8	$4.18n^{-1/3}$	$(0.6903n^{-1/3})^8$

3.6.2 Other L_P Criteria

The error analysis of histograms for any even p is straightforward using higher order binomial moments and proceeding as before. For example, using L_4 error, the asymptotic mean integrated fourth-power error is (see Problem 3.27)

$$\text{AMI4E} = \frac{3}{n^2h^2}\int f(x)^2 + \frac{h}{2n}\int f'(x)^2 f(x) + \frac{h^4}{80}\int f'(x)^4. \qquad (3.80)$$

Table 3.9 summarizes some known optimal bandwidths for $N(0,1)$ data. The L_∞ optimal bandwidth is $O(\log(n)n^{-1/3})$.

Again, any fixed-bandwidth criterion will pay most attention to regions where the density is rough; that region is not necessarily in the tails. The coefficients in the table suggest that L_4 may have some special attraction, but that is an open problem.

PROBLEMS

3.1 Show how the normal bin width rule can be modified if f is skewed or leptokurtotic, as discussed in the introduction to Section 3.3 using other reference densities. Examine the effect of bimodality. Compare your rules to Doane's (1976) extensions of Sturges' rule.

3.2 Perform a dimensional analysis for the quantities $f, f', f'', R(f), R(f')$, and $R(f'')$. Check your results by computing these quantities for the case $f = N(\mu, \sigma^2)$ by tracking the factor σ.

3.3 An approximation to the error of a Riemannian sum:

$$\left| \sum_{n=-\infty}^{\infty} g(nh)h - \int_{-\infty}^{\infty} g(x)dx \right| = \left| \sum_{n=-\infty}^{\infty} \int_{nh}^{nh+h} [g(nh) - g(x)]dx \right|$$

$$\leq \sum_{n=-\infty}^{\infty} \int_{nh}^{nh+h} |g(nh) - g(x)|dx \leq \sum_{n=-\infty}^{\infty} hV_g(nh, nh+h) \leq hV_g(\Re^1),$$

where $V_g(a,b)$ is the total variation of g on $[a, b]$ defined by the $\sup\{\sum_{i=1}^{n} |g(x_i) - g(x_{i-1})|\}$ over all partitions on $[a, b]$, including $(a,b) = (-\infty, \infty)$. Conclude that if $f'(\cdot)^2$ has finite total variation, then the remainder term in the bias (3.14) is $O(h^3)$.

3.4 Compute the roughness of several parametric densities: Cauchy, Student's t, Beta, lognormal. For each compute the optimal bin width. Express the optimal bin width in terms of the variance, if it exists. Compare the coefficients in these formulas to the normal rule in Equation (3.16).

3.5 Show that when $h = h^*$ for the histogram, the contribution to AMISE of the IV and ISB terms is asymptotically in the ratio 2:1.

3.6 Verify Equation (3.25). *Hint*: Substitute $h = ch^*$ into Equation (3.23).

3.7 Compare the sensitivity of the AMISE(ch^*) in Equation (3.25) for various combinations of d, p, and r.

3.8 Prove the exact IV and ISB expressions in Equation (3.26).

3.9 Show that the ISB in a bin containing the origin of the double exponential density, $f(x) = \exp(-|x|)/2$, is $O(h^3)$; hence, the discontinuity in the derivative of f does not have any asymptotic effect on consistency. Compare the choice $t_0 = 0$ to the choice $t_0 = h/2$ as the bin boundary. Formally, if f is the ordinary negative exponential density, $R(f')$ is infinite because of the jump at zero (integral of the square of the Dirac delta function at 0), but $R(f')$ is well-defined for the double exponential.

3.10 Consider the exact MISE of a histogram when $f = U(0,1)$.
  (a) If the mesh $t_k = kh$ is chosen where $h = 1/m$, show that MISE$(m,n) = (m-1)/n$. Trivially, $m^* = 1$ and $h^* = 1$, and the error is 0.
  (b) If $t_k = (k + \frac{1}{2})h$ with $h = 1/m$, show that

$$\text{MISE}(m,n) = \frac{1 - 2m + 2m^2 + n}{2mn}.$$

  Conclude that $h^* = \sqrt{2/(n+1)}$ and

$$\text{MISE}^* = \frac{\sqrt{2}\sqrt{n+1} - 1}{n} \approx \sqrt{\frac{2}{n}} - \frac{1}{n} + \cdots,$$

  which is a slower rate than the usual histogram with a smooth density.

3.11 Verify the inequalities for the adaptive MISE in Equation (3.32).

3.12 Find the optimal adaptive meshes for a skewed Beta density using a numerical optimization program.

3.13 Investigate the use of fixed and percentile meshes when applying chi-squared goodness-of-fit hypothesis tests.

3.14 Apply the oversmoothing procedure to the LRL dataset. Compare the results of using the range and variance as measures of scale.

3.15 Find an oversmoothed rule based on the interquartile range as a measure of scale.

3.16 Use Taylor series and Riemannian integral approximations to verify Equation (3.49).

3.17 Carefully evaluate and verify the formulas in Equation (3.53). *Hint*: $\hat{f}_{-i}(x_i) = (\nu_k - 1)/[(n-1)h]$.

3.18 What are the UCV and BCV bin widths for the LRL and snowfall datasets? Since the LRL data are pre-binned, try various bin widths that are multiples of 10 MeV, and various bin origins.

3.19 The UCV rule seems to estimate something between ISE and MISE [the middle quantity, $-2\int \hat{f}(x)f(x)dx$, being the focus]. Using the exact MISE formula with normal and lognormal data, investigate the behavior of the terms in the UCV approximation.

3.20 Take a normal sample of size 1000. Apply increasing degrees of rounding to the data and compare the resulting UCV curves. At what point does the minimizer of UCV become $h = 0$?

3.21 In BCV, the bias in the roughness was reduced by subtraction. Alternatively, the roughness could have been multiplied by 3/4. Examine the effect of this idea by example. Where does the factor of 3/4 originate?

3.22 Develop a BCV formulation based on a central difference estimator of f' given by $(\nu_{k+1}/nh - \nu_{k-1}/nh)/2h$ (see Scott and Terrell (1987)).

3.23 Verify the expressions in Equations (3.65) and (3.69) for bin widths and errors when the density is bivariate normal.

3.24 For bivariate normal data, examine the inefficiency of using square bins relative to rectangular bins. Examine combinations of situations where the correlation is or is not 0, and situations where the ratio of the marginal variances is or is not equal to 1.

3.25 Verify the AMISE expressions given in Equations (3.70) and (3.71) for the three regular bivariate meshes. *Hint*: Carefully integrate the bias over the individual bin shape, and then aggregate by using the Riemannian integral approximation.

3.26 Consider a two-component normal mixture density, $f(x) = 0.5N \cdot (-\mu, 1) + 0.5N(\mu, 1)$. How large must μ be so that the density exhibits two modes and bumps?

3.27 Compute one of the error criteria based on L_p for some even $p > 2$.

4

FREQUENCY POLYGONS

The discontinuities in the histogram limit its usefulness as a graphical tool for multivariate data. The *frequency polygon* (FP) is a continuous density estimator based on the histogram, with some form of linear interpolation. For example, the rootgram of the lateral root length (LRL) data in Figure 3.22 is actually a linearly interpolated histogram. With 172 bins in the data, the vertical lines representing the raw histogram overlap, an undesirable feature that can be seen in the 200-bin histogram of the German income data in Figure 3.18. Scott (1985a) examined the theoretical properties of univariate and bivariate frequency polygons and found them to have surprising improvements over histograms. Fisher (1932, p. 37) disapproved of the frequency polygon, ironically for graphical reasons:

> The advantage is illusory, for not only is the form of the curve thus indicated somewhat misleading, but the utmost care should always be taken to distinguish the infinitely large hypothetical population from which our sample of observations is drawn, from the actual sample of observations which we possess; the conception of a continuous frequency curve is applicable only to the former, and in illustrating the latter no attempt should be made to slur over this distinction.

Fisher was unaware of any theoretical differences between histograms and frequency polygons and was thinking only of univariate histograms when he wrote this passage. His objection to using a continuous nonparametric density estimator is no longer justified, but his concern about using techniques that totally obscure the statistical noise with mathematical sophistication is worth re-emphasizing. Finally, as a matter

Multivariate Density Estimation, First Edition. David W. Scott.
© 2015 John Wiley & Sons, Inc. Published 2015 by John Wiley & Sons, Inc.

of terminology, the distinction between the histogram and frequency polygon in the scientific literature is being blurred, with the histogram label being applied to both.

4.1 UNIVARIATE FREQUENCY POLYGONS

In one dimension, the frequency polygon is the linear interpolant of the mid-points of an equally spaced histogram. As such, the frequency polygon extends beyond the histogram into an empty bin on each extreme. The frequency polygon is easily verified to be a *bona fide* density function, that is, nonnegative with integral equal to 1 (see Problem 4.1).

4.1.1 Mean Integrated Squared Error

The asymptotic MISE (AMISE) is easily computed on a bin-by-bin basis, by considering a typical pair of histogram bins displayed in Figure 4.1. The frequency polygon connects the two adjacent histogram values, $\hat{f}_0$ and $\hat{f}_1$, between the bin centers, as shown. The FP is described by the equation

$$\hat{f}(x) = \left(\frac{1}{2} - \frac{x}{h}\right)\hat{f}_0 + \left(\frac{1}{2} + \frac{x}{h}\right)\hat{f}_1, \qquad -\frac{h}{2} \le x < \frac{h}{2}. \tag{4.1}$$

The randomness in the frequency polygon comes entirely from the randomness in the histogram levels, $\hat{f}_i = \nu_i/(nh)$. The "x" in $\hat{f}(x)$ is *not random* but is fixed.

As before, using the Taylor's series

$$f(x) = f(0) + xf'(0) + \frac{1}{2}x^2 f''(0) + \cdots, \tag{4.2}$$

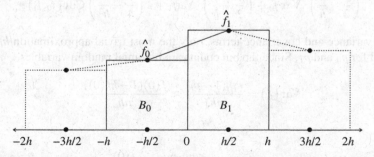

FIGURE 4.1 The frequency polygon in a typical bin, $(-h/2, h/2)$, which is derived from two adjacent histogram bins B_0 and B_1.

approximations for p_0 and p_1 can be obtained:

$$p_0 = \int_{-h}^{0} f(s)\, ds \approx hf(0) - h^2 f'(0)/2 + h^3 f''(0)/6$$

$$(4.3)$$

$$p_1 = \int_{0}^{h} f(s)\, ds \approx hf(0) + h^2 f'(0)/2 + h^3 f''(0)/6.$$

The bias is computed by noting that the pointwise expectation of the FP is a linear combination of the expectations of the two histogram values. As $E(\hat{f}_i) = p_i/h$, then from (4.1) and (4.3), and again noting that "x" is not random, we have

$$E\hat{f}(x) = \left(\frac{1}{2} - \frac{x}{h}\right)\frac{p_0}{h} + \left(\frac{1}{2} + \frac{x}{h}\right)\frac{p_1}{h} \approx f(0) + xf'(0) + h^2 f''(0)/6.$$

Subtracting (4.2) gives $\text{Bias}\{\hat{f}(x)\} \approx (h^2 - 3x^2)f''(0)/6$. The integral of the squared bias (ISB) over the FP bin $(-h/2, h/2)$ equals $[49h^4 f''(0)^2/2880] \times h$, with a similar expression for other FP bins. Summing over all of the bins and using standard Riemannian approximation, yields

$$\text{ISB} \approx \sum_k \frac{49}{2880} h^4 f''(kh) \times h = \frac{49}{2880} h^4 R(f'') + O(h^6).$$

Evidently the squared bias of the frequency polygon is of significantly higher order than the order $O(h^2)$ of the histogram. For purposes of cross-validation, the bias is driven by the unknown roughness $R(f'')$ rather than by $R(f')$. Recalling Equation (3.12), the FP extends the good property of the histogram at its bin centers, the elimination of $O(h)$ effects, to the entire estimator. The bias is a function of the curvature in the density function rather than the slope as with the histogram.

The variance calculation is similar. From the FP definition in (4.1), the variance of $\hat{f}(x)$ equals

$$\left(\frac{1}{2} - \frac{x}{h}\right)^2 \text{Var}\,\hat{f}_0 + \left(\frac{1}{2} + \frac{x}{h}\right)^2 \text{Var}\,\hat{f}_1 + 2\left(\frac{1}{4} - \frac{x^2}{h^2}\right)\text{Cov}(\hat{f}_0, \hat{f}_1). \qquad (4.4)$$

For the variance and covariance terms, only the most trivial approximation $hf(0)$ is required for p_0 and p_1. Since the bin counts are binomial random variables,

$$\text{Var}\,(\hat{f}_i) = \frac{np_i(1-p_i)}{(nh)^2} \approx \frac{f(0)(1-hf(0))}{nh}$$

and

$$\text{Cov}(\hat{f}_0, \hat{f}_1) = \frac{-np_0 p_1}{(nh)^2} \approx -\frac{f(0)^2}{n}.$$

Substituting these approximations into (4.4) gives

$$\mathrm{Var}\hat{f}(x) = \left(\frac{2x^2}{nh^3} + \frac{1}{2nh}\right)f(0) - \frac{f(0)^2}{n} + o(n^{-1}).$$

Integrating over the FP bin $(-h/2, h/2)$ yields $[2f(0)/(3nh) - f(0)^2/n] \times h$. Summing the corresponding expression for all bins and noting that $\int f = 1$ gives

$$\mathrm{IV} \approx \sum_k \left[\frac{2f(kh)}{3nh} - \frac{f(kh)^2}{n}\right] \times h = \frac{2}{3nh} - \frac{1}{n}R(f) + o(n^{-1}).$$

If the optimal histogram bin width were used with a frequency polygon, the asymptotic effect would be to eliminate the bias entirely relative to the variance in the MISE, the orders being $O(n^{-4/3})$ and $O(n^{-2/3})$, respectively. Since the ISB comprises a third of the MISE for a histogram, the reduction would be substantial. But a better FP can be constructed. The improved order in the bias suggests that a larger bin width could be used to reduce the variance, but still with smaller bias than the histogram. In fact, the bin width $h = O(n^{-1/5})$ turns out to be just right. The improvement is substantial, as the following theorem reveals (Scott, 1985a).

Theorem 4.1: *Suppose f'' is absolutely continuous and $R(f''') < \infty$. Then*

$$\mathrm{AMISE}(h) = \frac{2}{3nh} + \frac{49}{2880}h^4 R(f''); \quad \text{hence,} \qquad (4.5)$$

$$h^* = 2[15/(49R(f''))]^{1/5} n^{-1/5}$$

$$\mathrm{AMISE}^* = (5/12)[49R(f'')/15]^{1/5} n^{-4/5}. \qquad (4.6)$$

For example, with 800 normal data points, the optimal bin width for the FP is 50% wider than the corresponding histogram bin width given in Theorem 3.1. Apparently, in order for the discontinuous histogram to approximate a continuous density, the histogram must be quite rough to track the density function in regions where its level is changing rapidly. The FP is inherently continuous and can approximate the continuous density better with piecewise linear fits over wider bins. The FP does most poorly near peaks where the second derivative and the density are both large in magnitude. The improvement in MISE is reflected not only by a decrease in the constant in front of $n^{-2/3}$ but also by a real decrease in the exponent.

The one situation where FPs are at a disadvantage occurs when the underlying density is discontinuous. A histogram is unaffected by such points if they are known and placed at bin boundaries. A FP cannot avoid overlapping such points, and the asymptotic theory above does not apply (see Problem 4.3).

Finally, as was shown in Table 3.3 in the column with $p = 2$, frequency polygons are more sensitive than histograms with respect to errors in choice of bin width, particularly when $h > h^*$. On the other hand, quite a large error in bin width for the FP is required before its MISE is worse than the best histogram MISE (see Figure 4.2).

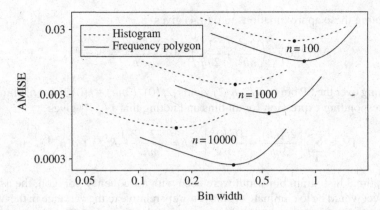

FIGURE 4.2   AMISE for histogram and frequency polygon for standard normal density.

EXAMPLE: For the normal density, $R(\phi'') = 3/(8\sqrt{\pi}\sigma^5)$; hence, from Theorem 4.1,

$$h^* = 2.15\,\sigma\,n^{-1/5} \quad \text{and} \quad \text{AMISE}^* = 0.3870\,\sigma^{-1}n^{-4/5}. \tag{4.7}$$

To understand the practical consequences and to see where the FP fits among the parametric estimators and the histogram with respect to sample size, examine Table 4.1, which extends Table 3.2. Clearly, the frequency polygon is not just a theoretical curiosity. The FP is even more data efficient relative to the histogram as the sample size grows. Of course, both nonparametric estimates will be increasingly inferior to the correct parametric fit.

Another way to see the difference between the histogram and the FP for normal data is shown in Figure 4.2. On a log–log scale, not only are the different rates of convergence easily seen but also the differences in the optimal bin widths. Continuing on to a million normal points, the optimal bin widths for the histogram and FP are 0.035 and 0.136, respectively. These are in the ratio 4:1, an example of which appears in Figure 3.7 labeled $h = 4h^*$. The stability (small variance) of the histogram with $h = 4h^*$ is evident; however, so too is the bias resulting from the staircase shape. The FP retains the stability of this histogram, while the linear interpolation dramatically reduces the bias. The ISE of the FP in Figure 3.7 is approximately equal to 5.40×10^{-6}, which is 14% of the ISE of the best histogram.

4.1.2   Practical FP Bin Width Rules

To highlight the differences with the corresponding histogram results, some bin width rules for the FP will be presented. The plug-in rule based on (4.7) is

$$\boxed{\text{FP Normal reference rule:} \quad \hat{h} = 2.15\,\hat{\sigma}\,n^{-1/5},} \tag{4.8}$$

TABLE 4.1 Sample sizes required for $N(0,1)$ data so that AMISE* $\approx 1/400$ and $1/4000$

Estimator	Equivalent sample sizes	
$N(\bar{x},1)$	57	571
$N(\bar{x},s^2)$	100	1,000
Optimal FP	546	9,866
Optimal histogram	2297	72,634

here $\hat{\sigma}$ is an estimate, perhaps robust, of the standard deviation. One robust choice appropriate based on the interquartile range is $\hat{\sigma} = \text{IQR}/1.348$, where 1.348 is $\Phi^{-1}(0.75) - \Phi^{-1}(0.25)$. The factors modifying rule (4.8) based on sample skewness and kurtosis were shown in Figure 3.5. The factors are based on the relationship

$$\frac{h_y^*}{h_N} = \left[\frac{R\left(\phi'' | 0, \sigma_y^2\right)}{R(g''(y))} \right]^{1/5},$$

corresponding to (3.19) and Theorem 4.1. Now $R(\phi'') = 3/(8\sqrt{\pi}\sigma_y^5)$ and the roughness $R(g'')$ of the lognormal and t_ν densities are

$$\frac{(9\sigma^4 + 20\sigma^2 + 12)e^{25\sigma^2/4}}{32\sqrt{\pi}\sigma^5} \quad \text{and} \quad \frac{12\Gamma\left(\nu + \frac{5}{2}\right)\Gamma\left(\frac{\nu+5}{2}\right)^2}{\sqrt{\pi}\nu^{5/2}\Gamma(\nu+5)\Gamma\left(\frac{\nu}{2}\right)^2},$$

respectively. Following the notation in Section 3.2.3,

$$\text{Skewness factor}\{\beta_1(\sigma)\} = \frac{12^{1/5}\sigma}{e^{7\sigma^2/4}(e^{\sigma^2} - 1)^{1/2}(9\sigma^4 + 20\sigma^2 + 12)^{1/5}}$$

$$\text{Kurtosis factor}\left\{\tilde{\beta}_2(\nu)\right\} = \frac{\sqrt{\nu-2}}{2}\left[\frac{\Gamma(\nu+5)\Gamma\left(\frac{\nu}{2}\right)^2}{\Gamma\left(\nu + \frac{5}{2}\right)\Gamma\left(\frac{\nu+5}{2}\right)^2}\right]^{1/5},$$

where $\tilde{\beta}_2 = 6/(\nu - 4)$, which are plotted in Figure 3.5 (see Problem 4.5).

Biased and unbiased cross-validation algorithms are only slightly more complicated to implement for the frequency polygon. For BCV, the following estimate of $R(f'')$ was proposed by Scott and Terrell (1987):

$$\hat{R}(f'') = \frac{1}{n^2 h^5}\sum_k (\nu_{k+1} - 2\nu_k + \nu_{k-1})^2 - \frac{6}{nh^5}. \tag{4.9}$$

Plugging this estimate into the AMISE expression (4.6) results in

$$\text{BCV}(h) = \frac{271}{480nh} + \frac{49}{2880n^2h} \sum_k (\nu_{k+1} - 2\nu_k + \nu_{k-1})^2.$$

The unbiased cross-validation formula is left as an exercise (see Problem 4.7).

As an example, consider the German income data displayed in Figures 3.18 and 3.19. The BCV(h) estimates for the histogram and frequency polygon are shown in Figure 4.3. The BCV estimate of MISE for the FP is 71% lower than that for the histogram. The BCV-optimal FP is constructed from a histogram with 51 bins, which is displayed in Figure 3.18. Examine the shapes of the two BCV curves more closely. For small bin widths, the curves are parallel with slope -1 on the log–log scale, since the integrated variances for the histogram and FP are $1/(nh)$ and $2/(3nh)$, respectively. For large bin widths, the difference in the slopes reflects the different orders in the bias.

Upper bounds for the bin width for a frequency polygon may be obtained by variational methods similar to those used with the histogram (see Scott and Terrell (1987) and Terrell (1990)). Examining the expression for the AMISE* in Theorem 4.1, the objective function becomes $R(f'')$, rather than $R(f')$ as with the histogram. Subject to the constraint that the range is the interval $[-0.5, 0.5]$, the smoothest density is

$$f_3(x) = \frac{15}{8}(1 - 4x^2)^2 \, I_{[-0.5, 0.5]}(x) \quad \text{so that} \quad R(f_{\prime}'') \geq \frac{720}{(b-a)^5}$$

when the support interval is the more general (a, b). Replacing $R(f'')$ in Equation (4.6) for h^* leads to

$$\boxed{\text{Number of bins} = \frac{b-a}{h^*} \geq \left(\frac{147}{2}n\right)^{1/5}.} \tag{4.10}$$

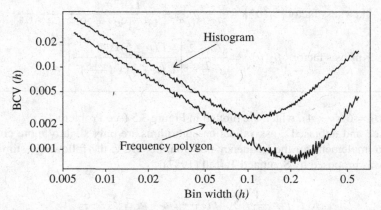

FIGURE 4.3   BCV for histogram and frequency polygon for German income data.

For example, with the large LRL dataset of 25,752 points, the optimal FP requires at least 18 bins, while the optimal histogram requires at least 37 bins. Given the amount of structure in the LRL data, these are conservative bounds.

A different version of the oversmoothed problem leads to a bin width rule. Among all densities with variance σ^2, the smoothest density is

$$f_4(x) = \frac{35}{96\sigma} \left(1 - \frac{x^2}{9\sigma^2}\right)^3 I_{[-3\sigma,3\sigma]}(x) \quad \text{so that} \quad R(f'') \geq \frac{35}{243\sigma^5}.$$

Substituting this inequality into the expression for h^* in Theorem 4.1 leads to the oversmoothed bin width rule:

$$h \leq \left(\frac{23,328}{343}\right)^{1/5} \sigma n^{-1/5} = 2.33\,\sigma\,n^{-1/5} \equiv h_{OS}. \tag{4.11}$$

This bin width is only 108% of the normal rule, which suggests that using the normal-based rule in Equation (4.8) will also oversmooth in most practical data situations. In general, a normal rule may be substituted for an oversmoothed rule whenever the variational problem is too difficult to solve explicitly.

4.1.3 Optimally Adaptive Meshes

Consider the theoretical improvement possible when applying frequency polygons to *adaptive* histogram meshes. Note, however, that connecting histogram midpoints in an adaptive mesh does not lead to an estimate that integrates to 1 except asymptotically. With that caveat, the following results are a consequence of Equation (4.6).

Theorem 4.2:   *The asymptotic properties of the optimal adaptive frequency polygon constructed by connecting midpoints of an adaptive histogram are*

$$\text{AMSE}(x) = \frac{2f(x)}{3nh} + \frac{49}{2880}h^4 f''(x)^2 \tag{4.12}$$

from which it follows that

$$h^*(x) = 2[15f(x)/49f''(x)^2]^{1/5}n^{-1/5}$$
$$\text{AMSE}^*(x) = (5/12)[49/15]^{1/5}[f''(x)^2 f(x)^4]^{1/5}n^{-4/5} \tag{4.13}$$
$$\text{AAMISE}^* = (5/12)[49/15]^{1/5}\left\{\int [f''(x)^2 f(x)^4]^{1/5}dx\right\}n^{-1/5}.$$

Comparing Equations (4.6) and (4.13), we see that

$$\text{AAMISE}^* \leq \text{AMISE}^* \Leftrightarrow \int \left[f''(x)^2 f(x)^4 \right]^{1/5} dx \leq \left[\int f''(x)^2 dx \right]^{1/5},$$

which is equivalent to the following inequality (which is true by Jensen's inequality):

$$E \left[\frac{f''(x)^2}{f(x)} \right]^{1/5} \leq \left[E \frac{f''(x)^2}{f(x)} \right]^{1/5}.$$

Thus, asymptotically, the MISE of an adaptive FP is only 91.5 and 76.7% of the MISE of a fixed-bin-width FP for normal and Cauchy data, respectively (see Problem 4.8).

The MISE for the FP of an adaptive histogram can be computed exactly. Since the resulting adaptive FP does not integrate to 1, judgment is reserved as to its practical value; however, there is much of interest to examine in the structure of the optimal mesh. Note that asymptotically, the optimal adaptive FP will integrate to 1 since the underlying adaptive histogram exactly integrates to 1.

The general pattern in an adaptive FP mesh may be inferred from Theorem 4.2. The FP mesh seems out of phase with the optimal adaptive histogram mesh at critical points. The FP bins are widest where the second derivative is small, which is at points of inflection and, to a lesser extent, in the tails. In between, the bins can be quite narrow depending on the magnitude of $f''(x)$. Consider the optimal adaptive mesh of the normal-like scaled Beta(5,5) density in Figure 4.4. In the tails, the optimal bins are not much wider. In fact, the pattern is relatively difficult to see except for the largest sample size. Given not only the complexity of an optimally adaptive mesh but also the relatively modest reduction in MISE, practical adaptive algorithms have been slow to appear. An intermediate strategy would be to perform data transformations to minimize skewness or to handle widely separated clusters individually.

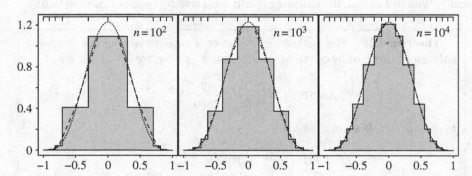

FIGURE 4.4 Optimal adaptive frequency polygon meshes for the scaled Beta(5,5) density. The histogram is drawn from which the FP (dotted line) is derived. The tick marks for the adaptive mesh are shown above each figure.

4.1.4 Modes and Bumps in a Frequency Polygon

In Chapter 3, the optimal MISE smoothing for a histogram was observed as unable to provide a reliable estimate of modes or bumps. Since the optimal FP bin widths are wider and of order $O(n^{-1/5})$, the discussion in Section 3.5 shows that the sample modes and bumps in an optimal frequency polygon are more reliable than those found in an optimal histogram.

Continuing the analysis of Section 3.5 and assuming that $x = 0$ is a mode, write

$$h^* = cn^{-1/5} \quad \text{and define} \quad \beta \equiv -\frac{1}{2}c^{5/2}\frac{f''(0)}{\sqrt{f(0)}}. \tag{4.14}$$

Using this h^* and noting that $(h^*)^{5/2}\sqrt{n} = c^{5/2}$ in Equation (3.76), we have that

$$\lim_{n\to\infty} \Pr\left(\nu_0 = \arg\max_{|j|\le k} \nu_j\right) = \int_y \prod_{\substack{j=-k \\ j\ne 0}}^{k} \Phi(y+j^2\beta)\,\phi(y)\,dy. \tag{4.15}$$

This probability is a constant that depends only on β (and a bit on k) but *not on the sample size*. A graph is shown in Figure 4.5 for the choice $k = 4$. When $\beta = 0$ [i.e., $f''(0) = f'(0) = 0$ so that the density function is very flat], then the probability that ν_0 is the mode is $1/(2k+1)$ since the density is locally uniform and all $2k+1$ bins are equally likely to contain the largest count. As β increases, the probability that ν_0 is the greatest count increases to 1. β, which is dimensionless, measures the "strength" of the mode at $x = 0$. A similar expression may be computed for the probability that each of $\{\nu_\ell, 1 \le |\ell| \le k\}$ is the largest bin count (an "error" since the mode is at the center of B_0). The probabilities are symmetric in ℓ. Each is also a function of β alone (see Figure 4.5).

A "confidence interval" for the true mode may be derived from this graph given an estimate of β. For example, if $\beta = 2.7$, then the probability the ν_0 is the greatest

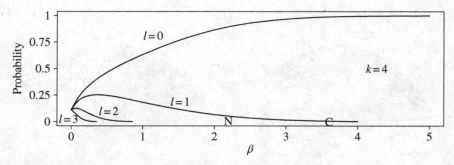

FIGURE 4.5 Probability distribution of the location of the sample mode as a function of β for the choice $k = 4$. The values of β for Normal and Cauchy densities are 2.15 and 3.6, respectively, and are indicated by the letters N and C.

count is 95%. Thus the sample bin interval $(-h/2, h/2)$ is a 95% confidence interval for the mode. If $\beta = 0.46$, then the probability is 95% that the sample mode is in $(-3h/2, 3h/2)$, that is, bins B_{-1}, B_0, B_1. (However, the derivation assumed that the true mode was at the center of a bin. Hence, to use this result in practice, the mesh should be shifted so that the values of the FP in the bins adjacent to the sample mode are approximately equal in height. This ensures that the sample mode is approximately at the center of a bin.) The sample estimate of β is sure to be underestimated because of the downward bias at modes, so the confidence interval is conservative. Are the estimates consistent if the probabilities are not changing? The answer is yes, because the confidence intervals are stated in terms of multiples of the bin width, which is shrinking towards zero at the rate $n^{-1/5}$. Notice that the calculation does not preclude other smaller sample modes (e.g., $\nu_3 > \nu_2$) in this neighborhood, but the interpretation should be acceptable in most situations. A figure similar to Figure 4.5 but with a larger value of k is virtually identical except near $\beta = 0$. For normal and Cauchy data, $\beta \approx 2.15$ and 3.6, respectively. There is an 89% chance that the sample mode for normal data is in bin B_0 for normal data (99% for Cauchy). Locating the modes for these densities is relatively easy.

4.2 MULTIVARIATE FREQUENCY POLYGONS

There are two important ways of defining a linear interpolant of a multivariate histogram with hyper-rectangular bins, where $\mathbf{x} \in \Re^d$. The first, considered by Scott (1985a, b), is to interpolate the values at the centers of $d+1$ adjacent histogram bins in a "triangular" or, more generally, a simplex-like configuration. The resulting collection of triangular pieces of a hyperplane defines a continuous but not differentiable surface in $\Re^{d+1}$. The definition is not unique since several reflections of the basic pattern work (see Figure 4.6).

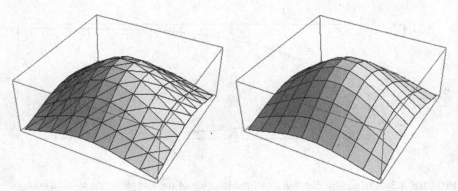

FIGURE 4.6 An example of the construction of a bivariate frequency polygon using triangular meshes (left) and linear blend elements (right).

The second definition for a multivariate FP, which was investigated independently by Terrell (1983) and Hjort (1986), is known as the *linear blend* in the computer graphics literature. For example, in two dimensions, the formula for a linear blend on the unit square is simply

$$f(x,y) = a + bx + cy + dxy,$$

which contains four of the six terms in a full quadratic model, omitting the two pure quadratic terms involving x^2 and y^2. For fixed $x = x_0$, the linear blend is $f(x_0, y) = (a + bx_0) + (c + dx_0)y$, which is linear in y. Likewise, for fixed $y = y_0$, the linear blend is $f(x, _0 y) = (a + cy_0) + (b + dy_0)x$, which is linear in x. Thus the surface is linear parallel to the coordinate axes. However, along the diagonal $x = y$, $f(x, y) = a + (b + c)x + dx^2$, which is quadratic. In general dimension, a single portion of a linear blend extends over a hyper-rectangle with 2^d vertices, defined by the centers of the 2^d adjacent histogram bins. Any cut of the surface parallel to a coordinate axis gives a linear fit (see the right frame in Figure 4.6). Certainly, this definition of a multivariate FP is smoother than the first, but the primary advantage of this formulation is the beautifully simple AMISE result.

The linear blend frequency polygon (LBFP) is only slightly more complicated to define than the triangular mesh. Consider a typical LBFP bin,

$$B_{k_1,\ldots,k_d} = \prod_{i=1}^{d} [t_{k_i}, t_{k_i} + h_i).$$

Then for $\mathbf{x} \in B_{k_1,\ldots,k_d}$, the LBFP is defined as

$$\hat{f}(\mathbf{x}) = \frac{1}{nh_1 \cdots h_d} \sum_{j_1,\ldots,j_d \in \{0,1\}^d} c_{j_1,\ldots,j_d} \, \nu_{k_1+j_1,\ldots,k_d+j_d}, \tag{4.16}$$

where

$$c_{j_1,\ldots,j_d} = \prod_{i=1}^{d} u_i^{j_i} (1 - u_i)^{1-j_i} \quad \text{and} \quad u_i = \frac{x_i - t_{k_i}}{h_i}.$$

Hjort (1986) showed that the LBFP integrates to 1 and that

$$\text{AMISE}(\mathbf{h}) = \frac{2^d}{3^d n h_1 \cdots h_d} + \frac{49}{2880} \sum_{i=1}^{d} h_i^4 R(f_{ii}) + \frac{1}{32} \sum_{i<j} h_i^2 h_j^2 R\left(\sqrt{f_{ii} f_{jj}}\right),$$

where f_{ij} is the mixed second-order partial derivative. Although this cannot be optimized in closed form except in special cases, it is easy to show that

$$h_i^* = O(n^{-1/(4+d)}) \quad \text{and} \quad \text{AMISE}^* = O(n^{-4/(4+d)}). \tag{4.17}$$

Not only are frequency polygons more efficient than histograms, but the difference in the order of convergence rates across dimensions is significant. If the notion that

TABLE 4.2   Asymptotic order of MISE for multivariate histogram and frequency polygon density estimators.

d	Histogram	Frequency polygon
1	$n^{-2/3}$	$n^{-4/5}$
2	$n^{-2/4}$	$n^{-4/6}$
3	$n^{-2/5}$	$n^{-4/7}$
4	$n^{-2/6}$	$n^{-4/8}$
5	$n^{-2/7}$	$n^{-4/9}$
6	$n^{-2/8}$	$n^{-4/10}$
7	$n^{-2/9}$	$n^{-4/11}$
8	$n^{-2/10}$	$n^{-4/12}$

The rate of convergence decreases as the dimension, d, increases. The arrows indicate identical rates for the histogram and frequency polygon.

bivariate histograms "work" is correct, then Table 4.2 suggests that quadrivariate frequency polygons should work equally well. Some authors have claimed good results for quadrivariate histograms, which would correspond to an eight-dimensional FP in terms of exponent order. On the other hand, the level of detail demanded in higher dimensions must diminish. Working in more than four or five dimensions is usually done for convenience of interpretation, not for reasons of structure (such as interactions in higher dimensions).

For graphical reasons, the first definition of an FP is simpler to work with because the resulting contours are comprised of piecewise polygonal sections, which can depicted with many CAD–CAM graphics packages. There is little practical difference in the approximation quality of the two estimators and the binning structure is easily apparent in the former and not in the latter. Advanced surface visualization algorithms require the value of the function on a 3-D mesh. The simplifying idea here is that the visualization mesh can be identical to the FP mesh. Usually, there are several interpolation options in visualization programs, including linear blends and piecewise triangular. Thus the choice of interpolation can be thought of as primarily an aesthetic issue of visualization smoothness, and secondarily, as a choice of density quality.

Using the triangular mesh with a bivariate normal data, Scott (1985a) showed that the optimal bin widths are approximately equal to

$$h_i^* = 2.105 \left(1 - \frac{107}{208}\rho^2 + \cdots\right) \sigma_i n^{-1/6}, \quad i = 1, 2.$$

For multivariate normal data with $\Sigma = I_d$, the optimal smoothing parameters in each dimension are equal with the constant close to 2. Thus Scott also proposed using

$$\boxed{\text{Approximate Normal FP reference rule:} \quad h_i = 2\hat{\sigma}_i n^{-1/(4+d)}.} \tag{4.18}$$

4.3 BIN EDGE PROBLEMS

As was mentioned in Chapter 3, the mesh is completely determined by the pair (h, t_0). The asymptotic theory indicates that the choice of bin origin is asymptotically negligible. Consider the Buffalo snowfall dataset set given in Table B.9. The annual snowfall during 63 winters was recorded from 1910/1911–1972/1973. Some have argued (Scott, 1980) that the data appear to be trimodal, but Parzen (1979) has suggested the evidence leans toward a unimodal density. One might imagine that the choice of the bin width and not the bin origin would be critical for understanding this issue. Indeed, in Figure 4.7 the histogram with 15 bins of width 10 inches suggests trimodality, while the histogram with 10 bins of width 15 inches suggests unimodality. But in Figure 4.8, the effect of bin origin choice is clearly revealed not to be negligible. The first histogram is almost unimodal. Four histograms are bimodal, but with secondary mode either on the left and right. And remarkably, only one histogram is trimodal. Continuing to the multivariate setting, the effect of the bin origin is more pronounced.

The statement that the choice of t_0 is asymptotically negligible with respect to MISE is also true for the frequency polygon. But the optimal bin widths for a FP are substantially wider than for the histogram. Thus there are many more possible choices for the bin origin. For a particular choice of bin width, one possible recommendation is to choose the bin origin so that the estimate is "smoothest." If this could always be done, then the amount of roughness in the estimate would always be determined by the bin width as much as possible, and not by the bin origin, t_0, which may be

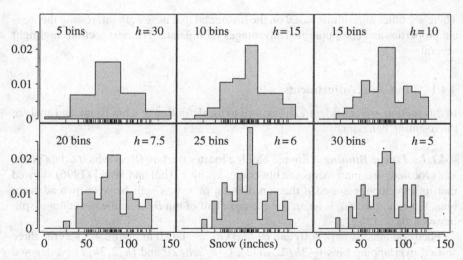

FIGURE 4.7 Histograms of the Buffalo snowfall data with bin origin $t_0 = 0$, and bin widths of 30, 15, 10, 7.5, 6, and 5 inches over the interval $(0, 150)$.

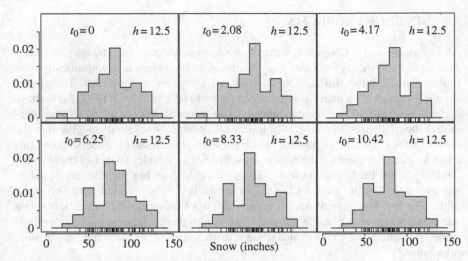

FIGURE 4.8 Six shifted histograms of the Buffalo snowfall data. All have a bin width of 12.5 inches, but different bin origins $t_0 = h/m$, $m = 1, \ldots, 6$.

thought of as a nuisance parameter. The next chapter introduces an ingenious device that eliminates the effect of this nuisance parameter entirely.

4.4 OTHER MODIFICATIONS OF HISTOGRAMS

There are other algorithms based on the histogram that have both interesting theoretical properties and computational advantages for big data. The next sections highlight several.

4.4.1 Bin Count Adjustments

In this section, we consider modifications to the definition of a bin count, and analyze the resulting behavior.

4.4.1.1 Linear Binning   Simple binning locates the bin, B_k, in which a data point, x_i, is located, and increments the bin count, ν_k, by 1. Hall and Wand (1996) showed that improvement resulted if the contribution of x_i was split between two adjacent bins. For example, if x_i is located in the right half of bin B_k, then the bin count is split between B_k and B_{k+1}.

Focus on bins $B_0 = (-h, 0)$ and $B_1 = (0, h)$, as shown in Figure 4.1. Let the three shifted overlapping bins $(-3h/2, -h/2)$, $(-h/2, h/2)$, and $(h/2, 3h/2)$ be denoted by J_{-1}, J_0, and J_1, respectively. If $x_i \in J_1$, then the bin "count" in B_1 is incremented by $(3/2 - x_i/h)$. Note the increment is 1, $\frac{1}{2}$, or 0 if $x_i = \frac{h}{2}$, h, or $\frac{3h}{2}$, respectively.

Let the linear binning counts be denoted by $\tilde{\nu}_0$ and $\tilde{\nu}_1$. Then

$$\tilde{\nu}_0 = \sum_{i=1}^{n} \left[\left(\tfrac{3}{2} + \tfrac{x_i}{h} \right) I(x_i \in J_{-1}) + \left(\tfrac{1}{2} - \tfrac{x_i}{h} \right) I(x_i \in J_0) \right]$$

$$\tilde{\nu}_1 = \sum_{i=1}^{n} \left[\left(\tfrac{1}{2} + \tfrac{x_i}{h} \right) I(x_i \in J_0) + \left(\tfrac{3}{2} - \tfrac{x_i}{h} \right) I(x_i \in J_1) \right].$$

Note the frequency polygon defined as in Equation (4.1) now uses these adjusted bin counts for $x \in J_0$. Let LB denote linear binning. Then the expectation is

$$\mathrm{E}\hat{f}_{LB}(x) = \left(\frac{1}{2} - \frac{x}{h} \right) \frac{\mathrm{E}\tilde{\nu}_0}{nh} + \left(\frac{1}{2} + \frac{x}{h} \right) \frac{\mathrm{E}\tilde{\nu}_1}{nh},$$

where

$$\mathrm{E}\tilde{\nu}_0 = n \left[\int_{J_{-1}} \left(\frac{3}{2} + \frac{x}{h} \right) f(x)\,dx + \int_{J_0} \left(\frac{1}{2} - \frac{x}{h} \right) f(x)\,dx \right]$$

$$\mathrm{E}\tilde{\nu}_1 = n \left[\int_{J_0} \left(\frac{1}{2} + \frac{x}{h} \right) f(x)\,dx + \int_{J_1} \left(\frac{3}{2} - \frac{x}{h} \right) f(x)\,dx \right].$$

Using the Taylor's series Equation (4.2) in Mathematica, we find

$$\mathrm{E}\hat{f}_{LB}(x) = f(0) + x f'(0) + \frac{5}{24} h^2 f''(0) + O(h^3). \qquad (4.19)$$

Subtracting Equation (4.2) from Equation (4.19), the leading term in the bias is $\frac{5}{24} h^2 f''(0) - \frac{1}{2} x^2 f''(0)$. Integrating the squared bias over the interval $J_0 = (-\frac{h}{2}, \frac{h}{2})$ gives $\frac{7}{240} h^4 f''(0)^2 \times h$. For $x \in J_k$, replace $f''(0)^2$ with $f''(kh)^2$. Thus the asymptotic ISB is given by adding up such terms over all the bins, $\{J_k\}$, giving

$$\mathrm{AISB}(h) = \sum_{k=-\infty}^{\infty} \frac{7}{240} h^4 f''(kh)^2 \times h = \frac{7}{240} h^4 R(f''). \qquad (4.20)$$

The computation of the variance of $\hat{f}_{LB}(x)$ is much more challenging. Rewrite $\hat{f}_{LB}(x)$ as

$$\left(\tfrac{1}{2} - \tfrac{x}{h} \right) \frac{1}{nh} \sum_{i=1}^{n} \left[\left(\tfrac{3}{2} + \tfrac{x_i}{h} \right) I(x_i \in J_{-1}) + \left(\tfrac{1}{2} - \tfrac{x_i}{h} \right) I(x_i \in J_0) \right]$$

$$+ \left(\tfrac{1}{2} + \tfrac{x}{h} \right) \frac{1}{nh} \sum_{i=1}^{n} \left[\left(\tfrac{1}{2} + \tfrac{x_i}{h} \right) I(x_i \in J_0) + \left(\tfrac{3}{2} - \tfrac{x_i}{h} \right) I(x_i \in J_1) \right].$$

Let $I_k(x_i)$ denote $I(x_i \in J_k)$. Bringing all the terms under one sum and defining

$$a_{-1} = \tfrac{3h-6x}{4h^2}; \ b_{-1} = \tfrac{h-2x}{2h^3}; \ a_0 = \tfrac{1}{2h}; \ b_0 = \tfrac{2x}{h^3}; \ a_1 = \tfrac{3h+6x}{4h^2}; \ b_1 = -\tfrac{h+2x}{2h^3},$$

we have that $\hat{f}_{LB}(x)$ may be written as

$$\frac{1}{n} \sum_{i=1}^{n} \left[(a_{-1} + b_{-1}x_i)I_{-1}(x_i) + (a_0 + b_0 x_i)I_0(x_i) + (a_1 + b_1 x_i)I_1(x_i) \right].$$

Since this is the mean of an *iid* set of random variables, the variance of $\hat{f}_{LB}(x)$ will be the variance of the first term in the summand (i.e., when $i = 1$) divided by n. This leads to six terms, three of the form

$$\text{Var}\left[(a_k + b_k X_1)I(X_1 \in J_k) \right], \quad \text{for } k = 0, 1, 2,$$

and three of the form, twice

$$\text{Cov}\left[(a_k + b_k X_1)I(X_1 \in J_k), (a_j + b_j X_1)I(X_1 \in J_j) \right],$$

where $(k,j) = (-1,0), (-1,1)$, or $(0,1)$. Define the local "moments"

$$m_\ell(k) = \int_{J_k} x^\ell f(x)\, dx;$$

note that the bin probability $p_k \equiv m_0(k)$. A little calculation shows that the three variance terms are given by

$$a_k^2 p_k(1 - p_k) + b_k^2 \left[m_2(k) - m_1(k)^2 \right] + 2a_k b_k m_1(k)(1 - p_k),$$

and the three covariance terms are given by

$$-\left[a_k p_k + b_k m_1(k) \right] \cdot \left[a_j p_j + b_j m_1(j) \right].$$

Plugging all these in (it is sufficient to use the one-term Taylor's Series $f(x) = f(0) + \cdots$) and integrating the variance over the interval $J_0 = (-\tfrac{h}{2}, \tfrac{h}{2})$ gives $f(0)/(2nh) \times h - f(0)^2/n \times h + \cdots$. Thus the asymptotic integrated variance is given by

$$\text{AIV} = \sum_{k=-\infty}^{\infty} \left[\frac{f(kh)}{2nh} \times h - \frac{f(kh)^2}{n} \times h \right] = \frac{1}{2nh} - \frac{R(f)}{n}.$$

Assembling these findings, we have

Theorem 4.3:   *Suppose f'' is absolutely continuous and $R(f''') < \infty$. Then for the linear binning frequency polygon density estimator, $\hat{f}_{LB}(x)$,*

$$\text{AMISE}(h) = \frac{1}{2nh} + \frac{7}{240}h^4 R(f''); \text{ hence,} \tag{4.21}$$
$$h^* = (30/7)^{1/5} R(f'')^{-1/5} n^{-1/5}$$
$$\text{AMISE}^* = \frac{5^{4/5} 7^{1/5}}{8 \cdot 6^{1/5}} R(f'')^{1/5} n^{-4/5}. \tag{4.22}$$

The leading AMISE* constants in Theorems 4.1 and 4.3 are 0.5280 and 0.4672, respectively. Thus linear binning reduces the AMISE of the frequency polygon for any density f by 11.5%. The leading h^* constants are 1.578 and 1.338, respectively. Thus the linear binning bandwidth is always 15.2% narrower. This is to be expected since the adjusted bin counts include data from wider intervals.

While the universal gain of 11.5% is most impressive, it should be noted that one practical limitation is that the bin width h must be known before the linear binning can be applied. Often the usual bin counts are computed using a much finer mesh, and then those counts aggregated into an appropriate choice for h by cross-validation. In the application imagined by the authors for kernel estimators, discussed later in Section 6.7.2.2, h can be computed after the fact by using linear binning on a much finer mesh.

4.4.1.2   *Adjusting FP Bin Counts to Match Histogram Areas*   Minnotte (1996) noted that the ordinary frequency polygon does not preserve the bin histogram areas. He observed that if the raw bin counts, $\{\nu_k\}$ were perturbed by the formula

$$\tilde{\nu}_k = \sum_{j=-\infty}^{\infty} c_j \nu_{k+j}, \quad \text{where } c_j = 2^{\frac{1}{2}} (2^{\frac{3}{2}} - 3)^{|j|} = 2^{\frac{1}{2}} (-0.1716)^{|j|},$$

then the frequency polygon constructed using $\{\tilde{\nu}_k\}$ would have bin areas exactly matching those of the histogram. The resulting FP reduces AIMSE* by 4.4%.

In a follow-up paper, Minnotte (1998) investigated the use of higher order spline density estimates that also matched the histogram bin areas. The result was that high-order approximations could be obtained. For example, the 0th- and 1st-order splines are the histogram and the adjusted frequency polygon with convergence rates of $O(n^{-2/3})$ and $O(n^{-4/5})$, respectively. The 2nd- and 3rd-order splines have convergence rates of $O(n^{-6/7})$ and $O(n^{-8/9})$, approaching the theoretical upper bound of $O(\log(n) n^{-1})$. Minnotte does observe that all of the higher order estimates will *not* be nonnegative in general, but argues that is not a serious deficiency.

4.4.2   Polynomial Histograms

A completely different approach is to compute not just the bin count in B_k, but also the mean of the points in each bin (and perhaps the variance, etc). Intuitively, if the bin mean is to the right of the bin center, then the density should be increasing in the bin, and vice versa. In a series of papers, Scott and Sagae (1997), Sagae et al. (2006, 2009), Papkov and Scott (2010), and Jing et al. (2012) considered this idea formally.

Consider fitting a piecewise linear equation for the density, as with the frequency polygon, but within the histogram bin. For convenience, let the 0th bin $B_0 = (-\frac{h}{2}, \frac{h}{2})$ and write for the *linear polynomial histogram* (LPH)

$$\hat{f}(x) = \hat{f}_{LPH}(x) = a + bx \quad \text{for } x \in B_0. \tag{4.23}$$

Now the area of $\hat{f}(x)$ over B_0 is simply $\int_{B_0} f(x)dx = ah$; hence, the conditional density is given by

$$\hat{f}_{B_0}(x) = \hat{f}(x|x \in B_0) = \frac{a + bx}{ah} \quad x \in B_0.$$

Note $\int_{B_0} \hat{f}_{B_0}(x)dx = 1$. The idea is to pick the two constants a and b in $\hat{f}(x)$ to match the first two moment constraints (area and conditional mean):

$$\int_{B_0} \hat{f}(x)dx = \frac{\nu_0}{n} \quad \text{and}$$

$$\int_{B_0} x\hat{f}_{B_0}(x)dx = \bar{x}_0 \quad \text{where} \quad \bar{x}_0 = \frac{1}{\nu_0}\sum_{x_i \in B_0} x_i.$$

The unique solution to these constraints is

$$a = \frac{\nu_0}{nh} \quad \text{and} \quad b = \frac{12\nu_0\bar{x}_0}{nh^3}.$$

Let $I_i = I(X_i \in B_0)$ denote the event that the random sample X_i is in B_0. Then we may write the explicit solution for $\hat{f}(x)$ in the convenient form for analysis as

$$\hat{f}_{LPH}(x) = \frac{1}{nh}\sum_{i=1}^{n} I_i + \frac{12x}{nh^3}\sum_{i=1}^{n} I_i X_i$$

$$= \frac{1}{n}\sum_{i=1}^{n}\left[\frac{1}{h}I_i + \frac{12x}{h^3}I_i X_i\right]. \tag{4.24}$$

Define the local moments

$$m_\ell = E\left[I_i X_i^\ell\right] = \int_{B_0} x^\ell f(x)\,dx,$$

using the Taylor's Series (4.2). (Again, $p_0 \equiv m_0$). Then the expected value of the linear polynomial histogram is

$$E\hat{f}_{LPH}(x) = \frac{1}{h}E(I_1) + \frac{12x}{h^3}E(I_1X_1) = \frac{1}{h}m_0 + \frac{12x}{h^3}m_1$$

$$= f(0) + xf'(0) + \frac{1}{24}h^2f''(0) + O(h^3).$$

Subtracting the Taylor's Series from Equation (4.2) gives the bias

$$\text{Bias}(x) = E\hat{f}(x) - f(x) = \left(\frac{h^2}{24} - \frac{x^2}{2}\right)f''(0) + O(h^3); \text{ hence,}$$

$$\int_{B_0} \text{Bias}(x)^2 dx = \frac{1}{720}h^4f''(0)^2 \times h + O(h^7) \quad \text{and}$$

$$\text{AISB}(h) = \sum_{k=-\infty}^{\infty} \frac{1}{720}h^4f''(kh)^2 \times h = \frac{1}{720}h^4R(f'').$$

The variance of $\hat{f}_{LPH}(x)$ in Equation (4.24) is

$$\text{Var}\hat{f}(x) = \frac{1}{n}\left[\frac{1}{h^2}\text{Var}(I_1) + \frac{24x}{h^4}\text{Cov}(I_1, I_1X_1) + \frac{144x^2}{h^6}\text{Var}(I_1X_1)\right]$$

$$= \frac{1}{n}\left[\frac{1}{h^2}p_0(1-p_0) + \frac{24x}{h^4}m_1(1-p_0) + \frac{144x^2}{h^6}(m_2 - m_1^2)\right]$$

$$= \frac{f(0)}{nh} + \frac{12x^2f(0)}{nh^3} - \frac{f(0)^2}{n} + \cdots.$$

Integrating over B_0 gives $\frac{2f(0)}{nh} \times h - \frac{f(0)^2}{n} \times h$; hence, the asymptotic integrated variance is

$$\text{AIV}(h) = \sum_{k=-\infty}^{\infty} \frac{2f(kh)}{nh} \times h - \frac{f(0)^2}{n} \times h = \frac{2}{nh} - \frac{R(f)}{n}.$$

Thus we have shown

Theorem 4.4: *Suppose f'' is absolutely continuous and $R(f''') < \infty$. Then for the linear polynomial histogram estimator, $\hat{f}_{LPH}(x)$,*

$$\text{AMISE}(h) = \frac{2}{nh} + \frac{1}{720}h^4R(f''); \text{ hence,} \qquad (4.25)$$

$$h^* = 360^{1/5}R(f'')^{-1/5}n^{-1/5}$$

$$\text{AMISE}^* = (625/2304)^{1/5}R(f'')^{1/5}n^{-4/5}. \qquad (4.26)$$

Now the leading coefficient of AMISE* is 0.565, which is 45.9% greater than for the frequency polygon. However, the optimal bin width is over twice as wide (2.056).

One strong advantage of this method is that it is well suited for analyzing massive streaming data, as the bin counts and bin moments are easily updated without retaining the raw data. Furthermore, for local estimation and boundary situations, this approach avoids having to cross bin boundaries.

In the first column of Figure 4.9, three LPH estimates are displayed with samples from the mixture density (3.78) for $n = 10^4, 10^5$, and 10^6 using the optimal bin width. Since the LPH is computed bin by bin, the estimate is not continuous, although for sufficiently large samples, it almost appears to be continuous.

The second column in Figure 4.9 displays a continuous LPH spline estimate. The spline is constrained to match the area and conditional mean in each bin, and have a continuous second derivative everywhere. The extra parameters are determined by minimizing the roughness of the spline (as in $R(f'')$). The positivity constraint is not enforced, but the estimates look very good.

Finally, the third column in Figure 4.9 displays the piecewise quadratic polynomial histogram (QPH), which adds the constraint that the conditional variance be matched. The estimate for $x \in B_0$ is quadratic, $\hat{f}_{QPH} = a + bx + cx^2$, and the additional constraint requires

$$\int_{B_0} x^2 \hat{f}_{B_0}(x)dx = \int_{B_0} x^2 \frac{a+bx+cx^2}{ah+ch^3/12}\, dx = \frac{1}{n} \sum_{x_i \in B_0} x_i^2 .$$

A little algebra reveals that

$$a = \frac{9\nu_0}{4nh} - \frac{15}{nh^3} \sum_{i=1}^{n} I_i X_i^2$$

$$b = \frac{12}{nh^3} \sum_{i=1}^{n} I_i X_i$$

$$c = -\frac{15\nu_0}{nh^3} + \frac{180}{nh^5} \sum_{i=1}^{n} I_i X_i^2 .$$

This estimate can be shown to be of higher order $O(n^{-6/7})$, but is noisier for small sample sizes.

To summarize, the histogram may be improved by constructing the piecewise linear interpolator, adjusting the bin counts, or collecting local moment information in each bin.

4.4.3   How Much Information Is There in a Few Bins?

In this section, we review a re-analysis of a dataset from a phone survey ($n = 1207$) of a histogram with only four irregularly spaced bins, but of relatively high accuracy.

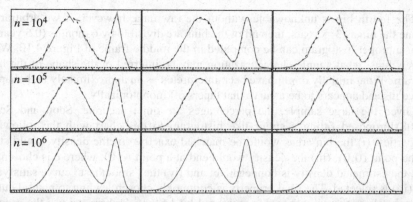

FIGURE 4.9 For three sample sizes from the mixture density, examples of the piecewise LPH, the continuous LPH, and the piecewise QPH estimates.

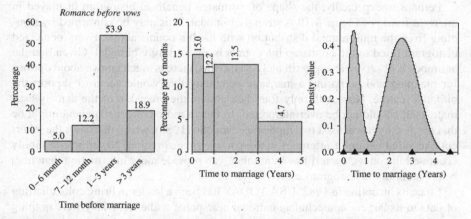

FIGURE 4.10 (Left) Barchart of raw improperly normalized binned marriage survey data. (Middle) Example of a barchart that is properly normalized. (Right) A penalized histogram of the data matching the four bin proportions.

A barchart of the data was published on October 13, 2006, in the front-page daily graphical feature *USA TODAY Snapshots®*. Men were asked how long they were romantically involved before marriage. The basic barchart that appeared is recreated in Figure 4.10. For such a simple chart based on only four counts, it is perhaps remarkable that there are two significant errors. From the statistician's point of view, the bin widths are not of equal width; hence, plotting raw percentages as rectangles of equal width is a serious graphical error, as it overemphasizes the counts in wider intervals. But a more fundamental error is that the percentages add up to only 90.0%. A check of the data source (Glenn, 2005) reveals that the first bin is actually 15.0%, as the bin counts are 181, 147, 651, and 228.

The fourth bin is unknowable without the raw data; however, if we arbitrarily define the bin as 3–5 years, then all of the bins are divisible by 6 months (0.5 years). Thus a proper histogram can be computed in the middle frame of Figure 4.10. With so few bins, what simple characterization may be inferred of the shape of the histogram? What initially might have been interpreted as an approximately bell-shaped curve instead appears to be a curve that tapers off monotonically.

Given the large sample, the percentages are quite accurate. Scott and Scott (2008) proposed fitting a spline-like histogram to this chart with the following properties: (i) the bin areas would be matched exactly; (ii) the density would start at the point $(0,0)$; (iii) the density would end at a point $(c,0)$, where c is chosen so that the estimated density is nonnegative; and (iv) the "smoothest" curve satisfying (i)–(iii) is selected. The final property is commonly chosen leading to natural cubic splines, where the smoothness is measured by $\int f''(x)^2 dx$; here we use the second difference in the place of the second derivative. The linear algebra formulation that solves these four properties may be found in Scott and Scott (2008). For these data, the value $c = 4.\overline{7}$ was appropriate to enforce nonnegativity.

Perhaps unexpectedly, the shape of estimated penalized histogram displayed in the right frame of Figure 4.10 is strongly bimodal. This may be confirmed by sampling from the multinomial distribution with the bin counts above. Every penalized histogram fitted to the bootstrap bin counts is also strongly bimodal. Given that the antimode is observed at approximately 14 months, there is much known about options for planning and preparing a marriage ceremony that would seem to support the bimodal feature. Now with only four data points, the precision of the density estimate itself should not be overinterpreted. In fact, a more correct statement might be that the density has at least two modes see Donoho (1988), which discusses the inherent one-sided nature of inference in nonparametric estimation. Nevertheless, highly accurate bin counts, even if few in number, can provide more information than may be apparent in the histogram itself.

Since its inception in 1982, USA TODAY has been a leader in bring color charting of data to its largely nontechnical national newspaper audience. Many were nothing more than pie charts, but it was common to find statisticians criticizing the quality of the diagrams. Today, however, the quality of such charts is quite high. Probably the leading popular newspaper for professionally produced graphics is the *New York Times*, which maintains a large number of graphical editors who construct the usually engrossing and data-rich graphics that are often interactive and map-based.

PROBLEMS

4.1 Demonstrate that the frequency polygon that interpolates the histogram at the midpoints of an equally spaced bins integrates to 1. Investigate alternative definitions of a FP derived from an adaptive or unequally spaced histogram.

4.2 Verify the derivations of the bias and variance of the FP in Section 4.1.1.

4.3 Consider the FP when the data come from the negative exponential density function, $f(x) = e^{-x}, x \geq 0$. Using the histogram mesh $(-h, 0, h, 2h, \ldots)$, compute the contribution to the bias from the bins adjacent to $x = 0$ and show that the total integrated squared bias over $(-h/2, h/2)$ is no longer $O(h^4)$, but rather $h/12 + O(h^2)$. Compare this result to the corresponding result for the histogram. *Hint*: Use a symbolic computer package to compute the probability exactly and take a Taylor's series at the end.

4.4 One suggested fix to the boundary problem is to reflect the data around 0, that is, compute the FP using the data $-x_n, \ldots, -x_1, x_1, \ldots, x_n$ and then doubling the estimate for $x \geq 0$. Consider again the negative exponential density.

  (a) Show that using the same mesh as in Problem 3.3 gives a "flat" histogram-like estimate over $(0, h/2)$, which contributes a term of order h^2 to the integrated squared bias.

  (b) Show that the histogram mesh $(-3h/2, -h/2, h/2, 3h/2, \ldots)$ with reflected data leads to a contribution towards the ISB from the bin $(0, h)$ equal to $h^3/48 + O(h^4)$, which is between the usual histogram and FP exponent orders.

4.5 Find some simple approximation to the skewness and kurtosis factors in Section 4.1.2. Try them on some simulated data.

4.6 Consider the FP roughness estimate given in Equation (4.9).

  (a) Show that it is unbiased to first order at $h = h^*$.

  (b) Alternatively, show that

$$[80/(129n^2h^5)] \sum_k (\nu_{k+1} - 2\nu_k + \nu_{k-1})^2$$

  is also unbiased to first order.

  (c) Construct the two BCV estimators that follow from these two estimators of $R(f'')$ and compare them empirically on simulated data.

4.7 Find the UCV formula for an FP and try it on the snowfall dataset. Can you use the LRL dataset? What happens if you blur LRL the data uniformly over the bins?

4.8 Compute the asymptotic efficiency of an optimal adaptive mesh relative to a fixed mesh for normal and Cauchy data.

4.9 How do kth-nearest-neighbor meshes (equal number of points in each bin) perform for the FP? Make a figure when $f(x) = \text{Beta}(5,5)$.

4.10 Consider the problem of estimating the curvature using a frequency polygon (or a histogram) by the finite second difference estimator

$$\hat{f}''(x) = \frac{\hat{f}_1 - 2\hat{f}_0 + \hat{f}_{-1}}{h^2}.$$

Show that the AISB is $h^2 R(f''')/12$, the AIV is $6/(nh^5)$, and hence $h^* = (180/nR(f'''))^{1/7}$. What is the normal reference rule? Hint: Use the mesh $(-3h/2, -h/2, h/2, 3h/2)$ for the "typical bin" $(-h/2, h/2)$.

4.11   We have seen the probability the bin containing the true mode is a sample mode varies if $h = O(n^{-1/3})$ or $h = O(n^{-1/5})$ in Equations (3.77) and (4.15), respectively. What is this probability if you use even wider bins of order $O(n^{-1/7})$, redefining h^* appropriately in Equation (4.14)? What do you conclude?

4.12   Perform a simulation study that confirms the predictions of Theorems 4.3 and 4.4, Minnotte's area-matching frequency polygon, and one of the several versions of the polynomial histogram.

5

AVERAGED SHIFTED HISTOGRAMS

A simple device has been proposed for eliminating the bin edge problem of the frequency polygon while retaining many of the computational advantages of a density estimate based on bin counts. Scott (1983, 1985b) considered the problem of choosing among the collection of multivariate frequency polygons, each with the same smoothing parameter but differing bin origins. Rather than choosing the "smoothest" such curve or surface, he proposed averaging several of the shifted frequency polygons. As the average of piecewise linear curves is also piecewise linear, the resulting curve appears to be a frequency polygon as well. If the weights are nonnegative and sum to 1, the resulting "averaged shifted frequency polygon" (ASFP) is nonnegative and integrates to 1.

A nearly equivalent device is to average several shifted histograms, which is just as general but simpler to describe and analyze. The result is the "averaged shifted histogram" (ASH). Since the average of piecewise constant functions such as the histogram is also piecewise constant, the ASH appears to be a histogram as well. In practice, the ASH is made continuous using either of the linear interpolation schemes described for the frequency polygon in Chapter 4, and will be referred to as the frequency polygon (FP) of the (ASH). The ASH is the practical choice for computationally and statistically efficient density estimation. Algorithms for its evaluation are described in detail.

Multivariate Density Estimation, First Edition. David W. Scott.
© 2015 John Wiley & Sons, Inc. Published 2015 by John Wiley & Sons, Inc.

5.1   CONSTRUCTION

Consider a collection of m histograms, $\hat{f}_1, \hat{f}_2, \ldots, \hat{f}_m$, each with bin width h, but with bin origins

$$t_0 = 0, \frac{h}{m}, \frac{2h}{m}, \ldots, \frac{(m-1)h}{m}, \tag{5.1}$$

respectively. The (*naive* or unweighted) ASH is defined as follows:

$$\hat{f}(\cdot) = \hat{f}_{\text{ASH}}(\cdot) = \frac{1}{m} \sum_{i=1}^{m} \hat{f}_i(\cdot). \tag{5.2}$$

Observe that the ASH is piecewise constant over intervals of width $\delta \equiv h/m$, as the bin origins in (5.1) differ by this amount.

Reexamine the sequence of shifted histograms of the Buffalo snowfall data shown in Figure 4.8. Each shifted histogram has bin width $h = 12.5$. In Figure 5.1, a series of ASHs using this same bin width is shown for an increasing sequence in the parameter m. Although the ordinary histogram (ASH with $m = 1$) displays a second bump to the right of the mode, every ASH with $m > 1$ reveals the presence of larger third bump to the left of the mode. The third bump was masked by the larger bump at the mode. The appearance of these additional bumps is not an artifact of the ASH algorithm, but rather the result of a significantly improved signal-to-noise ratio obtained by averaging out the nuisance parameter t_0. In a sense, the parameter t_0 has been replaced by

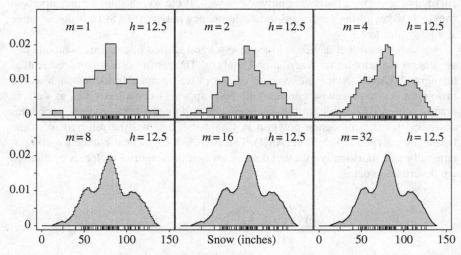

FIGURE 5.1   Naive averaged shifted histograms of the Buffalo snowfall data with bin width $h = 12.5$ inches.

a different parameter m that must be specified; however, the improvement over the ordinary histogram justifies any additional work.

Multivariate ASHs are constructed by averaging shifted multivariate histograms, each with bins of dimension $h_1 \times h_2 \times \cdots \times h_d$. If every possible multivariate histogram is constructed by coordinate shifts that are multiples of $\delta_i \equiv h_i/m_i, i = 1,\ldots,d$, then the multivariate ASH is the average of $m_1 \times m_2 \times \cdots \times m_d$ shifted histograms. In the bivariate case, the ASH is given by

$$\hat{f}(\cdot,\cdot) = \frac{1}{m_1 m_2} \sum_{i=1}^{m_1} \sum_{j=1}^{m_2} \hat{f}_{ij}(\cdot,\cdot), \tag{5.3}$$

where the bin origin for the bivariate shifted histogram $\hat{f}_{ij}$ is the point $(x,y) = ((i-1)\delta_1, (j-1)\delta_2)$. Figure 5.2 displays several bivariate ASH's of the plasma lipid dataset (see Table B.3) with $m_i = 1, 2, 3$. Only a few shifts along each axis are required to provide a smoother estimate. The underlying histogram bin size is the same for all three estimates, although the apparent number of bins in the ASH grows from 8^2 to 16^2 to 24^2. A contour plot of the linear interpolant of the ASH with $m_1 = m_2 = 3$ suggests a multimodal structure not apparent in the original histogram. Only the bivariate glyph histogram in Figure 3.21 hints at the structure in this dataset.

The recognition of the need to smooth a bivariate histogram is not new. In 1886, Galton performed bivariate bin smoothing on a cross-tabulation of 928 adult children and their parents' average height (Stigler, 1986, p. 285). Galton further adjusted all female heights upward by a factor of 1.08 to account for male–female height

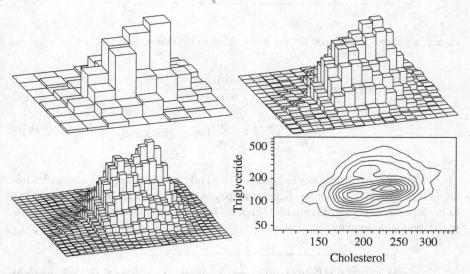

FIGURE 5.2 Bivariate averaged shifted histograms of the lipid dataset for 320 diseased males; see text for ASH parameter values.

differences. Stigler quotes Galton's description of how he then smoothed his raw bin counts:

> ... by writing at each intersection of a horizontal column with a vertical one, the sum of the entries in the four adjacent squares, and using these to work upon.

This smoothing sharpened the elliptical shape of the contours of the data. Galton's smoothing corresponds roughly to the bivariate ASH with $m_1 = m_2 = 2$ (see Problem 5.1).

5.2 ASYMPTOTIC PROPERTIES

As the univariate ASH is piecewise constant over the intervals $[k\delta, (k+1)\delta)$ where $\delta \equiv h/m$, it is convenient to refer to this *narrower interval* as the bin B_k, and let

$$v_k = \text{bin count in bin } B_k, \quad \text{where } B_k \equiv [k\delta, (k+1)\delta).$$

With this new definition of the bin intervals, the bin count for an ordinary histogram may be obtained by adding m of the adjacent bin counts $\{v_k\}$ from the finer grid.

Consider the ASH estimate in bin B_0. The height of the ASH for x in B_0 is the average of the heights of the m shifted histograms, each of width $h = m\delta$, that all include the bin count v_0 and the bin B_0 in their span:

$$\frac{v_{1-m} + \cdots + v_0}{nh}, \frac{v_{2-m} + \cdots + v_0 + v_1}{nh}, \ldots, \frac{v_0 + \cdots + v_{m-1}}{nh}.$$

Hence, a general expression for the naive ASH in Equation (5.2) is

$$\hat{f}(x; m) = \frac{1}{m} \sum_{i=1-m}^{m-1} \frac{(m - |i|)v_{k+i}}{nh}$$

$$= \frac{1}{nh} \sum_{i=1-m}^{m-1} \left(1 - \frac{|i|}{m}\right) v_{k+i} \quad \text{for } x \in B_k. \tag{5.4}$$

The weights on the bin counts in Equation (5.4) take on the shape of an isosceles triangle with base $(-1, 1)$. Other shapes may be contemplated, such as uniform weights or perhaps smoother (differentiable) shapes. The general ASH uses arbitrary weights, $w_m(i)$, and is defined by

$$\boxed{\text{General ASH:} \quad \hat{f}(x; m) = \frac{1}{nh} \sum_{|i| < m} w_m(i) \, v_{k+i} \quad \text{for } x \in B_k.} \tag{5.5}$$

In order that $\int \hat{f}(x; m)dx = 1$, the weights must sum to m (see Problem 5.2). An easy way to define general weights is

$$w_m(i) = m \times \frac{K(i/m)}{\sum_{j=1-m}^{m-1} K(j/m)} \quad i = 1 - m, \ldots, m - 1, \tag{5.6}$$

where K is a continuous function defined on $(-1, 1)$. K is often chosen to be a probability density function, such as

$$K(t) = \frac{15}{16}(1 - t^2)_+^2 = \frac{15}{16}(1 - t^2)^2 I_{[-1,1]}(t), \tag{5.7}$$

which is called the *biweight kernel* or quartic kernel.

The computational algorithm for the generalized ASH is quite simple. Construct an equally spaced mesh of width δ over the interval (a, b), and compute the corresponding bin counts $\{v_k, k = 1, \ldots, nbin\}$ for the n data points. Typically, $\delta \ll h$, and *nbin* refers to the number of bins width δ. This computation is accomplished by the BIN1 algorithm given in the box.

BIN1$(x, n, a, b, nbin)$ **Algorithm**:   (* Bin univariate data *)

$\delta = (b - a)/nbin$
for $k = 1, nbin \{v_k = 0\}$
for $i = 1, n\{$
$\quad k = (x_i - a)/\delta + 1$ (* integer part *)
$\quad$ if $(k \in [1, nbin])$ $v_k = v_k + 1\}$
return $(\{v_k\})$

Next, compute the weight vector, $\{w_m(i)\}$, as in Equation (5.6). Then the univariate ASH estimates, $\{f_k, k = 1, \ldots, nbin\}$, over the *nbin* intervals may be computed in an efficient manner by reordering the operations indicated in Formula (5.5). Rather than computing the ASH estimates individually in each bin by sweeping through the $2m - 1$ adjacent bin counts, a single pass is made through the bin counts, with a weighted count applied to the $2m - 1$ adjacent ASH estimates. This modification avoids repeated weighting of empty bins; see the ASH1 algorithm given in the box. The algorithm assumes that there are at least $m - 1$ empty bins on each end. Observe that the amount of work is determined by m and by the number of nonempty bins. The algorithm is quite efficient even when $n > 10^6$, in which case most of the work involves tabulating the several hundred bin counts.

ASH1 $(m, v, nbin, a, b, n, w_m)$ **Algorithm**:   (* Univariate ASH *)

$$\delta = (b - a)/nbin$$
$$h = m\delta$$
for $k = 1, nbin\{f_k = 0\}$
for $k = 1, nbin\{$
    if$(v_k = 0)$ next k
    for $i = \max(1, k - m + 1), \min(nbin, k + m - 1)\{$
        $f_i = f_i + v_k w_m(i - k)\}\}$
for $k = 1, nbin\{f_k = f_k/(nh); t_k = a + (k - 0.5)\delta\}$
return $(\mathbf{x} = \{t_k\}, \mathbf{y} = \{f_k\})$ (* Bin centers and ASH heights *)

In Figure 5.3, examples of the ASH using the biweight kernel are shown. For the Buffalo snowfall data, observe how the use of the biweight kernel weights rather than the isosceles triangle weights results in a visually smoother curve, with less local noise in the estimate. (As the variances of the triangle and biweight kernels are 1/6 and 1/7, respectively, a bin width of $h = 13.5 \times \sqrt{7/6} = 14.6$ inches was applied with the biweight kernel. This rescaling is justified in Section 6.2.3.3.) For a large dataset such as the German household income data, the additional visual smoothness is still apparent, even when the smoothing parameters are small enough to reveal any possible feature (compare with Figure 3.18).

In practice, the narrow bin width δ is usually fixed first by choosing between 50 and 500 bins over the sample range (extended by 5–10% to include some empty bins on both sides). Since $h = m\delta$, only values of the smoothing parameter h that are integer

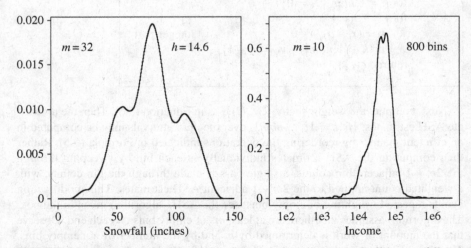

FIGURE 5.3 Examples of ASH with biweight kernel applied to the Buffalo snowfall and German household income datasets.

multiples of δ may be considered, although it is easy to remove this restriction (see Problem 5.4). On the other hand, if h is known, then δ may be computed as $h/5$ or $h/10$. This case is rare. Many large datasets are prebinned; that is, the raw data are not recorded, only the bin counts. If the width of those bins is called δ and h^* turns out to be close to δ, then no additional smoothing can be applied since $m = 1$ is the only option. Careful planning can avoid such an unfortunate outcome. For example, using the oversmoothed FP bin width rule in Equation (4.11), choose δ sufficiently small or n sufficiently large so that $\delta < h_{OS}/25$ or $\delta < h_{OS}/50$. Only a small pilot study is required to estimate the variance of the data to use in the bin width rule for oversmoothed frequency polygon.

The derivation of the AMISE for the naive (isosceles triangle weight function) ASH is similar to previous calculations and is not given here. The corresponding result for the general weighted ASH is much more complicated. Scott (1985b) proved the following result.

Theorem 5.1: *For the naive ASH with the isosceles triangle kernel,*

$$\text{AMISE} = \frac{2}{3nh}\left(1 + \frac{1}{2m^2}\right) + \frac{h^2}{12m^2}R(f')$$
$$+ \frac{h^4}{144}\left(1 - \frac{2}{m^2} + \frac{3}{5m^4}\right)R(f''). \tag{5.8}$$

The first term in the AMISE gives the error due to the integrated variance. The ISB or bias portion of the AMISE combines terms involving $R(f')$ and $R(f'')$, which were found in the ISB of the histogram and frequency polygon, respectively.

It is easy to check that the first two terms in this result match the ordinary histogram result in Theorem 3.1 when $m = 1$. On the other hand, as $m \to \infty$, the second histogram-like bias term disappears and the bias is similar to that for a frequency polygon in Theorem 4.1. Usually, for $m \geq 10$, the middle term is negligible compared to the last term, which may be taken to equal $h^4/144$. Comparing Equations (4.6) and (5.8), the IV terms are identical, while the ISB for the ASH is 41% of the ISB for the FP. The optimal bin width for the naive ASH as $m \to \infty$ is simply

$$h^*_{m=\infty} = \left[24/(nR(f''))\right]^{1/5} \quad \left(= 2.576\sigma n^{-1/5} \ \text{if} f(x) = N(\mu, \sigma^2)\right).$$

The sample sizes in Table 5.1 summarize the efficiency of the ASH and other estimators with normal data. The ASH requires 80% of the samples required by the FP to achieve the same MISE. In fact, this figure of 80% holds true for any sampling density, asymptotically (compare Theorems 4.1 and 5.1). In some difficult situations, such as a small sample from a rough density, the histogram may actually be competitive with the ASH. But asymptotically the efficiency of the histogram will be 0

TABLE 5.1   Equivalent Sample Sizes Required for
AMISE $\approx$ 1/400 for $N(0,1)$ Data

Estimator	$N(\bar{x}, s^2)$	ASH	FP-ASH	FP	Histogram
Sample size	100	436	436	546	2297

relative to the ASH or FP, because of the different rates of convergence of MISE. Of course, the improvement of the ASH relative to the FP is not as dramatic as the improvement of the FP relative to the histogram, as diminishing returns begin to take effect.

The expression for the asymptotic L_2 error of the FP-ASH or linear interpolant of the naive ASH is much simpler than for the naive ASH itself.

Theorem 5.2:   *For the frequency polygon interpolant of the naive ASH,*

$$\text{AMISE} = \frac{2}{3nh} + \frac{h^4}{144}\left(1 + \frac{1}{m^2} + \frac{9}{20m^4}\right)R(f''). \qquad (5.9)$$

Notice that the histogram-like bias term involving $R(f')$ has vanished. Further, the dependence of the remaining terms on the choice of m is greatly reduced. Usually, $m \geq 3$ is sufficient to achieve the 20% improvement in AMISE over the frequency polygon, and not $m \geq 10$ as recommended for the ASH itself.

The multivariate FP-ASH has been studied by Scott (1985b) using a triangular mesh, but the linear blend results of Hjort (1986) are more elegant and are reported here. Let subscripts on f denote partial derivatives.

Theorem 5.3:   *The AMISE of the multivariate linear blend of the naive ASH equals*

$$\frac{2^d}{3^d n\, h_1 \cdots h_d} + \frac{1}{720}\sum_{i=1}^{d}\delta_i^4 R(f_{ii}) + \frac{1}{144}\int_{\Re^d}\left[\sum_{i=1}^{d}h_i^2\left(1 + \frac{1}{2m_i^2}\right)f_{ii}\right]^2. \qquad (5.10)$$

Except in special circumstances, closed-form expressions for the optimal smoothing parameters are not available. Rather they must be obtained by solving a system of nonlinear equations. If $\delta_i \approx 0$ in (5.10), then $h_i^* = O(n^{-1/(4+d)})$ and $\text{AMISE}^* = O(n^{-4/(4+d)})$, which are comparable to the results for the multivariate frequency polygon in Equation (4.17). While the rates are the same, the multivariate FP is inferior by a fixed amount.

The BIN2 and ASH2 algorithms for $d = 2$ are given later. Note that the parameters in the univariate ASH become vectors in the bivariate algorithm. The BIN2 and ASH2 algorithms are easily extended to the cases $d = 3$ and 4 by increasing the dimensions on the vectors and matrices. For dimensions greater than 4, it is generally not possible to fit arrays of sufficient dimension directly in computer memory. In those cases, the ASH algorithm may be modified to compute only two- or three-dimensional slices of the higher dimensional ASH.

BIN2$(x, n, a, b, nbin)$ **Algorithm**: (* Bin bivariate data *)

> for $j = 1, 2\{\delta_j = (b_j - a_j)/nbin_j\}$
> for $k_1 = 1, nbin_1\{\text{for} k_2 = 1, nbin_2\{\nu_{k_1 k_2} = 0\}\}$
> for $i = 1, n\{$
> > for $j = 1, 2\{k_j = 1 + (x_{ij} - a_j)/\delta_j\}$ (* integer part *)
> > $\nu_{k_1 k_2} = \nu_{k_1 k_2} + 1\}$
> return $(\{\nu_{k\ell}\})$

ASH2$(m, v, nbin, a, b, n, w_{m_1}, w_{m_2})$ **Algorithm**: (* Bivariate ASH *)

> for $i = 1 - m_1, m_1 - 1\{\text{for} j = 1 - m_2, m_2 - 1\{$
> $w_{ij} = w_{m_1}(i)w_{m_2}(j)\}\}$
> for $j = 1, 2\{\delta_j = (b_j - a_j)/nbin_j; h_j = m_j\delta_j\}$
> for $k = 1, nbin_1\{\text{for} \ell = 1, nbin_2\{f_{k\ell} = 0\}\}$
> for $k = 1, nbin_1\{\text{for} \ell = 1, nbin_2\{$
> > if $(\nu_{k\ell} = 0)$nextℓ
> > for $i = \max(1, k - m_1 + 1), \min(nbin, k + m_1 - 1)\{$
> > > for $j = \max(1, \ell - m_2 + 1), \min(nbin, \ell + m_2 - 1)\{$
> > > $f_{ij} = f_{ij} + \nu_{k\ell}w_{(i-k)(j-\ell)}\}\}\}\}$
> for $k = 1, nbin_1\{\text{for} \ell = 1, nbin_2\{f_{k\ell} = f_{k\ell}/(nh_1 h_2)\}\}$
> for $k = 1, nbin_1\{t_{1k} = a_1 + (k - 0.5)\delta_1\}$ (* Bin centers x-axis *)
> for $k = 1, nbin_2\{t_{2k} = a_2 + (k - 0.5)\delta_2\}$ (* Bin centers y-axis *)
> return $(\mathbf{x} = \{t_{1k}\}, \mathbf{y} = \{t_{2k}\}, \mathbf{z} = \{f_{k\ell}\})$ (* $\mathbf{z}$ = ASH *)

5.3 THE LIMITING ASH AS A KERNEL ESTIMATOR

The parameter m in the ASH is a nuisance parameter, but much less so than the bin origin. The precise choice of m is unimportant as long as it is greater than 2 and h is well-chosen. Then why study the limiting behavior of the ASH as $m \to \infty$, where the ASH loses computational efficiency? The limit is in a class of nonparametric estimators that has been extensively studied since the pioneering works of Fix and Hodges (1951), Rosenblatt (1956), and Parzen (1962).

With h and n fixed and m increasing, it is easy to isolate the effect of a single data point x_j on the ASH estimate $\hat{f}(x)$, at a fixed point x. If $x \in B_k$ and $x_j \in B_{k+i}$, where the index labeling of the bins changes as m increases, then from Equation (5.4) the influence of x_j on x is proportional to

$$1 - \frac{|i|}{m} = 1 - \frac{|i| \cdot \delta}{m \cdot \delta} = 1 - \frac{|x - x_j|}{h} + O\left(\frac{\delta}{h}\right), \quad \text{if} |x - x_j| < h. \tag{5.11}$$

If x_j is not in the interval $(x - h, x + h)$, then the influence is 0. Note that the number of bins between x and x_j is approximately i, since these points are in bins B_k and B_{k+i}, respectively; hence, $|x - x_j| \approx |i| \cdot \delta$. Equation (5.4) may be reexpressed as follows:

$$\lim_{m \to \infty} \hat{f}(x; m) = \frac{1}{nh} \sum_{j=1}^{n} \left(1 - \frac{|x - x_j|}{h}\right) I_{[-1,1]}\left(\frac{x - x_j}{h}\right), \tag{5.12}$$

where the sum is over the number of data points rather than the number of bins. Defining a *kernel function* $K(\cdot)$ to be an isosceles triangle density,

$$K(t) = (1 - |t|)I_{[-1,1]}(t), \tag{5.13}$$

the limiting ASH may be written as follows:

$$\hat{f}(x) = \frac{1}{nh} \sum_{i=1}^{n} K\left(\frac{x - x_i}{h}\right). \tag{5.14}$$

Formula (5.14) also defines the *general kernel density estimator* with kernel K, corresponding to the generalized ASH in Equation (5.5). Apparently, the kernel estimate is simply a mixture density, which has n identical component densities centered on the data points. The component densities are the kernel functions. Any probability density may be chosen for the kernel, and sometimes kernels that are not densities are used. The ASH kernel always has finite support, but an infinite-support kernel such as the normal density is often chosen in (5.14). The isosceles triangle kernel density estimator could be described as an *indifferent histogram*, where the reference is to the uniform weighting over all possible choices for the bin origin of a histogram. Kernel estimators are studied in detail in Chapter 6.

Graphically, the kernel estimate places a probability mass of size $1/n$ in the shape of the kernel, which has been scaled by the smoothing parameter h, centered on each data point. These probability masses are then added vertically to give the kernel estimate. By contrast, the histogram uses a rectangular kernel but does not center these kernels on the data points; rather, these kernels are placed in a rigid mesh. In Figure 5.4, this process is illustrated with the silica dataset (see Table B.7) for several choices of the smoothing parameter and the isosceles triangle kernel. The 22 kernels for the individual data points are shown correctly scaled in each panel.

Of particular interest is the multivariate kernel corresponding to the multivariate naive ASH. Some algebra reveals that as $m_i \to \infty$,

$$\hat{f}(\mathbf{x}) = \frac{1}{nh_1 h_2 \cdots h_d} \sum_{i=1}^{n} \left\{ \prod_{j=1}^{d} K\left(\frac{x_j - x_{ij}}{h_j}\right) \right\}, \tag{5.15}$$

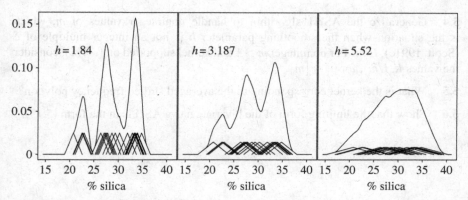

FIGURE 5.4   Triangle kernel estimates of the silica dataset showing the individual kernels.

where K is the univariate isosceles triangle kernel (5.13). This special form of the multivariate kernel function is called the *product kernel* and the estimate (5.15) the *product kernel estimator*. Although the individual multivariate product kernel does factor (implying that the coordinates are independent), the resultant density estimate does not factor, as is apparent from the examples displayed in Figure 5.2.

Thus the ASH provides a direct link to the better known kernel methods. However, kernel estimators are notoriously slow to compute, and many faster numerical approximations have been considered. The ASH is a *bona fide* density estimator and a natural candidate for computation. The ASH uses a discrete convolution to perform smoothing, a device well-known in spectral density estimation. The ASH construction was described independently by Chamayou (1980). The ASH is a special case of a more general framework called WARPing (weighted average of shifted points) developed by Härdle and Scott (1988), where the computational efficiency of the ASH is discussed in more detail. Wegman (1990) has used the ASH to address the problem of too much ink in the parallel coordinates plot discussed in Chapter 1. He proposed plotting the line segments as a series of points on a fine vertical mesh and plotting the contours of a bivariate ASH of those points.

PROBLEMS

5.1   Consider Galton's bivariate smoothing scheme, which placed equal weights on the counts in only four of the eight bins surrounding the bin of interest, and no weight on the count in the central bin. What are the weights on these nine bins with the bivariate naive ASH with $m_1 = m_2 = 2$?

5.2   Prove that if the weights $\{w_m(i)\}$ in Equation (5.5) sum to m, then the ASH integrates to 1.

5.3   Prove Theorem 5.2.

5.4   Generalize the ASH1 algorithm to handle noninteger values of m, which is the situation when the smoothing parameter h is not an integer multiple of δ (Scott, 1991c). *Hint:* For noninteger $m > 1$ and kernel supported on $(-1, 1)$, consider the values $K(i/m)$, for $|i| \leq \lfloor m \rfloor$.

5.5   What is the kernel corresponding to the averaged shifted frequency polygon?

5.6   Show that the limiting form of the bivariate naive ASH is in the form (5.15).

6

KERNEL DENSITY ESTIMATORS

It is remarkable that the histogram stood as the only nonparametric density estimator until the 1950s, when substantial and simultaneous progress was made in density estimation and in spectral density estimation. In a little-known technical report, Fix and Hodges (1951) introduced the basic algorithm of nonparametric density estimation. They addressed the problem of statistical discrimination when the parametric form of the sampling density was not known. Fortunately, this paper has been reprinted with commentary by Silverman and Jones (1989). During the following decade, several general algorithms and alternative theoretical modes of analysis were introduced by Rosenblatt (1956), Parzen (1962), and Cencov (1962). There followed a second wave of important and primarily theoretical papers by Watson and Leadbetter (1963), Loftsgaarden and Quesenberry (1965), Schwartz (1967), Epanechnikov (1969), Tarter and Kronmal (1970), and Wahba (1971). The natural multivariate generalization was introduced by Cacoullos (1966). Finally, in the 1970s came the first papers focusing on the practical application of these methods: Scott et al. (1978) and Silverman (1978b). These and later multivariate applications awaited the computing revolution.

The basic kernel estimator may be written compactly as

$$\hat{f}(x) = \frac{1}{nh} \sum_{i=1}^{n} K\left(\frac{x-x_i}{h}\right) = \frac{1}{n} \sum_{i=1}^{n} K_h(x-x_i), \qquad (6.1)$$

Multivariate Density Estimation, First Edition. David W. Scott.
© 2015 John Wiley & Sons, Inc. Published 2015 by John Wiley & Sons, Inc.

where $K_h(t) = K(t/h)/h$, which is a notation introduced by Rosenblatt (1956). The kernel estimator can be motivated not only as the limiting case of the averaged shifted histogram as in (5.14) but also by other techniques demonstrated in Section 6.1. In fact, virtually all nonparametric algorithms are asymptotically kernel methods, a fact demonstrated empirically by Walter and Blum (1979) and proved rigorously by Terrell and Scott (1992). Woodroofe (1970) called the general class "delta sequences."

6.1   MOTIVATION FOR KERNEL ESTIMATORS

From the vantage point of a statistician or instructor, the averaging of shifted histograms seems the most natural motivation for kernel estimators. However, following other starting points in numerical analysis, time series, and signal processing provides, a deeper understanding of kernel methods. When trying to understand a particular theoretical or practical point concerning a nonparametric estimator, not all approaches are equally powerful. For example, Fourier analysis provides sophisticated tools for theoretical purposes. The bias-variance trade-off can be recast in terms of low-pass and high-pass filters in signal processing. Each is describing the same entity but with different mathematics.

6.1.1   Numerical Analysis and Finite Differences

The kernel estimator originated as a numeric approximation to the derivative of the cumulative distribution function (Rosenblatt, 1956). The empirical probability density function, which was defined in Equation (2.2) as the formal derivative of the empirical cdf $F_n(x)$ is a sum of Dirac delta functions, which is useless as an estimator of a smooth density function. Consider, however, a one-sided finite difference approximation to the derivative of $F_n(\cdot)$:

$$\hat{f}(x) = \frac{F_n(x) - F_n(x-h)}{h}$$

$$= \frac{1}{nh} \sum_{i=1}^{n} I_{[x-h,x)}(x_i) = \frac{1}{nh} \sum_{i=1}^{n} I_{(0,1]}\left(\frac{x-x_i}{h}\right), \qquad (6.2)$$

which from Equation (6.1) is clearly a kernel estimator with $K = U(0,1]$. As $E[F_n(x)] = F(x)$ for all x, then with the Taylor's series

$$F(x-h) = F(x) - hf(x) + \frac{1}{2}h^2 f'(x) - \frac{1}{6}h^3 f''(x) + \cdots,$$

the bias is easily computed as

$$\text{Bias}\{\hat{f}(x)\} = E[\hat{f}(x)] - f(x) = -\frac{1}{2}hf'(x) + O(h^2).$$

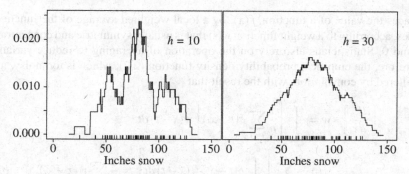

FIGURE 6.1 Central difference estimates of the Buffalo snowfall data.

Thus the integrated squared bias is $h^2 R(f')/4$, which is comparable to the order of the integrated squared bias (ISB) of the histogram in Theorem 3.1 rather than the $O(h^4)$ ISB of the frequency polygon (FP). Furthermore, the ISB of (6.2) is three times larger than the ISB of the histogram. (The integrated variances are identical; see Problem 6.1.) Thus, the one-sided kernel estimator (6.2) is inferior to a histogram.

Without comment, Rosenblatt proposed a two-sided or central difference estimator of f:

$$\hat{f}(x) = \frac{F_n\left(x + \frac{h}{2}\right) - F_n\left(x - \frac{h}{2}\right)}{h}. \tag{6.3}$$

The bias of (6.3) turns out to be $h^2 f''(x)/24$ (see Problem 6.2). Thus the squared bias is $O(h^4)$, matching that of the FP. The corresponding kernel is $K = U(-0.5, 0.5)$. Recall that the histogram placed a rectangular block into the bin where each data point fell. The one-sided estimator (6.2) places the left edge of a rectangular block at each data point, whereas the two-sided estimator (6.3) places the center of a rectangular block at each data point (see Tarter and Kronmal (1976)).

Figure 6.1 displays two central difference estimates of the Buffalo snowfall data. Compared with the FP, these estimates are inferior graphically, as the estimate contains $2n$ jumps that are not even equally spaced. However, most criteria such as MISE are not particularly sensitive to such local noisy behavior.

Quasi-Newton optimization codes routinely make use of numerical central difference estimates of derivatives. Some codes use even "higher order" approximations to the first derivative (see Section 6.2.3.1).

6.1.2 Smoothing by Convolution

An electrical engineer facing a noisy function will reach into a grab bag of convolution filters to find one which will smooth away the undesired high-frequency components. The convolution operation "$*$", which is defined as

$$(f * w)(x) = \int_{-\infty}^{\infty} f(t) w(x - t) \, dt,$$

replaces the value of a function, $f(x)$, by a local weighted average of the function's values, according to a weight function $w(\cdot)$ that is usually symmetric and concentrated around 0. Statisticians also rely on the operation of averaging to reduce variance. Therefore, the empirical probability density function (2.2), which is too noisy, may be filtered by convolution, with the result that

$$\left[\frac{dF_n}{dx}\right] * w = \int_{-\infty}^{\infty} \left[\frac{1}{n}\sum_{i=1}^{n}\delta(t-x_i)\right] w(x-t)\,dt$$

$$= \frac{1}{n}\sum_{i=1}^{n}\left[\int_{-\infty}^{\infty}\delta(t-x_i)\,w(x-t)\,dt\right] = \frac{1}{n}\sum_{i=1}^{n}w(x-x_i), \qquad (6.4)$$

which is the second kernel form given in (6.1) but without the smoothing parameter h appearing explicitly as, for example, $w(t) = K_h(t)$. In general, the shape and extent of the convolution filter weight function w will depend on the sample size. The kernel estimator (6.1) uses a single "shape" for all sample sizes, and the width of the kernel is explicitly controlled through the smoothing parameter h. The literature on filter design often uses different terminology. For example, the width of the filter w may be controlled by the half-power point, where the filter reaches half its value at the origin.

6.1.3 Orthogonal Series Approximations

The heuristic introduction to smoothing by convolution may be formalized by an orthogonal series approximation argument. For simplicity, suppose that the density function f is periodic on the interval $[0, 1]$ so that the ordinary Fourier series basis, $\phi_\nu(t) = \exp(2\pi i \nu t)$, is appropriate. Every function, even noisy functions, may be expressed in terms of these basis functions as

$$f(x) = \sum_{\nu=-\infty}^{\infty} f_\nu \phi_\nu(x) \quad \text{where} \quad f_\nu = <f, \phi_\nu> = \int_0^1 f(x)\phi_\nu^*(x)\,dx. \qquad (6.5)$$

The basis functions are orthonormal, that is, $\int \phi_\nu^*(x)\phi_\mu(x)dx = \delta_{\mu\nu}$, where $\delta_{\mu\nu}$ is the Kronecker delta function and ϕ^* denotes complex conjugate. As f is a density function, the coefficient f_ν in Equation (6.5) may be expressed in statistical terms as

$$f_\nu = E[\phi_\nu^*(X)]; \quad \text{hence} \quad \hat{f}_\nu = \frac{1}{n}\sum_{\ell=1}^{n}\phi_\nu^*(x_\ell) \qquad (6.6)$$

is an unbiased and consistent estimator of the Fourier coefficient f_ν. (Note that the sum is over ℓ to avoid confusion with $i = \sqrt{-1}$.) As an extreme example, consider the Fourier coefficients of the empirical probability density function (2.2):

$$f_\nu = \int_0^1 \left[\frac{1}{n}\sum_{\ell=1}^{n}\delta(x-x_\ell)\right] \phi_\nu^*(x)\,dx = \frac{1}{n}\sum_{\ell=1}^{n}\phi_\nu^*(x_\ell) = \hat{f}_\nu,$$

where $\hat{f}_\nu$ is defined in (6.6). Since $\hat{f}_\nu$ and f_ν for the empirical pdf are *identical*, the following is formally true for any sample $\{x_\ell\}$:

$$\sum_{\nu=-\infty}^{\infty} \hat{f}_\nu \, \phi_\nu(x) = \frac{1}{n} \sum_{\ell=1}^{n} \delta(x - x_\ell), \qquad (6.7)$$

which is the empirical probability density function.

Cencov (1962), Kronmal and Tarter (1968), and Watson (1969) suggested smoothing the empirical density function by including only a few selected terms from Equation (6.7). Excluding terms of the form $|\nu| > k$ corresponds to what the engineers call "boxcar" filter weights

$$w_\nu(k) = \begin{cases} 1 & |\nu| \le k \\ 0 & \text{otherwise.} \end{cases} \qquad (6.8)$$

As the Fourier transform of the boxcar function is the *sinc* function, $\sin(x)/(\pi x)$, the estimate will be rough and will experience "leakage"; that is, sample points relatively distant from a point x will influence $\hat{f}(x)$. Wahba (1977) suggests applying a smooth tapering window to this series, which provides more fine tuning of the resulting estimate. She introduces two parameters, λ and p, that control the shape and extent of the tapering window:

$$w_\nu(\lambda, p) = \frac{1}{1 + \lambda(2\pi\nu)^{2p}} \qquad \text{for } |\nu| \le n/2. \qquad (6.9)$$

Both forms of the weighted Fourier estimate may be written explicitly as

$$\hat{f}(x) = \sum_\nu w_\nu \left[\frac{1}{n} \sum_{\ell=1}^{n} \phi_\nu^*(x_\ell) \right] \phi_\nu(x) = \frac{1}{n} \sum_{\ell=1}^{n} \left[\sum_\nu w_\nu \, \phi_\nu^*(x_\ell) \, \phi_\nu(x) \right], \qquad (6.10)$$

where the order of summations has been exchanged. With the Fourier basis, the orthogonal series estimator (6.10) equals

$$\hat{f}(x) = \frac{1}{n} \sum_{\ell=1}^{n} \left[\sum_\nu w_\nu \, e^{2\pi i \nu (x - x_\ell)} \right].$$

This estimator is now in the convolution form (6.4) of a fixed kernel estimator, with the filter (or kernel) defined by the quantity in brackets. Some examples of these kernel functions are shown in Figure 6.2. Wahba's equivalent kernels are smoother and experience less leakage. The parameters λ and p can be shown to have interpretations corresponding to the smoothing parameter and to the order of the finite difference approximation, respectively.

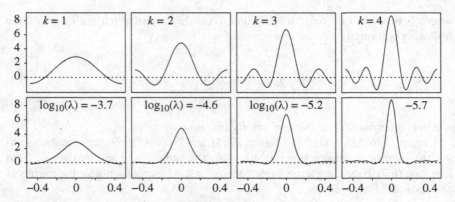

FIGURE 6.2 Examples of equivalent kernels for orthogonal series estimators. The four Wahba kernels (bottom row) have been selected to match the peak height of the corresponding Kronmal–Tarter–Watson kernels (top row). The Kronmal–Tarter–Watson kernels are independent of sample size; the Wahba examples are for $n = 16$.

6.2 THEORETICAL PROPERTIES: UNIVARIATE CASE

6.2.1 MISE Analysis

The statistical analysis of kernel estimators is much simpler than for histograms, as the kernel estimator (6.1) is the *arithmetic mean* of n independent and identically distributed random variables,

$$K_h(x, X_i) \equiv \frac{1}{h} K\left(\frac{x - X_i}{h}\right).$$

Therefore,

$$\mathrm{E}\{\hat{f}(x)\} = \mathrm{E}K_h(x, X) \quad \text{and} \quad \mathrm{Var}\{\hat{f}(x)\} = \frac{1}{n}\mathrm{Var}K_h(x, X). \tag{6.11}$$

The expectation equals

$$\mathrm{E}K_h(x, X) = \int \frac{1}{h} K\left(\frac{x - t}{h}\right) f(t)\, dt = \int K(w) f(x - hw)\, dw \tag{6.12}$$

$$= f(x) \int K(w) - h f'(x) \int w K(w) + \frac{1}{2} h^2 f''(x) \int w^2 K(w) + \cdots, \tag{6.13}$$

and the variance is given by

$$\mathrm{Var}K_h(x, X) = \mathrm{E}\left[\frac{1}{h} K\left(\frac{x - X}{h}\right)\right]^2 - \left[\mathrm{E}\frac{1}{h} K\left(\frac{x - X}{h}\right)\right]^2. \tag{6.14}$$

The second term in (6.14) was computed in (6.13) and is approximately equal to $[f(x) \int K(w) + \cdots]^2$, while the first term may be approximated by

$$\int \frac{1}{h^2} K\left(\frac{x-t}{h}\right)^2 f(t)\, dt = \int \frac{1}{h} K(w)^2 f(x-hw)\, dw \approx \frac{f(x)R(K)}{h}. \tag{6.15}$$

From Equation (6.13), if the kernel K satisfies

$$\int K(w) = 1, \quad \int wK(w) = 0, \quad \text{and} \quad \int w^2 K(w) \equiv \sigma_K^2 > 0,$$

then the expectation of $\hat{f}(x)$ will equal $f(x)$ to order $O(h^2)$. In fact,

$$\text{Bias}(x) = \frac{1}{2}\sigma_K^2 h^2 f''(x) + O(h^4) \quad \Rightarrow \quad \text{ISB} = \frac{1}{4}\sigma_K^4 h^4 R(f'') + O(h^6). \tag{6.16}$$

Similarly, from (6.14), (6.15), and (6.13),

$$\text{Var}(x) = \frac{f(x)R(K)}{nh} - \frac{f(x)^2}{n} + O\left(\frac{h}{n}\right) \Rightarrow$$

$$\text{IV} = \frac{R(K)}{nh} - \frac{R(f)}{n} + \cdots. \tag{6.17}$$

These results are summarized in the following theorem.

Theorem 6.1: *For a nonnegative univariate kernel density estimator,*

$$\text{AMISE} = \frac{R(K)}{nh} + \frac{1}{4}\sigma_K^4 h^4 R(f'')$$

$$h^* = \left[\frac{R(K)}{\sigma_K^4 R(f'')}\right]^{1/5} n^{-1/5} \tag{6.18}$$

$$\text{AMISE}^* = \frac{5}{4}[\sigma_K R(K)]^{4/5} R(f'')^{1/5} n^{-4/5}.$$

The conditions under which the theorem holds have been explored by many authors, including Parzen (1962). A simple set of conditions is that the kernel K be a continuous probability density function with finite support, $K \in L_2, \mu_K = 0, 0 < \sigma_K^2 < \infty$, and that f'' be absolutely continuous and $f''' \in L_2$ (Scott, 1985b).

It is easy to check that the ratio of asymptotic integrated variance (AIV) to asymptotic integrated squared bias (AISB) in the asymptotic mean integrated squared error (AMISE)* is 4:1. That is, the ISB comprises only 20% of the AMISE. The similarity to the FP results in Theorem 4.1 is clear. If K is an isosceles triangle, then the results in Theorem 6.1 match those for the naive average shifted histogram (ASH) with $m = \infty$ in Equation (5.8) (see Problem 6.3). Since $R(\phi''(x|0, \sigma^2)) = 3/(8\sqrt{\pi}\sigma^5)$, the normal reference rule bandwidth with a normal kernel is

$$\boxed{\text{normal reference rule:} \quad h = (4/3)^{1/5}\sigma n^{-1/5} \approx 1.06\hat{\sigma} n^{-1/5}.} \quad (6.19)$$

6.2.2 Estimation of Derivatives

Occasionally, there arises a need to estimate the derivatives of the density function; for example, when looking for modes and bumps. Derivatives of an ordinary kernel estimate behave consistently if the kernel is sufficiently differentiable and if wider bandwidths are selected. Larger smoothing parameters are required as the derivative of the estimated function is noisier than the estimated function itself. Take as an estimator of the rth derivative of f the rth derivative of the kernel estimate:

$$\hat{f}^{(r)}(x) = \frac{d^r}{dx^r}\frac{1}{nh}\sum_{i=1}^{n}K\left(\frac{x - x_i}{h}\right) = \frac{1}{nh^{r+1}}\sum_{i=1}^{n}K^{(r)}\left(\frac{x - x_i}{h}\right). \quad (6.20)$$

A calculation similar to that leading to Equation (6.15) shows that

$$\text{Var}\{\hat{f}^{(r)}(x)\} \approx \frac{n}{(nh^{r+1})^2}\text{E}\left[K^{(r)}\left(\frac{x - X}{h}\right)^2\right] \approx \frac{f(x)R(K^{(r)})}{nh^{2r+1}};$$

hence, the asymptotic integrated variance of $\hat{f}^{(r)}$ is $R(K^{(r)})/(nh^{2r+1})$. After an expansion similar to (6.13) to find the bias, the expectation of the first derivative estimator is

$$\text{E}\hat{f}'(x) = \frac{1}{h}\left[f_x \int K' - hf_x'\int wK' + \frac{h^2}{2}f_x''\int w^2K' - \frac{h^3}{6}f_x'''\int w^3K' + \cdots\right],$$

where $f_x^{(r)} \equiv f^{(r)}(x)$. Assuming K is symmetric, $\int w^r K' = 0$ for even r. Integrating by parts, $\int wK' = -1$ and $\int w^3 K' = -3\sigma_K^2$. Hence, the pointwise bias is of order h^2 and involves the *third* derivative of f. A general theorem is easily given (see Problem 6.6).

> **Theorem 6.2:** *Based on a nonnegative univariate kernel density estimator $\hat{f}$,*
>
> $$\text{AMISE}(\hat{f}^{(r)}) = \frac{R(K^{(r)})}{nh^{2r+1}} + \frac{1}{4}h^4\sigma_K^4 R(f^{(r+2)}), \tag{6.21}$$
>
> $$h_r^* = \left[\frac{(2r+1)R(K^{(r)})}{\sigma_K^4 R(f^{(r+2)})}\right]^{1/(2r+5)} n^{-1/(2r+5)}$$
>
> $$\text{AMISE}^*(\hat{f}^{(r)}) = \frac{2r+5}{4}R(K^{(r)})^{\frac{4}{2r+5}}\left[\sigma_K^4 R(f^{(r+2)})/(2r+1)\right]^{\frac{2r+1}{2r+5}} n^{\frac{-4}{2r+5}}.$$

While the order of the bias term remains $O(h^4)$, each additional derivative order introduces two extra powers of h in the variance. The optimal smoothing parameters h^* for the first and second derivatives are $O(n^{-1/7})$ and $O(n^{-1/9})$, respectively, while the AMISE* is $O(n^{-4/7})$ and $O(n^{-4/9})$. If the optimal density rate $h^* = O(n^{-1/5})$ is used in the estimate of the second derivative, the asymptotic IV in the AMISE does not vanish, since $nh^5 = O(1)$. The estimation of an additional derivative is more difficult than estimating an additional dimension. For example, the optimal AMISE rate for the second derivative is $O(n^{-4/9})$, which is the same (slower) rate as for the optimal AMISE of a 5-D multivariate frequency polygon density estimator.

6.2.3 Choice of Kernel

Much of the first decade of theoretical work focused on various aspects of estimation properties relating to the characteristics of a kernel. Within a particular class of kernel (e.g., the order of its first nonzero moment), the quality of a density estimate is now widely recognized to be primarily determined by the choice of smoothing parameter, and only in a minor way by the choice of kernel, as will become evident in Table 6.2. Thus the topic could be de-emphasized. However, there has been a recent spurt of useful research on kernel design in special situations. While many potential hazards face the user of density estimation (e.g., underestimating the smoothness of the unknown density), the specification of desired properties for the kernel is entirely at the disposal of the worker, who should have a good understanding of the following results.

6.2.3.1 Higher Order Kernels Bartlett (1963) considered the possibility of carefully choosing the kernel to further reduce the contribution of the bias to the MISE. If the requirement that the kernel estimate should itself be a true density is relaxed, then it is possible to achieve significant improvement in the MISE. Suppose a kernel of order p is chosen so that

$$\int K = 1; \quad \int w^i K = 0, \quad i = 1, \cdots, p-1; \quad \text{and} \quad \int w^p K \neq 0, \tag{6.22}$$

then continuing the expansion in Equation (6.13), the pointwise kernel bias becomes [letting $\mu_i \equiv \int w^i K(w) dw$]

$$\text{Bias}\{\hat{f}(x)\} = \frac{1}{p!} h^p \mu_p f^{(p)}(x) + \cdots.$$

Since the formulas for the pointwise and integrated variances are unchanged, the following theorem may be proved.

Theorem 6.3: *Assuming that f is sufficiently differentiable and that the kernel K is of order p,*

$$\text{AMISE}(h) = \frac{R(K)}{nh} + \frac{1}{(p!)^2} \mu_p^2 R(f^{(p)}) h^{2p}$$

$$h^* = \left[\frac{(p!)^2 R(K)}{2p \mu_p^2 R(f^{(p)})} \right]^{1/(2p+1)} n^{-1/(2p+1)} \qquad (6.23)$$

$$\text{AMISE}^* = \frac{2p+1}{2p} \left[2p\, \mu_p^2 R(K)^{2p} R(f^{(p)})/(p!)^2 \right]^{1/(2p+1)} n^{-2p/(2p+1)}.$$

Asymptotically, this result indicates it is possible to approach the usual parametric rate of $O(n^{-1})$ for the AMISE. However, the width of the optimal bandwidths increases as the order of the kernel p increases, suggesting that much of the benefit may be quite asymptotic for large p.

Table 6.1 shows some higher order kernels, together with the optimal AMISE for the case of standard normal data. These kernels are shown in Figure 6.3. Selecting representative higher order kernels is difficult, for reasons given in Section 6.2.3.2. Only *even* values of p are considered, because all *odd* "moments" of symmetric kernels vanish. Each increase of 2 in p adds another (even) moment constraint in Equation (6.22). If the kernels are polynomials, then the degree of the polynomial must also increase so that there are sufficient degrees of freedom to satisfy the constraints. The kernels in Table 6.1 begin with the so-called Epanechnikov kernel and are the unique continuous polynomial kernels of degree p that satisfy the constraints *and* have their support on the interval $[-1, 1]$.

TABLE 6.1   Some Simple Polynomial Higher Order Kernels

p	K_p on $(-1, 1)$	$N(0, 1)$ AMISE*
2	$\frac{3}{4}(1 - t^2)$	$0.320 n^{-4/5}$
4	$\frac{15}{32}(1 - t^2)(3 - 7t^2)$	$0.482 n^{-8/9}$
6	$\frac{105}{256}(1 - t^2)(5 - 30t^2 + 33t^4)$	$0.581 n^{-12/13}$
8	$\frac{315}{4,096}(1 - t^2)(35 - 385t^2 + 1001t^4 - 715t^6)$	$0.681 n^{-16/17}$

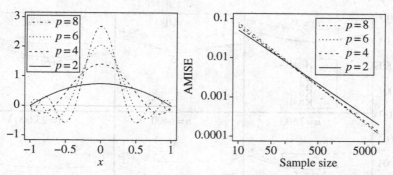

FIGURE 6.3 Examples of higher-order kernels that are low-order polynomials. The right panel shows the corresponding $N(0, 1)$ AMISE* curves on a log–log scale.

The plots of the AMISE for normal data in Figure 6.3 suggest that the higher-order kernels require several thousand data points before a substantial gain may be realized. For rougher data, the gains are even more asymptotic. The improvement made possible by going to a higher order kernel is not simply a constant multiplicative factor but rather an exponential change in the order of convergence of n. Of course, for small samples, the difference between MISE and AMISE may be substantial, particularly for higher order kernels. The exact MISE may be obtained by numerical integration of the bias and variance equations. For normal data, the exact MISE was obtained for several sample sizes with the kernels in Table 6.1 plus the histogram. The results are depicted in Figure 6.4. The individual MISE curves are plotted against h/h^*, so that the minimum is centered on 1. These figures suggest that in most practical situations, kernels of order 2 and 4 are sufficient. The largest gain in MISE is obtained when going from the histogram to the order-2 kernel, with diminishing returns beyond that. Higher order kernels also seem sensitive to oversmoothing. The order-8 kernels are inferior to the histogram if $h > 2h^*$.

These higher order kernels have negative components. They will be referred to as "negative kernels," although the more accurate phrase is "not nonnegative." The introduction of negative kernels does provide improvement in the MISE but at the cost of having to provide special explanations. This negativity is particularly a nuisance in multiple dimensions where the regions of negative estimate can be scattered all over the domain. Statisticians may be comfortable ignoring such features, but care should be taken in their actual use. In practice, negative portions of the estimate could be clipped at 0. Clipping introduces discontinuities in the derivative of the estimate, and the modified density estimate now integrates to slightly more than 1.

In Figure 6.5, ASH estimates from Equation (5.5) using the order 2 and 4 kernels in Table 6.1 for weights in Equation (5.6) are applied to the steel surface data (Bowyer, 1980; Silverman, 1986). The data are measurements from an arbitrary origin of the actual height of a machined flat surface at a grid of 15,000 points. The bandwidths were selected so that the values at the mode matched. This example clearly indicates the reason many statisticians are willing to use negative kernels with large datasets.

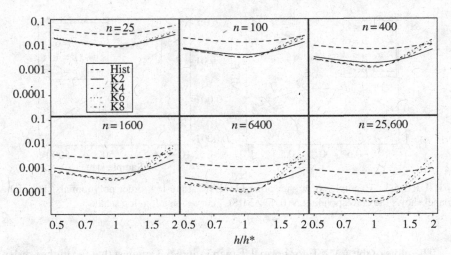

FIGURE 6.4 Exact MISE using higher order kernels with normal data for several sample sizes. The histogram MISE is included for reference.

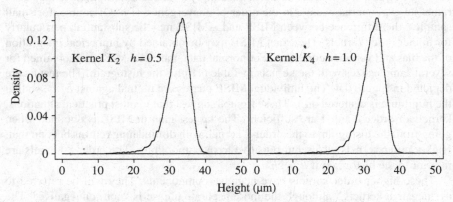

FIGURE 6.5 Positive and negative kernel estimates of the steel surface data. Kernels used were K_2 and K_4 from Table 6.1.

The negativity problem is quite marginal (minimum value -0.00008) and there is some improvement in the appearance of the estimate at the peak (simultaneously lower bias and variance).

A second problem introduced by the use of negative kernels involves the subjective choice of bandwidth. With negative kernels, the estimates can appear rough for moderately oversmoothed values of h, and not just for small, undersmoothed values, as with positive kernels. This dual roughness can be a problem for the novice, especially given the promise of higher order "better" estimates. This phenomenon is easy to demonstrate (see Figure 6.6 with the snowfall data).

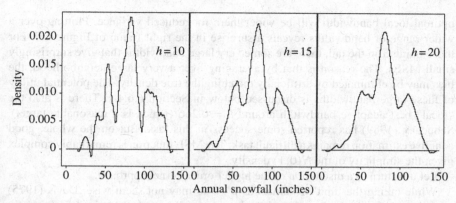

FIGURE 6.6 Kernel K_4 applied to the Buffalo snowfall data with three smoothing parameters. The ASH estimate is depicted in its histogram form.

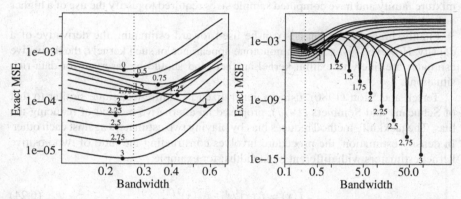

FIGURE 6.7 The exact MSE as a function of the bandwidth h for $f \sim N(0,1)$ and $n = 1000$ for selected values of x between 0 and 3. The globally optimal bandwidth $h = 0.266$ is indicated by the vertical dotted line. The bandwidth range is expanded in the right frame.

It is interesting to speculate about the relative merits of using a higher order kernel versus using a lower order kernel with an adaptive procedure. (An adaptive kernel behaves much like an adaptive frequency polygon, as in Theorem 4.2.) There is reason to believe that there is a role for both. Asymptotically, higher order kernels outperform adaptive procedures. The sensitivity to errors in the optimal bandwidth grows with the kernel order, suggesting that adaptivity is more important with higher order kernels.

Consider, for example, the exact $\text{MSE}(h, x)$ for the $N(0,1)$ density with a normal kernel. A closed form expression for $\text{MSE}(h, x)$ was given by Fryer (1976). A plot of MSE versus h for several values of x between 0 and 3 is shown in Figure 6.7. The global best bandwidth for $n = 1000$ is $h^* = 0.266$. In the left frame, the best bandwidth as x varies is close to this value, except when $x \approx 1$. Recall the bias is a function of $f''(x)$ and the second derivative for a normal density vanishes at $x = \pm 1$; hence, the optimal local bandwidth will be wider there for reduced variance. Plotting over a wider range of bandwidths reveals a surprise in the right frame of Figure 6.7. For those points x in the tail, there are some very large bandwidths that give surprisingly small MSEs. The reason is that by averaging over a very large neighborhood, the bias may be eliminated by artificially matching the true density. The potential utility of these large bandwidths is discussed below in Section 6.6.4.1. There is also the "local" best adaptive bandwidth around $h = 0.266$ (which is a reasonable target). Schucany (1989) has reported some success in this task. But on the whole, good adaptive estimation remains a difficult task. The MSE function is remarkably complex given the simplicity of the $N(0, 1)$ density.

Let us return to a discussion of the higher order kernel approach.

While taking the limit of kernels as $p \to \infty$ may not seem wise, Davis (1975) investigated the properties of a particular "$p = \infty$" kernel, the "sinc" function sin $(x)/(\pi x)$. She showed that the MISE$^* = O(n/\log n)$. Marron and Wand (1992) have examined the MISE of a variety of more complex densities and kernels in the normal mixture family and have computed sample sizes required to justify the use of a higher order kernel.

Finally, higher order kernels can be used toward estimating the derivative of a density. However, given the nonmonotone appearance of such kernels, the derivative estimates are likely to exhibit kernel artifacts and should be reserved for data-rich situations.

Terrell and Scott (1980), using a generalized jackknife technique similar to that of Schucany and Sommers (1977), proposed an alternative method of reducing the bias. The jackknife method reduces bias by playing two estimators against each other. In density estimation, the procedure involves constructing the ratio of two positive kernel estimators with different bandwidths; for example,

$$\hat{f}(x) = \hat{f}_h(x)^{4/3} \div \hat{f}_{2h}(x)^{1/3}. \tag{6.24}$$

The result follows from jackknifing the $O(h^2)$ bias term in a Taylor's series expansion of the log bias. The expectation E_h of the usual kernel estimate is $E_h \equiv \mathrm{E}[\hat{f}_h(x)] = f(x)[1 + c_2 h^2/f(x) + c_4 h^4/f(x) + \cdots]$, where $c_2 = h^2 \sigma_K^2 f''(x)/2$, etc. Then by Taylor's expansion,

$$\log E_h = \log f(x) + c_2 h^2/f(x) + \left[c_4 f(x) - c_2^2/2\right] h^4/f(x)^2 + \cdots$$
$$\log E_{2h} = \log f(x) + 4c_2 h^2/f(x) + 16\left[c_4 f(x) - c_2^2/2\right] h^4/f(x)^2 + \cdots.$$

Then $\frac{4}{3}\log E_h - \frac{1}{3}E_{2h} = \log f(x) + O(h^4)$, which, after taking exponentials, suggests the form given in Equation (6.24) (See Problem 6.14 for details). The authors show that $h^* = 1.42\,\sigma n^{-1/9}$ in the case $f = N(0, 1)$ and $K = N(0, \sigma^2)$. The resulting estimate is nonnegative and continuous, but its integral is usually slightly greater than 1. Generally, exceeding the rate $n^{-4/5}$ requires violating 1 of the 2 posits of a density function: nonnegativity and total probability mass of 1.

6.2.3.2 *Optimal Kernels* The kernel density estimate inherits all the properties of its kernel. Hence, it is important to note that the naive Rosenblatt kernel is discontinuous, the ASH triangle kernel has a discontinuous derivative, and the Cauchy kernel has no moments. A conservative recommendation is to choose a smooth, clearly unimodal kernel that is symmetric about the origin. However, strange kernel shapes are seldom visible in the final estimate, except perhaps in the tails, because of all the averaging.

The question of finding an optimal kernel for nonnegative estimates was considered by Epanechnikov (1969); the same variational problem was considered by Bartlett (1963), and in another context by Hodges and Lehmann (1956). From Equation (6.18), the kernel's contribution to the optimal AMISE is the following dimensionless factor:

$$\text{AMISE}^* \propto [\sigma_K R(K)]^{4/5}. \tag{6.25}$$

The problem of finding the smoothest density for the oversmoothed bandwidth problem is similar to the problem of minimizing (6.25), which may be written as

$$\min_K R(K) \quad \text{s/t} \quad \sigma_K^2 = \sigma^2.$$

The solution is a scaled version of the so-called Epanechnikov's kernel:

$$K_2^*(t) = \frac{3}{4}(1 - t^2)I_{[-1,1]}(t).$$

It is interesting that the optimal kernel has finite support. The optimal kernel is not differentiable at $t = \pm 1$.

The variance of K_2^* is $1/5$ and $R(K_2^*) = 3/5$ in (6.25). Since the AMISE is also proportional to $n^{-4/5}$, other kernels require (see Problem 6.8)

$$\frac{\sigma_K R(K)}{\sigma_{K_2^*} R(K_2^*)} = \frac{\sigma_K R(K)}{3/(5\sqrt{5})} \tag{6.26}$$

times as much data to achieve the same AMISE as the optimal Epanechnikov kernel. Table 6.2 lists many commonly used kernels and computes their asymptotic relative efficiency. The optimal kernel shows only modest improvement. Therefore, the kernel can be chosen for other reasons (ease of computation, differentiability, etc.) without undue concern for loss of efficiency. It is somewhat surprising that the popular normal kernel is so wasteful. Given the computational overhead computing exponentials, it is difficult to recommend the actual use of the normal kernel except as a point of reference.

In the last part of the table, a few absurd kernels are listed to illustrate that a very large inefficiency is still less than 2. From Theorem 4.1, the frequency polygon estimator belongs in the same grouping as the positive kernel estimators. The entries in the table by the FP were obtained by matching the AMISE expressions

TABLE 6.2 Some Common and Some Unusual Kernels and Their Relative Efficiencies. All kernels are supported on $[-1, 1]$ unless noted otherwise.

Kernel	Equation	$R(K)$	σ_K^2	$\sigma_K R(K)$	Efficiencies
Uniform	$U(-1, 1)$	1/2	1/3	0.2887	1.0758
Triangle	$(1 - |t|)_+$	2/3	1/6	0.2722	1.0143
Epanechnikov	$\frac{3}{4}(1 - t^2)_+$	3/5	1/5	0.2683	1
Biweight	$\frac{15}{16}(1 - t^2)_+^2$	5/7	1/7	0.2700	1.0061
Triweight	$\frac{35}{32}(1 - t^2)_+^3$	$\frac{350}{429}$	1/9	0.2720	1.0135
Normal	$N(0, 1)$	$1/2\sqrt{\pi}$	1	0.2821	1.0513
Cosine arch	$\frac{\pi}{4}\cos\frac{\pi}{2}t$	$\pi^2/16$	$1 - \frac{8}{\pi^2}$	0.2685	1.0005
Indifferent FP	See Problem 6.17	11/20	1/4	0.2750	1.0249
Dble. exp.	$\frac{1}{2}e^{-|t|}, |t| \le \infty$	1/4	2	0.3536	1.3176
Skewed	$2860(t + \frac{2}{7})_+^3(\frac{5}{7} - t)_+^9$	$\frac{7436}{3059}$	2/147	0.2835	1.0567
Dble. Epan.	$3|t|(1 - |t|)_+$	3/5	3/10	0.3286	1.2247
Shifted exp.	$e^{-(t+1)}, t > -1$	1/2	1	0.5743	1.8634
FP	See Theorem 4.1	2/3	$7/12\sqrt{5}$	0.3405	1.2690

in Theorems 4.1 and 6.1. The conclusion is that the FP is indeed in the same class, but inefficient. Finally, note that the limiting kernel of the averaged shifted histogram (isosceles triangle) is superior to the limiting kernel of the averaged shifted frequency polygon (indifferent FP), although the FP itself is superior to the histogram.

The symmetric Beta density functions, when transformed to the interval $(-1, 1)$ so that the mean is 0, are a useful choice for a class of kernels:

$$K_k(t) = \frac{(2k + 1)!!}{2^{k+1}k!}(1 - x^2)_+^k, \tag{6.27}$$

where the double factorial notation means $(2k + 1)!! = (2k + 1)(2k - 1)\cdots 5 \cdot 3 \cdot 1$. The Epanechnikov and biweight kernels are in this class. So is the normal density as $k \to \infty$ (see Problem 6.18).

The search for optimal high-order kernels is quite different and not so fruitful. Suppose that $K_4^*, K_6^*, \ldots$ are the optimal order-4, order-6, $\ldots$ kernels, respectively. From Theorem 6.2, the kernel's contribution to the AMISE* for order-4 kernels is the following nonnegative and dimensionless quantity:

$$\text{AMISE}_4^* \propto [R(K)^8 \mu_4^2]^{1/9}. \tag{6.28}$$

Consider the following fourth-order kernel, which is a mixture of K_4^* and K_6^*:

$$K_\epsilon(t) = \epsilon K_4^*(t) + (1 - \epsilon)K_6^*(t), \qquad 0 \le \epsilon \le 1.$$

K_ϵ has finite roughness but its fourth moment vanishes as $\epsilon \to 0$. Thus, the fourth moment of K_4^* must be 0 and the criterion in (6.28) equals 0 at the solution. But by definition, then, K_4^* is no longer a fourth-order kernel. As many kernels have zero fourth "moment" but finite roughness, the lower bound of 0 is achieved by many kernels, none of which are in any sense order-4 or interesting in the sense of Epanechnikov. Of course, the AMISE would not in fact be 0, but would involve higher order terms.

Choosing among higher order kernels is quite complex and it is difficult to draw guidelines. In practice, second- and fourth-order methods are probably the most one should consider, as kernels beyond the order-4 provide little further reduction in MISE. For very large samples, where higher order methods do provide a substantial *fractional* reduction, the *absolute* MISE may already be so small that any practical advantage is lost.

The general advice on choosing a kernel based on these observations is to choose a symmetric kernel that is a low-order polynomial. Gasser et al. (1985) have developed a smooth hierarchy of higher order kernels. They show that their kernels are optimal but in a different sense: these kernels have minimum roughness subject to a fixed number of sign changes in the kernel. Such justification does not really warrant the label "optimal" in the sense of Epanechnikov. However, they have provided a valuable formula for low-order polynomial kernels appropriate for various combinations of the kernel order p, and for the rth derivative of the density,

$$K_{(k,p,r)}^* = \sum_{i=0}^{k+2(r-1)} \lambda_i \, t^i \, I_{[-1,1]}(t),$$

where $k \geq p+2$ and (k, p) are both odd or both even, with

$$\lambda_i = \frac{(-1)^{\frac{i+p}{2}}(k+p+2r)!(k-p)(k+2r-i)(k+i)!}{i!(i+p+1)2^{2(k+r)+1}\left(\frac{k-p}{2}\right)!\left(\frac{k+p+2r}{2}\right)!\left(\frac{k+2r-i}{2}\right)!\left(\frac{k+i}{2}\right)!}$$

if $k+i$ is even, and 0 otherwise (see Müller (1988)).

Given the availability of symbolic manipulation programs, it is probably sufficient to solve the set of linear equations governing the particular application simply. A "designer kernel" approach allows the addition of any linear conditions and results in a new kernel of higher polynomial degree. Again, the choice of kernel is not a critical matter.

6.2.3.3 Equivalent Kernels

For a variety of reasons, there is no single kernel that can be recommended for all circumstances. One serious candidate is the normal kernel; however, it is relatively inefficient and has infinite support. The optimal Epanechnikov kernel is not continuously differentiable and cannot be used to estimate derivatives. In practice, the ability to switch between different kernels without having to reconsider the calibration problem at every turn is convenient. This task is easy to accomplish, *but only for kernels of the same order*. As Scott (1976) noted, if

TABLE 6.3   Factors for Equivalent Smoothing Among Popular Kernels[a]

From\To	Normal	Uniform	Epan.	Triangle	Biwt.	Triwt.
Normal	1	1.740	2.214	2.432	2.623	2.978
Uniform	0.575	1	1.272	1.398	1.507	1.711
Epanech.	0.452	0.786	1	1.099	1.185	1.345
Triangle	0.411	0.715	0.910	1	1.078	1.225
Biwt.	0.381	0.663	0.844	0.927	1	1.136
Triwt.	0.336	0.584	0.743	0.817	0.881	1

[a]To go from h_1 to h_2, multiply h_1 by the factor in the table in the row labeled K_1 and in the column labeled K_2.

h_1 and h_2 are smoothing parameters to be used with kernels K_1 and K_2, respectively, then Theorem 6.1 implies that asymptotically

$$\frac{h_1^*}{h_2^*} = \left[\frac{R(K_1)/\sigma_{K_1}^4}{R(K_2)/\sigma_{K_2}^4} \right]^{1/5} = \frac{\sigma_{K_2}}{\sigma_{K_1}} \left[\frac{\sigma_{K_1} R(K_1)}{\sigma_{K_2} R(K_2)} \right]^{1/5}. \tag{6.29}$$

Table 6.3 gives a summary of factors for equivalent smoothing bandwidths among popular kernels.

The term in brackets on the right in Equation (6.29) is the ratio of dimensionless quantities for each kernel. Those quantities $\sigma_k R(K)$ are almost equal to each other, as may be seen from Equation (6.26) and in Table 6.2. Thus the term in Equation (6.29) in brackets can be set to 1, so that the task of choosing equivalent smoothing parameters for different kernels can be accomplished by scaling according to the standard deviations:

$$\boxed{\text{Equivalent kernel rescaling:} \qquad h_2^* \approx \frac{\sigma_{K_1}}{\sigma_{K_2}} h_1^*.} \tag{6.30}$$

For example, the "exact" factor going from a normal to triweight bandwidth in Table 6.3 is 2.978, while the approximate rule (6.30) gives the factor 3.

Equivalent bandwidth scaling provides nearly identical estimates not only for optimal smoothing parameters but also for nonoptimal values. This rescaling is often used when computing and plotting a biweight kernel estimate, but using a smoothing parameter derived from a normal kernel cross-validation rule.

If all kernels were presented with equal variances, then no changes in smoothing parameters would be required. However, it would be extremely difficult to remember the formulas of those kernels, as the variances would be incorporated into the kernel forms. On balance, it seems easier to write kernels in a parsimonious form. Marron and Nolan (1988) have proposed scaling all kernels to their "canonical" form having equivalent bandwidths to a normal kernel. This proposal is slightly different than the variance rescaling proposal, since the kernel roughness is also taken into account.

For higher order kernels K_{p1} and K_{p2} of order p, a similar rescaling follows from Theorem 6.3 based on the appropriate higher order "moment" (see Problem 6.20):

$$\frac{h_{p1}^*}{h_{p2}^*} = \left[\frac{\mu_{p2}}{\mu_{p1}}\right]^{\frac{1}{p}} \left\{\left[\frac{\mu_{p1}}{\mu_{p2}}\right]^{\frac{1}{p}} \frac{R(K_{p1})}{R(K_{p2})}\right\}^{\frac{1}{2p+1}} \approx \left[\frac{\mu_{p2}}{\mu_{p1}}\right]^{\frac{1}{p}}, \qquad (6.31)$$

since the quantity in brackets is dimensionless and approximately equal to 1 for most symmetric kernels. When $p = 2$, Equations (6.31) and (6.30) agree. Some of the higher order moments are *negative*, but their ratio is positive in (6.31).

6.2.3.4 Higher Order Kernels and Kernel Design

Between kernels of different order, there is no similar notion for choosing bandwidths that give "equivalent smoothing." Furthermore, the lack of a true "optimal kernel" beyond order 2 is troubling. In this section, higher order kernels are reintroduced from two other points of view. The results have some curious implications for bandwidth selection and kernel design, and suggest ways in which further work may be helpful.

The first alternative approach appeals to the numerical analysis argument first introduced by Rosenblatt (1956). Using forward and central difference approximations for the derivative, it was shown in Section 6.1.1 that the equivalent kernels are the order-1 and order-2 kernels $U(0,1)$ and $U(-0.5, 0.5)$, respectively. Using the notation

$$\Delta F_n(x,c) \equiv F_n(x+c) - F_n(x-c),$$

the two-point Rosenblatt estimator with kernel $U(-1,1)$ is $\hat{f}_2(x) = \Delta F_n(x,h)/(2h)$. Consider the following 4-, 6-, and 8-point derivative estimators (see Problem 6.21):

$$\hat{f}_4(x) = \frac{8\Delta F_n(x,h) - \Delta F_n(x,2h)}{12h} \qquad (6.32)$$

$$\hat{f}_6(x) = \frac{45\Delta F_n(x,h) - 9\Delta F_n(x,2h) + \Delta F_n(x,3h)}{60h}$$

$$\hat{f}_8(x) = \frac{224\Delta F_n(x,h) - 56\Delta F_n(x,2h) + \frac{32}{3}\Delta F_n(x,3h) - \Delta F_n(x,4h)}{280h}.$$

The biases for these three estimators are $-h^4 f^{(4)}(x)/30$, $-h^6 f^{(6)}(x)/140$, and $-h^8 f^{(8)}(x)/630$, respectively. The equivalent kernels are easily visualized by plotting $\hat{f}_r(x)$ with one data point at 0 with $h = 1$ as in Figure 6.8.

In Figure 6.9, the $p = 2$ naive kernel estimate is plotted for 320 cholesterol levels of patients with heart disease. As $\hat{\sigma} = 43$, the bandwidth chosen was $h = 25$, which is the equivalent bandwidth rule for a $U(-1,1)$ kernel to the normal reference rule (6.19). Specifically, the equivalent bandwidth is computed as $h \approx 3^{1/2}[1.06 \times 43 \times 320^{-1/5}]$,

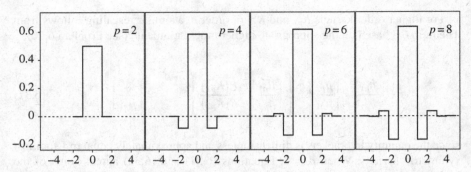

FIGURE 6.8   Higher order boxcar or naive kernels based on finite differences.

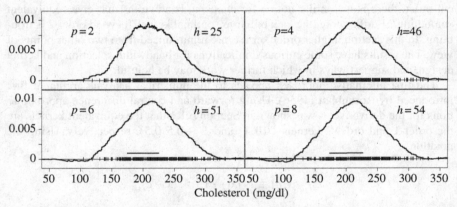

FIGURE 6.9   Higher order naive kernels applied to the cholesterol data for diseased males ($n = 320$). Bandwidths are indicated by the horizontal lines.

where $3^{1/2}$ is the standard deviation of the $U(-1, 1)$ kernel. The bandwidths for the $p =$ fourth-, sixth-, and eighth-order kernels were chosen so that the levels at the modes were equal. The negative side lobes are easily seen. The second- and fourth-order estimates are substantially different. The horizontal line shows the bandwidth h, which is half the width for the central lobe of the kernel. Thus for K_8, the support is eighth times wider than h; the influence of each data point extends 216 units (mg/dl) in both directions. For the order-2, 4-, and 6- kernels, the extent is 25, 92, and 153, respectively. Higher order kernels even with finite support are not local.

An empirical observation is that when a higher order kernel fails to provide further reduction in ISE, then the optimal higher order density estimates are all very similar (except for small local noise due to the added roughness in the tails of the kernel). This similarity occurs when the central lobe of the kernel [over the interval $(-h, h)$] remains of fixed width even as the order of the kernel grows. The sequence of bandwidths is 25, 46, 51, and 54 for the cholesterol data. Thus $p = 4$ seems a plausible choice for these data, possibly with a narrower bandwidth. As p increases, the negative lobes have an unfortunate tendency to grow and spread out.

Alternatively, the bandwidths for the naive higher order kernels could have been chosen to increase the estimate at the mode by an estimate of the bias there; for example, for the order-2 kernel

$$\text{Bias}\{\hat{f}(x)\} = \frac{1}{2}h^2\sigma_K^2 f''(x). \tag{6.33}$$

If the resulting estimate (with the order-4 kernel) is much rougher or if a bandwidth cannot be found that accomplishes the desired increase, then a reasonable conclusion is that the maximum feasible kernel order has been exceeded. The resulting bandwidth for the naive higher order kernel may be transformed by (6.31) to a smoother higher order kernel in Table 6.1.

The second alternative introduction to negative kernels suggests that the problem of bandwidth selection is even more complicated than is apparent. Beginning with a positive kernel estimate and the bias estimate given earlier, the idea is to estimate and remove the bias pointwise by using a kernel estimate of the second derivative. For clarity, a possibly different bandwidth g is used to estimate $f''(x)$:

$$\hat{f}''(x) = \frac{1}{ng^3}\sum_{i=1}^{n} K''\left(\frac{x-x_i}{g}\right). \tag{6.34}$$

Consider the "bias-corrected" kernel estimate, which is obtained by combining Equations (6.33) and (6.34):

$$\hat{f}(x) = \frac{1}{nh}\sum_{i=1}^{n} K\left(\frac{x-x_i}{h}\right) - \frac{1}{2}h^2\sigma_k^2 \times \frac{1}{ng^3}\sum_{i=1}^{n} K''\left(\frac{x-x_i}{g}\right). \tag{6.35}$$

As $\hat{f}''(x)$ integrates to 0, this modified kernel estimate still integrates to 1. By inspection, (6.35) is itself in the form of a kernel estimator with kernel

$$K_{h,g}(t) = K_h(t) - \frac{h^2\sigma_k^2}{2g^2}K_g''(t). \tag{6.36}$$

A straightforward calculation verifies that this new kernel has a vanishing second moment, so that this constructive procedure in fact results in a order-4 kernel (see Problem 6.22). However, this approach suggests that $g \neq h$, since the bandwidth for the second derivative should be wider than for the density estimate itself. If so, then a well-constructed higher order kernel should slowly "expand" as the sample size increases. Given the relatively slow changes in the bandwidths, there may not be any practical improvement over allowing $g = h$.

6.2.3.5 Boundary Kernels When the unknown density is discontinuous, kernel estimates suffer the same dramatic loss of MISE efficiency as for the frequency polygon. In the case where the location of the discontinuity is known, an elegant fix is

available. Without loss of generality, suppose that the discontinuity occurs at zero
and that the density vanishes for $x < 0$. Careful examination of the theoretical argument which led to the elimination of the $O(h)$ bias term for a kernel estimate reveals
that the critical requirement is that the kernel satisfy $\int tK(t) = 0$. The fundamental
requirement is not that the kernel be symmetric; symmetry is only a simple way that
the kernel may satisfy the integral equation.

The task, then, is to design finite-support kernels for use with samples in the
boundary region $x_i \in (0, h)$. In order that the kernel estimate vanish for $x < 0$, the
kernel for a sample point $x_i \in [0, h)$ should cover the interval $[0, x_i + h)$ rather than
the interval $(x_i - h, x_i + h)$. As the interval $[0, x_i + h)$ is narrower than $2h$, the roughness of the kernels (and hence the IV) will increase rather dramatically. In a regression
setting, Gasser and Müller (1979) and Rice (1984a) have suggested using the wider
interval $(0, 2h)$ for every $x_i \in [0, h)$. This suggestion is equivalent to choosing kernels
supported on the interval $(c, c+2)$, for $-1 \le c \le 0$, that satisfy $\int_c^{c+2} tK(t)\, dt = 0$. An
attempt to allow the interval width to vary so as to achieve "equivalent smoothing"
using the full rule (6.29) seems doomed to failure because a wider interval cannot usually be found with equivalent smoothing. Thus the simple choice of $2h$ is a reasonable
compromise.

A designer boundary kernel is described. Assume that the desired boundary kernel
is to be a modification of the ordinary biweight kernel, with similar properties at the
right-hand endpoint $x = c + 2$. This suggests looking at a designer boundary kernel
of the form

$$K_c(t) = [c_1 + c_2(t-c)^2] \cdot [t - (c+2)]^2, \qquad -1 \le c \le 0,$$

where the constants c_1 and c_2 are determined by the constraints $\int K_c(t)\, dt = 1$ and
$\int tK_c(t)\, dt = 0$. The form for $K_c(\cdot)$ ensures that the two constraints are linear in the
unknowns c_1, c_2. Solving those equations gives

$$K_c(t) = \frac{3}{4}\left[(c+1) - \frac{5}{4}(1+2c)(t-c)^2 \right] [t - (c+2)]^2 I_{[c,c+2]}(t). \qquad (6.37)$$

Figure 6.10 shows some examples of these kernels in two different ways. First, the
kernels are shown centered on the sample (taken to be the origin). Second, the kernels
are shown as they would be placed on top of the samples, so that they *begin* at zero.
Note that if $x_i \in [0, h)$, then the kernel K_c with $c = (0 - x_i)/h$ should be used instead
of the ordinary biweight kernel.

Clearly, Figure 6.10 indicates that these are *floating boundary kernels*, mean-
ing that the value of the kernel floats at the left boundary. An example of the use
of these kernels with a sample of 100 points from the negative exponential density illustrates the effectiveness of the floating boundary kernels (see Figure 6.11).
Notice how the unmodified biweight kernel estimate is quite biased in the interval
$(0, h)$ and spills into the $x < 0$ region. However, without any checking or indica-
tion of a boundary problem, the unmodified kernel estimate appears quite smooth.
(This smoothness should serve as a warning to check for errors resulting from the

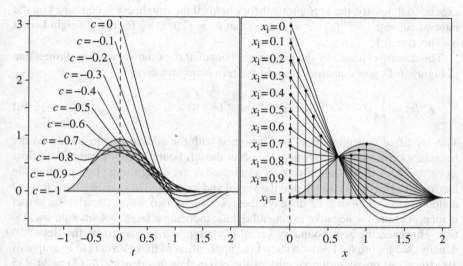

FIGURE 6.10   (Left frame) Examples of the "floating" boundary kernels $K_c(t)$, where $-1 < c < 0$. (Right frame) Assuming the boundary $x \geq 0$, each kernel $K_c(t)$ is drawn centered on the data point, $x_i = -c$, which is indicated by the dashed vertical line.

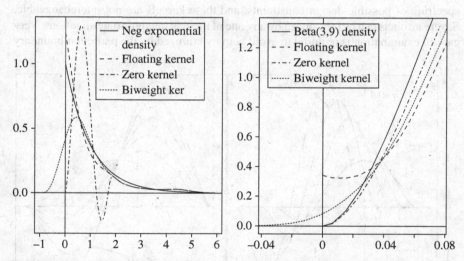

FIGURE 6.11   (Left frame) Example with negative exponential data—with and without boundary modification for $n = 100$ and $h = 0.93$. The "floating" and "zero" boundary kernels are defined in Equations (6.37) and (6.38), respectively. (Right frame) Example with Beta(3,9) density in a neighborhood of 0 for $n = 100$ and $h = 0.11$.

lack of prior knowledge of the existence of a boundary problem.) For the negative exponential density, the asymptotic theory holds if the roughness is computed on the interval $(0, \infty)$: $\int_0^\infty f''(x)^2 dx = 1/2$, so that $h^* = (70/n)^{1/5}$ for the biweight kernel by Theorem 6.1.

The density estimate of the negative exponential data shown as a big-dashed line in Figure 6.11 was constructed using the *zero boundary kernel*,

$$K_c^0(t) = \frac{15}{16}(t-c)^2(2+c-t)^2[(7c^2+14c+8)-7t(c+1)]I_{[c,c+2]}(t). \qquad (6.38)$$

This modified biweight kernel was designed with the additional constraint that the boundary kernel and its derivative vanish at the left boundary $x = c$ rather than float as before. This modification was accomplished at the design stage by including the factor $(t-c)^2$, which ensures that the kernel and its derivative vanish at the left-hand endpoint $t = c$. Figure 6.12 displays these kernels in two ways. Clearly, the kernel is inappropriate for negative exponential data, inducing a large, but smooth, oscillation. However, the zero boundary kernel is appropriate for data from the Beta(3,9) density; see Figure 6.11, which shows the application of these kernels to a sample of 100 Beta(3,9) points in the vicinity of the origin. For this density, $R(f'') \approx 24,835$ and $h^* \approx (0.269/n)^{1/5}$. The ordinary estimate spills into the negative region and the "floating" kernel estimate lives up to its name.

While boundary kernels can be very useful, there are potentially serious problems with real data. There are an infinite number of boundary kernels reflecting the spectrum of possible design constraints, and these kernels are not interchangeable. Severe artifacts can be introduced by any one of them in inappropriate situations. Very careful examination is required to avoid being victimized by the particular boundary

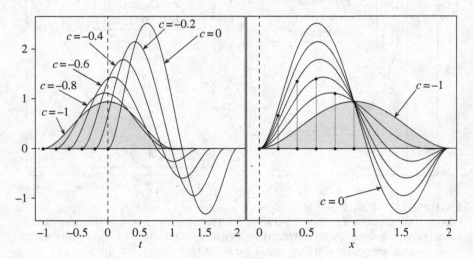

FIGURE 6.12 Examples of "zero" boundary kernels as in Figure 6.10.

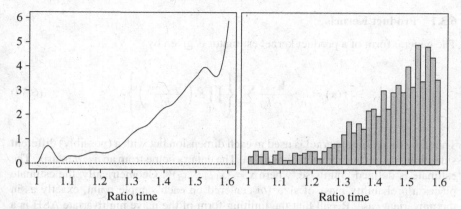

FIGURE 6.13 Density estimate of 857 fastest times in the 1991 Houston Tenneco Marathon. The data are the ratio to the leader's time for the race. Different boundary kernels were used on each extreme. A histogram is shown for comparison.

kernel chosen. Artifacts can unfortunately be introduced by the choice of the support interval for the boundary kernel. Little is known about the best way to avoid this situation, but the Rice–Müller solution seems the best of several possible alternatives that have been attempted. Finally, while the boundary kernels seem to cover a wide region of the density estimate, the effect is generally limited to the interval $(0, h/2)$ for appropriately smoothed estimates.

Some data may require a different boundary kernel at each end. For example, the fastest 857 times in the January 20, 1991, Houston Tenneco Marathon were recorded as a ratio to the winning time. Clearly, the features in the density are expected to differ at the two boundaries. An estimate was constructed using K_c^0 on the left and the floating K_c on the right with $h = 0.05$ (see Figure 6.13). The clump among the leaders is real; however, the extra bump on the right appears to be more of an artifact.

Reflection Boundary Technique There is a more conservative technique that can replace the "floating" kernel. If the data are nonnegative and the discontinuity is at $x = 0$, an ordinary kernel estimate is computed but on the augmented data $(-x_n, \ldots, -x_1, x_1, \ldots, x_n)$. The final estimate is obtained by doubling this estimate for $x \geq 0$. The bandwidth should be based on the sample size n and not $2n$. This technique avoids the pitfalls of negative boundary kernels, but is generally of lower order consistency (see Problem 4.4 in the context of the frequency polygon).

6.3 THEORETICAL PROPERTIES: MULTIVARIATE CASE

The theoretical analysis of multivariate kernel estimators is the same as for frequency polygons save for a few details. The initial discussion will be limited to product kernel density estimators. The general kernel analysis will be considered afterward.

6.3.1   Product Kernels

The general form of a product kernel estimator is given by

$$\hat{f}(\mathbf{x}) = \frac{1}{nh_1 \cdots h_d} \sum_{i=1}^{n} \left\{ \prod_{j=1}^{d} K\left(\frac{x_j - x_{ij}}{h_j}\right) \right\}. \tag{6.39}$$

The same (univariate) kernel is used in each dimension but with a (possibly) different smoothing parameter for each dimension. The data x_{ij} come from an $n \times d$ matrix. The estimate is defined pointwise, where $\mathbf{x} = (x_1, \dots, x_d)^T$. Geometrically, the estimate places a probability mass of size $1/n$ centered on each sample point, exactly as in the univariate case. Recall that the limiting form of the naive multivariate ASH is a product triangle kernel estimator. Several bivariate product kernels are displayed in Figure 6.14.

Consider the pointwise bias of the multivariate estimator. Clearly,

$$E\hat{f}(\mathbf{x}) = E \prod_{j=1}^{d} \frac{1}{h_j} K\left(\frac{x_j - X_j}{h_j}\right) = \int_{\mathfrak{R}^d} \prod_{j=1}^{d} \frac{1}{h_j} K\left(\frac{x_j - t_j}{h_j}\right) f(\mathbf{t}) \, d\mathbf{t}$$

$$= \int_{\mathfrak{R}^d} \prod_{j=1}^{d} K(w_j) f(x_1 - h_1 w_1, \dots, x_n - h_n w_n) \, d\mathbf{w}$$

$$\approx \int_{\mathfrak{R}^d} \prod_{j=1}^{d} K(w_j) \left[f(\mathbf{x}) - \sum_{r=1}^{d} h_r w_r f_r(\mathbf{x}) + \sum_{r,s=1}^{d} \frac{h_r h_s}{2} w_r w_s f_{rs}(\mathbf{x}) \right] d\mathbf{w}$$

$$= f(\mathbf{x}) + \frac{1}{2} \sigma_K^2 \sum_{j=1}^{d} h_j^2 f_{jj}(\mathbf{x}) + O(h^4). \tag{6.40}$$

As before, the $O(h)$ bias terms vanish if the univariate kernels have zero mean. Similarly, the $h_r h_s$ terms vanish (see Problem 6.24). It follows that the integrated squared bias is as given in Theorem 6.4. The pointwise variance is $f(x)R(K)^d / (nh_1 h_2 \cdots h_n)$, from which the integrated variance follows easily.

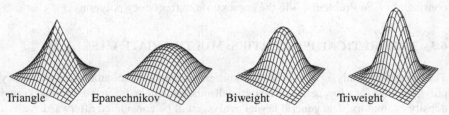

Triangle   Epanechnikov   Biweight   Triweight

FIGURE 6.14   Product kernel examples for four kernels.

> **Theorem 6.4:**   *For a multivariate product kernel estimator, the components of the AMISE are*
>
> $$\text{AISB} = \frac{1}{4}\sigma_K^4 \left[\sum_{i=1}^{d} h_i^4 R(f_{ii}) + \sum_{i \neq j} h_i^2 h_j^2 \int_{\Re^d} f_{ii}f_{jj}\, d\mathbf{x} \right]$$
>
> $$\text{AIV} = \frac{R(K)^d}{nh_1 h_2 \cdots h_d} - \frac{R(f)}{n} + O\left(\frac{h}{n}\right). \tag{6.41}$$

The order of the optimal smoothing parameters is precisely the same as for the multivariate FP: $h_i^* = O(n^{-1/(4+d)})$ and $\text{AMISE}^* = O(n^{-4/(4+d)})$.

It is a minor inconvenience, but no general closed-form expression for the optimal smoothing parameters exists, save as the solution to d nonlinear equations. Solutions may be found in special cases, for example, when $d \leq 2$ or if $h_i = h$ for all i. For example, with general bivariate normal data and a normal kernel, straightforward integration shows that

$$R(f_{11}) = 3 \left[16\pi(1 - \rho^2)^{5/2}\sigma_1^5\sigma_2 \right]^{-1},$$

$$R(f_{22}) = 3 \left[16\pi(1 - \rho^2)^{5/2}\sigma_1\sigma_2^5 \right]^{-1},$$

$$\int_{\Re^2} f_{11}f_{22}\, dx_1\, dx_2 = (1 + 2\rho^2) \left[16\pi(1 - \rho^2)^{5/2}\sigma_1^3\sigma_2^3 \right]^{-1}.$$

From Theorem 6.4, the AMISE is minimized when

$$h_i^* = \sigma_i(1 - \rho^2)^{5/12}(1 + \rho^2/2)^{-1/6}n^{-1/6}$$

$$\approx \sigma_i(1 - \rho^2/2 - \rho^4/16 - \cdots)n^{-1/6} \quad i = 1, 2, \tag{6.42}$$

for which

$$\text{AMISE}^* = \frac{3}{8\pi}(\sigma_1\sigma_2)^{-1}(1 - \rho^2)^{-5/6}(1 + \rho^2/2)^{1/3}n^{-2/3}.$$

Observe that the AMISE diverges to infinity when the data are perfectly correlated (the *real* curse of dimensionality). In comparison to other bivariate estimates, if $\rho = 0$ and $\sigma_i = 0$, then the bivariate AMISE is equal to 1/400 when $n = 302$. A bivariate FP and histogram require $n = 557$ and $n = 4244$, respectively.

A second example of a special case is the multivariate normal, where all the variables are independent. If a normal kernel is used, then a short calculation with Theorem 6.4 gives the

$$\text{normal reference rule:} \qquad h_i^* = \left(\frac{4}{d+2}\right)^{1/(d+4)} \sigma_i\, n^{-1/(d+4)}. \qquad (6.43)$$

As the dimension d varies, the constant in Equation (6.43) ranges over the interval (0.924, 1.059), with a limit equal to 1. The constant is exactly 1 in the bivariate case and smallest when $d = 11$. Hence, an easy-to-remember data-based rule is

$$\text{Scott's rule in } \Re^d: \qquad \hat{h}_i = \hat{\sigma}_i n^{-1/(d+4)}. \qquad (6.44)$$

For other kernels, the equivalent kernel smoothing parameter may be obtained by dividing by the standard deviation of that kernel. Just as in one dimension, these formulas can be used in place of more precise oversmoothing values as independent normal data are very smooth. Any special structure will require narrower bandwidths. For example, the modification based on skewness and kurtosis in $\Re^1$ are identical to the factors for the frequency polygon in Section 4.1.2. If the data are not full-rank, kernel methods perform poorly. Dimension reduction techniques will be considered in Chapter 7.

6.3.2 General Multivariate Kernel MISE

In practice, product kernels are recommended. However, for various theoretical studies, general multivariate kernels will be required. This section presents a brief summary of those studies.

The general multivariate kernel estimator will include not only an arbitrary multivariate density as a kernel but also an arbitrary linear transformation of the data. Let H be a $d \times d$ nonsingular matrix and $K : R^d \to R^1$ be a kernel satisfying conditions given below.

Then the general multivariate kernel estimator is

$$\hat{f}(\mathbf{x}) = \frac{1}{n|H|} \sum_{i=1}^{n} K(H^{-1}(\mathbf{x} - \mathbf{x}_i)). \qquad (6.45)$$

It should be apparent from Equation (6.45) that the linear transformation H could be incorporated into the kernel definition. For example, it is equivalent to choose K to be $N(\mathbf{0}_d, \Sigma)$ with $H = I_d$, or to choose K to be $N(\mathbf{0}_d, I_d)$ with $H = \Sigma^{1/2}$ (see Problem 6.25). Thus it is possible to choose a multivariate kernel with a simple covariance structure without loss of generality. It will not, however, be sufficient to consider only product kernels, as that would limit the discussion to multivariate kernels that are independent (and not just uncorrelated) and to kernels that are supported on a rectangular region.

The multivariate kernel will be assumed hereafter to satisfy three moment conditions (note these are matrix equations):

$$\int_{\Re^d} K(\mathbf{w})\, d\mathbf{w} = 1$$

$$\int_{\Re^d} \mathbf{w} K(\mathbf{w})\, d\mathbf{w} = \mathbf{0}_d \tag{6.46}$$

$$\int_{\Re^d} \mathbf{w}\mathbf{w}^T K(\mathbf{w})\, d\mathbf{w} = I_d.$$

If K is indeed a multivariate probability density, then the last two equations summarize many assumptions about the *marginal kernels*, $\{K_i(w_i), i = 1, \ldots, d\}$. The second equation says that the means of the marginal kernels are all zero. The third equation states that the marginal kernels are all pairwise uncorrelated and that each has unit variance. Thus any simple linear transformation is assumed to be captured entirely in the matrix H and not in the kernel.

In matrix notation, it is straightforward to compute the error of the multivariate kernel estimator. For letting $\mathbf{w} = H^{-1}(\mathbf{x} - \mathbf{y})$,

$$\mathrm{E}\hat{f}(\mathbf{x}) = \int_{\Re^d} K(H^{-1}(\mathbf{x} - \mathbf{y})) f(\mathbf{y})\, d\mathbf{y}/|H|$$

$$= \int_{\Re^d} K(\mathbf{w}) f(\mathbf{x} - H\mathbf{w})\, d\mathbf{w}$$

$$= \int_{\Re^d} K(\mathbf{w}) \left[f(\mathbf{x}) - \mathbf{w}^T H \nabla f(\mathbf{x}) + \frac{1}{2} \mathbf{w}^T H^T \nabla^2 f(\mathbf{x}) H \mathbf{w} \right] d\mathbf{w} \tag{6.47}$$

to second order. Further simplification is possible using the following property of the trace (tr) of a matrix: $\mathrm{tr}\{AB\} = \mathrm{tr}\{BA\}$, assuming that the matrices A and B have dimensions $r \times s$ and $s \times r$, respectively. Now the quadratic form in Equation (6.47) is a 1×1 matrix, which trivially equals its trace. Hence, using the trace identity and exchanging the trace and integral operations yields

$$\mathrm{E}\hat{f}(\mathbf{x}) = f(\mathbf{x}) - 0 + \frac{1}{2}\mathrm{tr}\left\{ \int_{\Re^d} \mathbf{w}\mathbf{w}^T K(\mathbf{w})\, d\mathbf{w} \cdot H^T \nabla^2 f(\mathbf{x}) H \right\}.$$

As the covariance matrix of K is I_d by assumption (6.46), the integral factor in the trace vanishes. Therefore,

$$\mathrm{Bias}\{\hat{f}(\mathbf{x})\} = \frac{1}{2}\mathrm{tr}\{H^T \nabla^2 f(\mathbf{x}) H\} = \frac{1}{2}\mathrm{tr}\{HH^T \nabla^2 f(\mathbf{x})\}.$$

Next, define the scalar $h > 0$ and the $d \times d$ matrix A to satisfy

$$H = hA \quad \text{where } |A| = 1.$$

Choosing A to have determinant equal to 1 means that the elliptical shape of the kernel is entirely controlled by the matrix AA^T and the size of the kernel is entirely controlled by the scalar h. Observe that this parameterization is entirely general and permits different smoothing parameters for each dimension. For example, if

$$H = \begin{pmatrix} h_1 & & 0 \\ & \ddots & \\ 0 & & h_d \end{pmatrix}; \quad \text{then} \quad H = h \cdot \begin{pmatrix} h_1/h & & 0 \\ & \ddots & \\ 0 & & h_d/h \end{pmatrix},$$

where $h = (h_1 h_2 \cdots h_d)^{1/d}$ is the geometric mean of the d smoothing parameters. Check that $|A| = 1$.

It follows that

$$\text{Bias}\{\hat{f}(\mathbf{x})\} = \frac{1}{2}h^2 \text{tr}\{AA^T\nabla^2 f(\mathbf{x})\}, \tag{6.48}$$

so that

$$\text{AISB} = \frac{1}{4}h^4 \int_{\Re^d} \left[\text{tr}\{AA^T\nabla^2 f(\mathbf{x})\}\right]^2 d\mathbf{x}.$$

As usual, the variance term is dominated by $EK_H(\mathbf{x} - \mathbf{x}_i)^2$; therefore,

$$\text{Var}\{\hat{f}(\mathbf{x})\} = \frac{f(\mathbf{x})}{n|H|} \int_{\Re^d} K(\mathbf{w})^2 d\mathbf{w} \quad \Rightarrow \quad \text{AIV} = \frac{R(K)}{nh^d}. \tag{6.49}$$

Together, these results may be summarized in a theorem.

Theorem 6.5: *For a general multivariate kernel estimator* (6.45) *parameterized by $H = hA$,*

$$\text{AMISE} = \frac{R(K)}{nh^d} + \frac{1}{4}h^4 \int_{\Re^d} \left[\text{tr}\{AA^T\nabla^2 f(\mathbf{x})\}\right]^2 d\mathbf{x}. \tag{6.50}$$

In spite of appearances, this is not using the same bandwidth in each dimension, but rather is applying a general elliptically shaped kernel at an arbitrary rotation.

The integral in Equation (6.50) will be quite complicated to evaluate unless the matrix A has a very simple structure. However, Wand and Jones (1995) provide a clever expression in their Section 4.3 that makes it much easier to evaluate this integral. Define the symmetric matrix $M = 2\nabla^2 f(\mathbf{x}) - \text{Diag}\left[\nabla^2 f(\mathbf{x})\right]$. Put the lower triangular portion of M into a vector $\mathbf{n}$ of length $d(d+1)/2$, a procedure referred to as the half-vectorization operation, $\mathbf{n} = \text{vech} M$. Their final expression is the scalar

$$\left(\text{vech} AA^T\right)^T \left[\int \mathbf{n}\mathbf{n}^T dx\right] \left(\text{vech} AA^T\right). \qquad (6.51)$$

The integral is applied to the elements of the $\frac{d(d+1)}{2} \times \frac{d(d+1)}{2}$ matrix $\mathbf{n}\mathbf{n}^T$. The integrals can be computed or estimated using integration by parts in many cases. Since AA^T is also a symmetric $d \times d$ matrix, $\text{vech} AA^T$ is a vector of length $\frac{d(d+1)}{2}$. See Wand and Jones (1995) and Problem 6.38 for further details.

6.3.3 Boundary Kernels for Irregular Regions

Staniswalis et al. (1993) showed that a boundary kernel for an arbitrarily compli-cated domain may be constructed by a simple device. Suppose that an estimate at $\mathbf{x}$ is desired. They propose using a kernel with spherical support, with radius h. Only sam-ples $\mathbf{x}_i$ in the sphere of radius h around $\mathbf{x}$ influence the estimate. For each sample $\mathbf{x}_i$ in that region, determine if the diameter on which it falls (the center being the estima-tion point $\mathbf{x}$) intersects the boundary. If it does, construct a one-dimensional boundary kernel along that diameter. Repeating this construction for all samples in the sphere, the authors prove that the resulting estimate retains the correct order of bias.

6.4 GENERALITY OF THE KERNEL METHOD

6.4.1 Delta Methods

Walter and Blum (1979) catalogued the common feature of the already growing list of different density estimators. Namely, each could be reexpressed as a kernel estima-tor. Such a demonstration for orthogonal series estimators was given in Section 6.1.3. Reexamining Equation (3.2), even the histogram can be thought of as a kernel estima-tor. Surprisingly, this result was shown to hold even for estimators that were solutions to optimization problems. For example, consider one of the several maximum penal-ized likelihood (MPL) criteria suggested by Good and Gaskins (1972):

$$\max_f \left[\sum_{i=1}^n \log f(x_i) - \alpha \int_{-\infty}^{\infty} f'(x)^2 dx\right] \qquad \text{for some } \alpha > 0. \qquad (6.52)$$

Without the *roughness penalty term* in (6.52), the solution would be the empirical pdf. The many MPL estimators were shown to be kernel estimators by de Montricher et al. (1975) and Klonias (1982). The form of the kernel solutions differs in that the weights on the kernels were not all equal to $1/n$. For some other density estimation algorithms, the equivalent kernel has weight $1/n$ but has an adaptive bandwidth. A simple example of this type is the kth nearest-neighbor (k-NN) estimator. The k-NN estimate at x is equivalent to a histogram estimate with a bin centered on x with bin width sufficiently large so that the bin contains k points (in two and three dimensions, the histogram bin shape is a circle and a sphere, respectively). Thus the equivalent kernel in $\Re^d$ is simply a uniform density over the unit ball in $\Re^d$, but with bin widths that adapt to x.

6.4.2 General Kernel Theorem

There is a theoretical basis for the empirical observations of Walter and Blum that many algorithms may be viewed as generalized kernel estimates. Terrell provided a theorem to this effect that contains a constructive algorithm for obtaining the generalized kernel of any density estimator (see Terrell and Scott (1992)). The construction is not limited to nonparametric estimators, a fact that is exploited later.

Theorem 6.6: *Any density estimator that is a continuous and Gâteaux differentiable functional on the empirical distribution function may be written as*

$$\hat{f}(x) = \frac{1}{n} \sum_{i=1}^{n} K(x, x_i, F_n), \tag{6.53}$$

where K is the Gâteaux derivative of $\hat{f}$ under variation of x_i.

The Gâteaux derivative of a functional T at the function ϕ in the direction of the function η is defined to be

$$DT(\phi)[\eta] = \lim_{\epsilon \to 0} \frac{1}{\epsilon} \left[T(\phi + \epsilon\,\eta) - T(\phi) \right]. \tag{6.54}$$

Theorem 6.6, which is proved below, has an analogous multivariate version (Terrell and Scott, 1992). The kernel K simply measures the influence of x_i on $\hat{f}(x)$. As F_n converges to F, then asymptotically, the form of K is independent of the remaining $n - 1$ observations. Thus, any continuous density estimator may be written (asymptotically) as

$$\hat{f}(x) = \frac{1}{n} \sum_{i=1}^{n} K(x, x_i, n) = \frac{1}{n} \sum_{i=1}^{n} K_n(x, x_i). \tag{6.55}$$

6.4.2.1 *Proof of General Kernel Result* The empirical cdf in Equation (2.1) can be written in the unusual form

$$F_n(\cdot) = \frac{1}{n} \sum_{i=1}^{n} I_{[x_i, \infty)}(\cdot). \tag{6.56}$$

Write the density estimator as an operator $\hat{f}(x) = T_x\{F_n\}$. Define

$$K(x,y,F_n) \equiv \lim_{\epsilon \to 0} \frac{1}{\epsilon} [T_x\{(1-\epsilon)F_n + \epsilon I_{[y,\infty)}\} - (1-\epsilon)T_x\{F_n\}] \qquad (6.57)$$

$$= \lim_{\epsilon \to 0} \frac{1}{\epsilon} [T_x\{F_n + \epsilon(I_{[y,\infty)} - F_n)\} - T_x\{F_n\}] + T_x\{F_n\}$$

$$= DT_x(F_n)[I_{[y,\infty)} - F_n] + \hat{f}(x),$$

where $DT(\phi)[\eta]$ is the Gâteaux derivative of T at ϕ in the direction η. Proposition (2.7) of Tapia (1971) shows that the Gâteaux derivative is linear in its second argument, so

$$\frac{1}{n}\sum_{i=1}^{n} K(x,x_i,F_n) = \frac{1}{n}\sum_{i=1}^{n} DT_x(F_n)[I_{[x_i,\infty)} - F_n] + \hat{f}(x)$$

$$= DT_x(F_n)\left[\frac{1}{n}\sum_{i=1}^{n} I_{[x_i,\infty)} - F_n\right] + \hat{f}(x)$$

$$= 0 + \hat{f}(x),$$

where the term in brackets is 0 by Equation (6.56). Note that by linearity, the Gâteaux variation in the direction 0, $DT(\phi)[0]$, is identically 0. This concludes the proof.

6.4.2.2 *Characterization of a Nonparametric Estimator* An estimator is defined to be nonparametric when it is consistent in the mean square for a large class of density functions. With a little effort, this definition translates into specific requirements for the equivalent kernel, $K_n(x,y)$. From the many previous examples, a nonparametric estimator that is consistent must be *local*; that is, the influence of sample points outside an ϵ–neighborhood of x must vanish as $n \to \infty$. As suggested by the term "delta function sequences" coined by Watson and Leadbetter (1963),

$$\lim_{n \to \infty} K_n(x,x_i,F_n) = \delta(x - x_i).$$

However, K_n must not equal $\delta(x - x_i)$ for finite n, as was the case in Section 6.1.3 with the orthogonal series estimator with all $\hat{f}_\nu$ coefficients included.

Beginning with the bias of the asymptotic kernel form in (6.55)

$$E\hat{f}(x) = EK_n(x,X) = \int K_n(x,y)f(y)\,dy$$

$$= \int K_n(x,y)\left[f(x) + (y-x)f'(x) + (y-x)^2 f''(\xi_y)/2\right]dy$$

by the exact version of Taylor's theorem where $\xi_y \in (x,y)$. To be asymptotically unbiased, all three terms in the following must vanish:

$$\text{Bias}\{\hat{f}(x)\} = f(x)\left[\int K_n(x,y)\,dy - 1\right] + f'(x)\int K_n(x,y)(y-x)\,dy$$

$$+ \frac{1}{2}\int K_n(x,y)(y-x)^2 f''(\xi_y)\,dy. \qquad (6.58)$$

Therefore, the first condition is that

$$\lim_{n\to\infty} \int K_n(x,y)\,dy = 1 \quad \forall x \in \Re^1.$$

Assume that the estimator has been rescaled so that the integral is exactly 1 for all n. Therefore, define the random variable Y to have pdf $K_n(x,\cdot)$. The second condition is that

$$\lim_{n\to\infty} \int K_n(x,y)\,y\,dy = \int K_n(x,y)\,x\,dy = x \quad \forall x$$
$$\implies \lim_{n\to\infty} K_n(x,y) = \delta(y-x)$$

as Watson and Leadbetter (1963) and Walter and Blum (1979) suggested. The precise behavior of the bias is determined by the rate at which this happens. For example, suppose that $\int K_n(x,y)\,y\,dy = x$ for all n and that

$$\sigma_{x,n}^2 = \int K_n(x,y)\,(y-x)^2\,dy \neq 0 \tag{6.59}$$

for finite n so that the first two moments of the random variables $Y \sim (x, \sigma_{x,n}^2)$ and $T \equiv (Y-x)/(\sigma_{x,n}) \sim (0,1)$. Suppose that the density function of T, which is a simple linear transformation of K_n, converges to a nice density:

$$\tilde{L}_n(x,t) = K_n(x, x+\sigma_{x,n}t)\,\sigma_{x,n} \to L(x,t) \quad \text{as } n \to \infty.$$

Then the last bias term in (6.58) may be approximated by

$$\frac{1}{2}f''(x)\int K_n(x,y)\,(y-x)^2\,dy = \frac{1}{2}f''(x)\int K_n(x,x+\sigma_{x,n}t)\,(\sigma_{x,n}t)^2\sigma_{x,n}\,dt$$
$$= \frac{1}{2}\sigma_{x,n}^2 f''(x)\int t^2\tilde{L}(x,t)\,dt$$
$$\approx \frac{1}{2}\sigma_{x,n}^2 f''(x)\int t^2 L(x,t)\,dt = \frac{1}{2}\sigma_{x,n}^2 f''(x)$$

since the $\mathrm{Var}(T) = 1$, so that the bias is $O(\sigma_{x,n}^2)$, the familiar rate for second-order kernels. Thus the third condition is that $\sigma_{x,n} \to 0$ as $n \to \infty$.

In order that the variance vanish asymptotically, consider

$$\mathrm{Var}\{\hat{f}(x)\} = \frac{1}{n}\mathrm{Var}\{K_n(x,X)\} \leq \frac{1}{n}\mathrm{E}[K_n(x,X)^2] = \frac{1}{n}\int K_n(x,y)^2 f(y)\,dy$$
$$= \frac{1}{n}\int K_n(x,x+\sigma_{x,n}t)^2 f(x+\sigma_{x,n}t)\,\sigma_{x,n}\,dt$$
$$= \frac{1}{n\sigma_{x,n}}\int \tilde{L}_n(x,t)^2\,[f(x)+\cdots]\,dt \approx \frac{f(x)\,R[L(x,\cdot)]}{n\sigma_{x,n}}.$$

Thus the fourth condition required is that the variance of the equivalent kernel satisfy $n\sigma_{x,n} \to \infty$ as $n \to \infty$.

These findings may be summarized in a theorem, the outline of which is presented in Terrell (1984).

Theorem 6.7: *Suppose $\hat{f}$ is a density estimator with asymptotic equivalent kernel $K_n(x,y)$ and that $\sigma_{x,n}^2$ defined in (6.59) is bounded and nonzero. Then $\hat{f}$ is a nonparametric density estimator if, for all $x \in \Re^1$,*

$$\lim_{n\to\infty} \int K_n(x,y)\,dy = 1$$

$$\lim_{n\to\infty} \int K_n(x,y)y\,dy = x \qquad (6.60)$$

$$\lim_{n\to\infty} \sigma_{x,n} = 0$$

$$\lim_{n\to\infty} n\sigma_{x,n} = \infty.$$

6.4.2.3 Equivalent Kernels of Parametric Estimators

Theorem 6.6 shows how to construct the kernel for any density estimator, parametric or nonparametric. For example, consider the parametric estimation of $f = N(\mu, 1) = \phi(x|\mu, 1)$ by $\hat{\mu} = \bar{x}$. Thus $T_x(F_n) = \phi(x|\bar{x}, 1)$. Examining the argument in the first line in Equation (6.57) and comparing it to the definition of the ecdf in (6.56), it becomes apparent that the empirical pdf $n^{-1}\sum_{i=1}^n \delta(x - x_i)$ is being replaced by

$$\frac{1-\epsilon}{n}\sum_{i=1}^n \delta(x - x_i) + \epsilon\,\delta(x - y),$$

which is the original empirical pdf with a small portion ϵ of the probability mass proportionally removed and placed at $x = y$. The sample mean of this perturbed epdf is $(1 - \epsilon)\bar{x} + \epsilon y$. Thus the kernel may be computed directly from Equation (6.57) by

$$\begin{aligned} K(x,y,F_n) &= \lim_{\epsilon \to 0} \frac{1}{\epsilon}\left[\phi(x|(1-\epsilon)\bar{x} + \epsilon\,y, 1) - (1-\epsilon)\,\phi(x|\bar{x}, 1)\right] \\ &= \frac{1 + (y-\bar{x})(x-\bar{x})}{\sqrt{2\pi}}\,e^{-\frac{1}{2}(x-\bar{x})^2} \end{aligned} \qquad (6.61)$$

by a Taylor's series (see Problem 6.28). The asymptotic equivalent kernel is

$$K(x,y) = \lim_{n\to\infty} K(x,y,F_n) = \frac{1 + (y-\mu)(x-\mu)}{\sqrt{2\pi}}\,e^{-\frac{1}{2}(x-\mu)^2}.$$

This kernel is never *local* and so the estimator is *not* nonparametric (if there was any doubt). Note that the *parametric kernel estimator* with kernel (6.61) is quite good, as

$$\hat{f}(x) = \frac{1}{n} \sum_{i=1}^{n} \frac{1 + (x_i - \bar{x})(x - \bar{x})}{\sqrt{2\pi}} e^{-(x - \bar{x})^2/2} = \frac{1}{\sqrt{2\pi}} e^{-(\hat{x} - \bar{x})^2/2}.$$

Of course, just as in the parametric setting, this "kernel" estimator will always be $\phi(x|\bar{x}, 1)$ for all datasets, no matter how non-normal the underlying density is.

6.5 CROSS-VALIDATION

6.5.1 Univariate Data

The goal is to go beyond the theoretical results for optimal bandwidth specification and to achieve practical data-based algorithms. The oversmoothed and normal reference rules provide reasonable initial choices, together with the simple modifications based on sample skewness and kurtosis given in Section 4.1.2. (The bandwidth modification factors for positive kernels and the frequency polygon are identical.) The unbiased and biased cross-validation algorithms for the histogram are easily extended to both kernel and ASH estimators. However, the development of bandwidth selection algorithms for kernel estimators has progressed beyond those for the histogram.

The importance of choosing the optimal bandwidth is easily overstated. From the sensitivity results in Table 3.3, any choice of h within 15–20% of h^* will often suffice. Terrell (1990) has even suggested that an oversmoothed rule be generally used. With real data, it is relatively easy to examine a sequence of estimates based on the sequence of smoothing parameters

$$h = \hat{h}_{OS}/1.05^k \qquad \text{for } k = 0, 1, 2, \ldots,$$

starting with the sample oversmoothed bandwidth $(\hat{h}_{OS})$, and stopping when the estimate displays some instability and very local noise near the peaks. Silverman (1978a) has characterized the expected amount of local noise present in $\hat{f}''(x)$ in the L_∞ norm (see Equation (6.65)). He suggested examining plots of $\hat{f}''(x)$ interactively in addition to $\hat{f}(x)$, a procedure referred to as the *test graph method*.

The biased cross-validation algorithm, which estimates $R(f'')$ from a kernel estimate, attempts to correct the overestimation by $R(\hat{f}''(\cdot))$ that is the direct result of presence of the local noise (see Section 6.5.1.3). However, the power of the interactive approach to bandwidth selection should not be underestimated. The interactive process can be quite painless in systems supporting animation, such as LISP-STAT (Tierney, 1990).

It should be emphasized that from an exploratory point of view, all choices of the bandwidth h lead to useful density estimates. Large bandwidths, on the one hand, provide a picture of the global structure in the unknown density, including general features such as skewness, outliers, clusters, and location. Small bandwidths, on the other hand, reveal local structure which may or may not be present in the true density. Furthermore, the optimality of h is dependent not only on the choice of metric L_p but also on the feature in the density to be emphasized $(F, f, f', f'', \ldots)$.

However, the desire for fully automatic and reliable bandwidth selection procedures has led inevitably to a series of novel algorithms. These objective (not subjective) procedures often have the stated goal of finding a bandwidth that minimizes the actual L_2 error rather than using the bandwidth that minimizes the expected L_2 error (MISE). In an early paper, Wahba (1981) expressed the expectation that her generalized cross-validation algorithm would accomplish this goal. The best L_2 bandwidth, h_{ISE}, remained the target in unbiased cross-validation for Hall and Marron (1987a, b). Scott and Factor (1981) had expressed the view that h_{MISE} was an appropriate target. The MISE-optimal bandwidth depends only on (f_n), whereas the ISE-optimal bandwidth depends on the sample as well, $(f, x, \{x_i\})$. However, it has been shown that the sample correlation between h_{ISE} and $\hat{\sigma}$ approached -0.70 for normal data (Scott and Terrell, 1987; Scott, 1988b). Given such a large negative correlation with the scale of the data, tracking $\hat{h}_{ISE}$ closely would require guessing whether $\hat{\sigma} > \sigma$, or vice versa, a difficult task.

Of some interest is the fact that while $\hat{h}_{ISE} \approx h_{MISE}$,

$$\frac{\sigma_{\hat{h}_{ISE}}}{h_{MISE}} = O(n^{-1/10}),$$

so that the ISE-optimal bandwidth is only slowly converging to h_{MISE}, as was shown by Hall and Marron (1987a). Scott and Terrell (1987) showed that unbiased cross-validation (UCV) and biased CV (BCV) bandwidths converged to h_{MISE} at the same slow rate. Some of the more recent extensions have been able to push the relative convergence rate all the way to $O(n^{-1/2})$, which is the best rate possible (Hall and Marron, 1991). These procedures require the introduction of one or two auxiliary smoothing parameters. Given the slow rate of convergence of the unattainable $\hat{h}_{ISE}$, it is perhaps unclear whether there is a practical advantage to be had in the faster rates. This question is examined further in Section 6.5.1.5.

In an empirical study of the performance of nine cross-validation algorithms and four sampling densities, Jones and Kappenman (1992) report the "broad equivalence of almost all" of these algorithms with respect to the observed ISE. Other simulations (Scott and Factor, 1981; Bowman, 1985; Scott and Terrell, 1987; Park and Marron, 1990) have reported less equivalence among the estimated smoothing parameter values themselves. Jones and Kappenman reported that the fixed AMISE bandwidth h^* outperformed all nine CV algorithms with respect to ISE. Their results reinforce the suggestion that h^* is an appropriate target and that any choice within 15–20% of h^* should be adequate. Most efforts now focus on h_{MISE} as the target bandwidth.

6.5.1.1 Early Efforts in Bandwidth Selection

The earliest data-based bandwidth selection ideas came in the context of orthogonal series estimators, which were discussed in Section 6.1.3. Using the Tarter–Kronmal weights (6.8) and the representation in Equation (6.5), the pointwise error is

$$\hat{f}(x) - f(x) = \sum_{v=-m}^{m} \hat{f}_v \phi_v(x) - \sum_{v=-\infty}^{\infty} f_v \phi_v(x).$$

As the basis functions are orthonormal,

$$\text{ISE} = \sum_{v=-m}^{m} ||\hat{f}_v - f_v||^2 + \sum_{v \notin [-m,m]} ||f_v||^2.$$

Recall that MISE = E(ISE). Tarter and Kronmal's selection procedure provided unbiased estimates of the increment in MISE in going from a series with $m - 1$ terms to one with m terms (noting the equality of the $\pm v$ MISE terms):

$$\text{MISE}(m) - \text{MISE}(m - 1) = 2\{\text{E}\,||\hat{f}_m - f_m||^2 - ||f_m||^2\}. \tag{6.62}$$

Unbiased estimates of the two terms on the right-hand side may be obtained for the Fourier estimator in Equation (6.6), as $\text{E}\hat{f}_v = f_v$ and $\text{E}\hat{f}_v\hat{f}_v^* = (1 - (n-1)|\hat{f}_v|^2)/n$, where $\hat{f}_v^*$ denotes the complex conjugate of $\hat{f}_v$; hence, $\text{Var}(\hat{f}_v) = \text{E}\hat{f}_v\hat{f}_v^* - |\hat{f}_v|^2 = (1 - |\hat{f}_v|^2)/n$. The data-based choice for m is achieved when the increment becomes *positive*. Notice that accepting the inclusion of the mth coefficient in the series estimator is the result of the judgment that the additional variance of $\hat{f}_m$ is less than the reduction in bias $||f_m||^2$. Usually, fewer than six terms are chosen, so that only relatively coarse adjustments can be made to the smoothness of the density estimator. Sometimes the algorithm misses higher order terms. But the real significance of this algorithm lies in its claim to be the first unbiased cross-validation algorithm for a density estimator.

Likewise, the credit for the first biased cross-validation algorithm goes to Wahba (1981) with her generalized cross-validation algorithm. She used the same unbiased estimates of the Fourier coefficients as Tarter and Kronmal, but with her smoother choice of weights, she lost the simplicity of examining the incremental changes in MISE. However, those same unbiased estimates of the Fourier coefficients lead to a good estimate of the AMISE. By ignoring all the unknown Fourier coefficients for $|v| > n/2$, a small bias is introduced. Both groups recommend plotting the estimated risk function in order to find the best data-based smoothing parameter rather than resorting to (blind) numerical optimization.

The earliest effort at choosing the kernel smoothing parameter in a fully automatic manner was a modified maximum likelihood algorithm due to Habbema et al. (1974) and Duin (1976). While it has not withstood the test of time, it is significant for having introduced a leave-one-out modification to the usual maximum likelihood (ML) criterion. Choosing the bandwidth h to maximize the usual ML criterion results in the (rough) empirical pdf:

$$0 = \arg\max_h \sum_{i=1}^n \log \hat{f}(x_i; h) \quad \Rightarrow \quad \hat{f}(x; h=0) = \frac{1}{n} \sum_{i=1}^n \delta(x - x_i),$$

a solution with "infinite" likelihood. The problem arises since, as $h \to 0$, the contribution to the likelihood at $x = x_i$ from the point x_i itself becomes infinite. The authors sought to eliminate that "self-contribution" by modifying the ML criterion:

$$\max_h \sum_{i=1}^n \log \hat{f}_{-i}(x_i; h),$$

where $\hat{f}_{-i}(x_i; h)$ is a kernel estimator based on the $n-1$ data points excluding x_i and then evaluated there. In spite of some promising empirical small sample and consistency results (Chow et al., 1983), the algorithm was shown to be overly influenced by outliers and tight clusters (Schuster and Gregory, 1981; Scott and Factor, 1981). With a finite support kernel, for example, the bandwidth cannot be less than $x_{(2)} - x_{(1)}$, which is the distance between the first two order statistics; for many densities the distance between these order statistics does not converge to zero and so the bandwidth does not converge to zero as required.

A simple fixed-point algorithm was suggested by Scott et al. (1977). For kernel estimates, the only unknown in h^* in (6.18) is $R(f'')$. If a normal kernel is used, then a straightforward calculation finds that

$$R(\hat{f}_h'') = \frac{3}{8\sqrt{\pi} n^2 h^5} \sum_{i=1}^n \sum_{j=1}^n \left(1 - \Delta_{ij}^2 + \frac{1}{12} \Delta_{ij}^4 \right) e^{-\frac{1}{4}\Delta_{ij}^2}, \tag{6.63}$$

where $\Delta_{ij} = (x_i - x_j)/h$. Following Equation (6.18), the search for a fixed-point value for h^* is achieved by iterating

$$h_{k+1} = \left[\frac{R(K)}{n \sigma_K^4 R(\hat{f}_{h_k}'')} \right]^{1/5},$$

with h_0 chosen to be the normal reference bandwidth. As the ratio of the optimal bandwidths for estimating f and f'' diverges as $n \to \infty$, it is clear that the algorithm is not consistent. That the algorithm worked well for small samples (Scott and Factor, 1981) is not surprising since the optimal bandwidths are reasonably close to each other for small samples. Note that this algorithm as stated provides no estimate of the MISE. It is a simple matter to use the roughness estimate (6.63) in Equation (6.18), following the lead of Wahba, to obtain

$$\widehat{\text{AMISE}}(h) = \frac{R(K)}{nh} + \frac{1}{4}\sigma_K^4 h^4 R(\hat{f}_h''). \tag{6.64}$$

Finding the minimizer of Equation (6.64) not only provides a data-based bandwidth estimate, but also an estimate of the MISE. This idea was resurrected with biased cross-validation using a consistent estimator for $R(f'')$ rather than Equation (6.63). Alternatively, a much wider bandwidth appropriate to f'' rather than f might be used in the iteration. Sheather (1983) proposed such a scheme while trying to estimate the density at the origin, providing an example of the plug-in (PI) estimators discussed by Woodroofe (1970). Sheather's motivation is discussed in Sheather and Jones (1991). In a talk in 1991, Gasser also reported success in this way for choosing a global bandwidth by inflating the bandwidth used in $R(\hat{f}_n'')$ by the factor $n^{-1/10}$.

For a normal kernel, Silverman (1978a) proved that

$$\frac{\sup |\hat{f}'' - E\hat{f}''|}{\sup |E\hat{f}''|} \approx 0.4. \tag{6.65}$$

He proposed choosing the bandwidth where it appears that the ratio of the noise to the signal is 0.4. This ratio is different for other kernels. The *test graph* procedure can be used in the bivariate case as well.

6.5.1.2 *Oversmoothing*

The derivation of the oversmoothed rule for kernel estimators will be constructive, unlike the earlier derivations of the histogram and frequency polygon oversmoothed rules. The preferred version uses the variance as a measure of scale. Other scale measures have been considered by Terrell and Scott (1985) and Terrell (1990).

Consider the variational problem

$$\min_f \int_{-\infty}^{\infty} f''(x)^2 dx \quad \text{s/t} \quad \int f = 1 \text{ and } \int x^2 f = 1. \tag{6.66}$$

Clearly, the solution will be symmetric. The associated Lagrangian is

$$L(f) = \int_{-\infty}^{\infty} f''(x)^2 dx + \lambda_1 \left(\int f - 1 \right) + \lambda_2 \left(\int x^2 f - 1 \right).$$

At a solution, the Gâteaux variation, defined in Equation (6.54), of the Lagrangian in any "direction" η must vanish. For example, the Gâteaux variation of $\Psi(f) = \int f''(x)^2$ is

$$\Psi'(f)[\eta] = \lim_{\epsilon \to 0} \frac{1}{\epsilon} \left[\int [(f + \epsilon \eta)'']^2 - \int [f'']^2 \right]$$

$$= \lim_{\epsilon \to 0} \frac{1}{\epsilon} \left[\int \left(f''^2 + 2\epsilon f'' \eta'' + \epsilon^2 \eta''^2 - f''^2 \right) \right] = \int 2f'' \eta''.$$

Computing the Gâteaux variation of $L(f)$, we have

$$0 = L'(f)[\eta] = \int 2f''(x)\eta''(x) + \lambda_1 \int \eta(x) + \lambda_2 \int x^2\eta(x)$$

$$= 2f''(x)\eta'(x) \big|_{-c}^{c} - 2f'''(x)\eta(x) \big|_{-c}^{c} + \int \left[2f^{iv}(x) + \lambda_1 + \lambda_2 x^2\right]\eta(x) \qquad (6.67)$$

after integrating by parts twice and where c is the boundary, possibly infinite, for f. Now $\eta(\pm c)$ must vanish so that there is not a discontinuity in the solution; therefore, the second term vanishes. The remaining two terms must vanish for all feasible choices for η; therefore, $f''(\pm c)$ must vanish, leaving only the integral. It follows that the integrand must vanish and that f is a sixth-order polynomial with only even powers (by symmetry). Therefore, the solution must take the form

$$f(x) = a(x-c)^3(x+c)^3$$

so that $f''(\pm c) = 0$. The two constraints in (6.66) impose two linear conditions on the unknowns (a, c), with the result that

$$f^*(x) = \frac{35}{69,984}(9-x^2)_+^3 \qquad \text{and} \qquad R[(f^*)''] = \frac{35}{243}.$$

A simple change of variables shows that $R(f'') \geq 35/(243\sigma^5)$; therefore,

$$h^* = \left[\frac{R(K)}{n\sigma_K^4 R(f'')}\right]^{1/5} \leq \left[\frac{243\sigma^5 R(K)}{35n\sigma_K^4}\right]^{1/5} \Rightarrow \qquad (6.68)$$

$$\boxed{\text{Oversmoothing rule:} \qquad h_{OS} = 3\left[\frac{R(K)}{35\sigma_K^4}\right]^{1/5}\sigma n^{-1/5}.} \qquad (6.69)$$

For the normal kernel, $h_{OS} = 1.144\,\sigma n^{-1/5}$. For the biweight kernel, it is *exactly* $h_{OS} = 3\,\sigma n^{-1/5}$. The rule is 1.08 times wider than the normal reference rule.

6.5.1.3 Unbiased and Biased Cross-Validation

The presentation in the section will rely heavily on the references for certain details. The overall flavor is the same as in the application to the histogram and frequency polygon.

The remarkable fact is that the UCV justification for the histogram is entirely general. Thus the definition in Equation (3.52) applies in the case of a kernel estimator. For the case of a normal kernel, Rudemo (1982) and Bowman (1984) showed that (replacing $n \pm 1$ with n for simplicity)

$$\boxed{\text{UCV}(h) = \frac{1}{2nh\sqrt{\pi}} + \frac{1}{n^2 h\sqrt{\pi}}\sum_{i<j}\left(e^{-\Delta_{ij}^2/4} - \sqrt{8}e^{-\Delta_{ij}^2/2}\right),} \qquad (6.70)$$

which is a special case of the general formula (Scott and Terrell, 1987)

$$\text{UCV}(h) = \frac{R(K)}{nh} + \frac{2}{n^2h} \sum_{i<j} \gamma(\Delta_{ij}),$$

where

$$\gamma(\Delta) = \int K(w) K(w + \Delta) \, dw.$$

The BCV algorithm follows from the result (Scott and Terrell, 1987) that

$$\text{ER}(\hat{f}_h'') = R(f'') + \frac{R(K'')}{nh^5} + O(h^2),$$

where $R(K'')/(nh^5)$ is asymptotically a constant, representing the fixed but finite noise that exists in the kernel estimate. Therefore, $R(\hat{f}_h'') - R(K'')/(nh^5)$ is an asymptotically unbiased estimator for the unknown roughness $R(f'')$. Substituting into Equation (6.18), the estimate of the MISE becomes

$$\text{BCV}(h) = \frac{R(K)}{nh} + \frac{\sigma_K^4}{2n^2h} \sum_{i<j} \phi(\Delta_{ij}), \tag{6.71}$$

where

$$\phi(\Delta) = \int K''(w) K''(w + \Delta) \, dw.$$

The similarity of the general UCV and BCV formulas is remarkable, given their quite different origins. In the case of a normal kernel,

$$\boxed{\text{BCV}(h) = \frac{1}{2nh\sqrt{\pi}} + \frac{1}{64n^2h\sqrt{\pi}} \sum_{i<j} \left(\Delta_{ij}^4 - 12\Delta_{ij}^2 + 12 \right) e^{-\Delta_{ij}^2/4}.} \tag{6.72}$$

As before, $\lim_{h\to\infty} \text{BCV}(h) = 0$; therefore, $\hat{h}_{\text{BCV}}$ is taken to be the largest local minimizer of $\text{BCV}(h)$ less than or equal to the oversmoothed bandwidth.

The asymptotic theory of the CV criteria is straightforward once it is recognized that the stochastic part is all contained in the so-called U-statistics, which are double sums of the form

$$U_n = \sum_{i<j} H_n(X_i, X_j).$$

Hall (1984) proved that if the function H_n is symmetric and the random variable $E[H_n(X, Y) \mid X] = 0$, then together with a certain moment condition,

$$U_n = \text{AN}\left(0, \tfrac{1}{2}n^2 \, \text{E}H_n^2\right).$$

The asymptotic normality of UCV and BCV is not obvious because the number of terms in the double sum effectively shrinks, as the bandwidth is a decreasing function of n. Hall used an argument from degenerate Martingale theory to prove this theorem.

Scott and Terrell (1987) then proved that for a fixed bandwidth, h, the UCV and BCV functions were (1) both asymptotically normal; (2) both converged to AMISE(h); and (3) the asymptotic (vertical) variances of UCV and BCV at h are

$$\frac{2R(\gamma)R(f)}{n^2 h} \quad \text{and} \quad \frac{\sigma_K^8 R(\phi)R(f)}{8n^2 h}, \tag{6.73}$$

respectively. For kernels in the symmetric Beta family of the form (6.27), the (vertical) variance of UCV is at least 80 times greater than for the BCV criterion. This smaller vertical variance suggests that the actual minimizer of the BCV criterion will have smaller variance than $\hat{h}_{UCV}$. The variances in Equation (6.73) are of order $O(n^{-9/5})$ if $h = O(n^{-1/5})$. However, this rapid rate of decrease can be explained by the fact that the CV functions are themselves going to 0 at the rate $O(n^{-4/5})$. Thus the relevant quantity is the coefficient of variation $(C.V.)$

$$C.V. = \frac{\sqrt{\text{Var UCV}(h)}}{O[\text{UCV}(h)]} = \frac{O(n^{-9/10})}{O(n^{-4/5})} = O(n^{-1/10}),$$

as was claimed earlier. The astute reader will note that the bias has not been counted in this error in the BCV case; however, the bias turns out to be $O(n^{-1})$, and hence the squared bias is $O(n^{-2})$, which is of lower order than the variance.

Using a delta argument outlined below, Scott and Terrell (1987) showed that $\hat{h}_{UCV}$ and $\hat{h}_{BCV}$ converged to h_{AMISE} and were asymptotically normal with respective variances given by

$$\frac{2R(f)R[\Delta\gamma'(\Delta)]}{25n^2(h^*)^7\sigma_K^4 R(f'')^2} \quad \text{and} \quad \frac{R(f)R[\Delta\phi'(\Delta)]}{200n^2(h^*)^7 R(f'')^2}. \tag{6.74}$$

Observe that if $h^* = O(n^{-1/5})$, then these variances are $O(n^{-3/5})$, from which the (horizontal) coefficient of variation is found to still be of $O(n^{-1/10})$:

$$C.V. = \frac{\sqrt{\text{Var}\,\hat{h}_{UCV}}}{O(\hat{h}_{UCV})} = \frac{O(n^{-3/10})}{O(n^{-1/5})} = O(n^{-1/10}).$$

Again, for densities in the symmetric Beta family, the UCV variance is at least 16 times that of the BCV. These results were confirmed in a simulation study. However, it was noted that BCV performed poorly for several difficult densities without a very large dataset. This finding is not surprising, given that the basis for the BCV formula is AMISE, while the exact MISE is the basis of UCV.

It is instructive to outline the derivation of Equation (6.74). Clearly, the BCV bandwidth satisfies

$$\frac{d}{dh}[\text{BCV}(h)]\Big|_{h=\hat{h}_{BCV}} = 0.$$

Noting that $\Delta'_{ij} = -(x_i - x_j)/h^2 = -\Delta_{ij}/h$, the derivative of BCV as defined in (6.71) equals

$$\frac{-R(K)}{nh^2} - \frac{\sigma_K^4}{2n^2h^2}\sum_{i<j}\phi(\Delta_{ij}) + \frac{\sigma_K^4}{2n^2h}\sum_{i<j}\phi'(\Delta_{ij})\frac{-\Delta_{ij}}{h} = 0$$

or

$$\sum_{i<j}[\phi(\Delta_{ij}) + \Delta_{ij}\phi'(\Delta_{ij})]\Big|_{h=\hat{h}_{BCV}} = -\frac{2nR(K)}{\sigma_K^4}.$$

Define $\psi(\Delta) = \Delta\,\phi'(\Delta)$; then computing approximations to the moments of ϕ and ψ (a nontrivial calculation given in Section 9 of Scott and Terrell (1987)), it follows that

$$\sum_{i<j}\left[\phi(\Delta_{ij}) + \psi(\Delta_{ij})\right] = \text{AN}\left(-2n^2h^5R(f''), n^2hR(f)R(\psi)/2\right).$$

Hence, combining these two results and rearranging,

$$-2n^2R(f'')\hat{h}_{BCV}^5 = \text{AN}\left(-2nR(K)/\sigma_K^4, n^2h^*R(f)R(\psi)/2\right)$$

or

$$\hat{h}_{BCV}^5 = \text{AN}\left(\frac{R(K)}{\sigma_K^4 nR(f'')}, \frac{h^*R(f)R(\psi)}{8n^2R(f'')^2}\right). \tag{6.75}$$

Clearly, the asymptotic mean of $\hat{h}_{BCV}^5$ is $(h^*)^5$. The random variable $\hat{h}_{BCV}$ is the 1/5 power of that given in (6.75). Applying the delta method, it may be concluded that $\hat{h}_{BCV}$ is AN with mean h^* and variance which may be computed by the formula

$$\text{Var}\{g(h)\} = \left(\frac{dg}{dh}\right)^2_{h=h^*}\text{Var}\{h\}.$$

Now $g(h) = h^5$ and $g'(h) = 5h^4$, so that

$$\text{Var}\{\hat{h}_{BCV}\} = \text{Var}\{\hat{h}_{BCV}^5\}/[25(h^*)^8]. \tag{6.76}$$

The variance (6.74) follows immediately combining (6.75) and (6.76).

Despite these apparently favorable findings, BCV does not qualify as a general replacement for UCV. UCV may be noisier but it tends to produce nearly unbiased smoothing parameters. However, there is a need for an auxiliary CV criterion since UCV is susceptible to certain problems. Clearly, BCV has its own set of limitations. But by carefully examining of the trio of smoothing parameters suggested by UCV, BCV, and OS as well as the shapes of the UCV and BCV curves, good bandwidths should be reliably available.

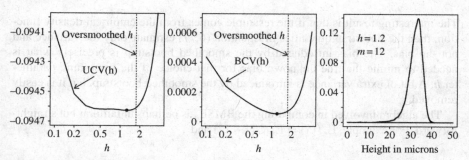

FIGURE 6.15 UCV and BCV estimates of the steel surface data ($n = 15,000$) using the triweight ASH. The UCV bandwidth was tied at 1.2 and 1.3, while the BCV bandwidth was 1.2 (shown): both estimates were virtually identical.

As a practical matter, both UCV and BCV involve $O(n^2)$ computation due to the double sums. If the normal kernel versions are used, the work can easily be prohibitive for $n > 500$. This work can be substantially reduced by using an ASH implementation. The computational details are given in Scott and Terrell (1987) and are not repeated here. Of course, code is available from several sources to perform these and other computations. The ASH implementation is illustrated in Figure 6.15 for the steel surface data. The standard deviation of these data is 3.513, so that the oversmoothed bandwidth for the triweight kernel is $\hat{h}_{OS} = 1.749$ from Equation (6.69). Observe that the UCV estimate is flat over a relatively wide interval even with such a large dataset. However, the minima of the two criteria are virtually identical, $h_{BCV} = 1.2$ and $h_{UCV} = 1.3$ for the triweight kernel. (The triweight estimate with equivalent smoothing parameter $h = 0.67$ is shown in Figure 6.5.) The original data were binned into 500 intervals, so that the use of the ASH implementation of the CV and estimation functions is natural.

6.5.1.4 Bootstrapping Cross-Validation
Taylor (1989) investigated a data-based algorithm for choosing h based on bootstrap estimates of the MSE$\{\hat{f}(x)\}$ and MISE$\{\hat{f}\}$. The bootstrap resample is not taken from the empirical pdf f_n as in the *ordinary bootstrap*, but rather the bootstrap sample $\{x_1^*, x_2^*, \ldots, x_n^*\}$ is a random sample from the candidate kernel density estimate $\hat{f}(x; h)$ itself. Such a resample is called a *smoothed bootstrap sample* and is discussed further in Chapter 9. Letting E_* represent expectation with respect to the smoothed bootstrap random sample, Taylor examined

$$\text{BMSE}_*(x; h) = E_*[\hat{f}_*(x; h) - \hat{f}(x; h)]^2$$

$$= E_* \left[\frac{1}{n} \sum_{i=1}^{n} K_h(x - x_i^*) - \frac{1}{n} \sum_{i=1}^{n} K_h(x - x_i) \right]^2 \quad (6.77)$$

$$\text{BMISE}_*(h) = \int_x \text{BMSE}_*(x; h) \, dx.$$

The interesting result is that if the resample comes from the empirical density function, then the bootstrap quantities in Equation (6.77) estimate only the variance and not the bias. The bias introduced by the smoothed bootstrap is precisely what is needed to mimic the true unknown bias for that choice of the smoothing parameter h. A bit of extra variance is introduced by the smoothed bootstrap, but it is easily removed.

The algebra involved in computing the $BMSE_*$ is perhaps unfamiliar but straightforward. For example, $E_* \hat{f}(x) = E_* K_h(x - x^*)$ and

$$E_* K_h(x - x^*) = \int K_h(x - y) \hat{f}(y) \, dy = \frac{1}{n} \sum_{i=1}^{n} \int K_h(x - y) K_h(y - x_i) \, dy.$$

The computation of $E_* K_h(x - x^*)^2$ is a trivial extension. For the particular case of the normal kernel, the convolutions indicated in the bootstrap expectation may be computed in closed form. After a few pages of work using the normal kernel and adjusting the variance, Taylor proposes minimizing

$$BMISE_*(h) = \frac{1 + \frac{\sqrt{2}}{n} \sum_{i<j} \left[\sqrt{2} \, e^{-\frac{\Delta_{ij}^2}{4}} - \frac{4}{\sqrt{3}} e^{-\frac{\Delta_{ij}^2}{6}} + e^{-\frac{\Delta_{ij}^2}{8}} \right]}{2nh\sqrt{\pi}} \qquad (6.78)$$

Taylor shows that $BMISE_*(h)$ has the same order variance as $UCV(h)$ and $BCV(h)$, but with a smaller constant asymptotically.

In the Figure 6.16, the three CV functions, with comparable vertical scalings, are plotted for the snowfall data along with the corresponding density estimates using the normal kernel. Many commonly observed empirical results are depicted in these graphs. The BCV and bootstrap curves are similar, although $BCV(h)$ has a sharper minimum. Both the biased and bootstrap CV functions have minima at bandwidths greater than the oversmoothed bandwidth $h_{OS} = 11.9$. This serves to emphasize how difficult estimating the bias is with small samples. The unbiased CV function better reflects the difficulty in precisely estimating the bias by presenting a curve that is flat over a wide interval near the minimum. There is some visual evidence of three modes in this data. Since these CV functions do not take account of the time series nature of these data and the first-order autocorrelation is -0.6, a smaller bandwidth is probably justifiable (see Chiu (1989) and Altman (1990)).

These computations are repeated for the Old Faithful geyser dataset, which is clearly bimodal, and the results depicted in Figure 6.17. It is interesting to note that both the biased and bootstrap CV functions have two local minima. Fortunately, only one local minima is smaller than the oversmoothed upper bound $h_{OS} = 0.47$, although the bootstrap curve barely exhibits the second local minimum. The UCV function leads to a narrow bandwidth and a density estimate that seems clearly undersmoothed given the sample size. These observations seem to recur with many "real" datasets

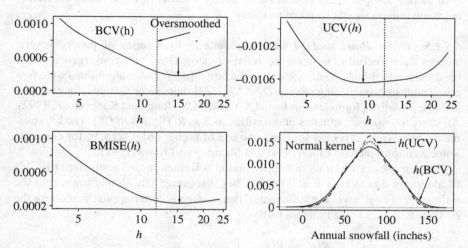

FIGURE 6.16 Normal kernel cross-validation algorithms and density estimates for the snowfall data ($n = 63$). The CV bandwidths are indicated by arrows and the oversmoothed bandwidth by the dashed line. The UCV, BCV, and oversmoothed density estimates are represented by the dashed, solid, and dotted lines, respectively.

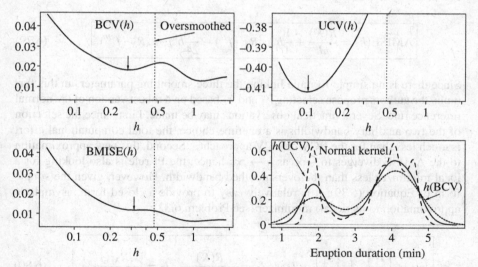

FIGURE 6.17 Normal kernel cross-validation algorithms and density estimates for the geyser dataset ($n = 107$). The CV bandwidths are indicated by arrows and the oversmoothed bandwidth by a solid line. The UCV, BCV, and oversmoothed density estimates are represented by the dashed, solid, and dotted lines, respectively.

with modest sample sizes. Concordance among a subset of these rather differently behaving criteria should be taken seriously.

6.5.1.5 Faster Rates and PI Cross-Validation

In a series of papers, several authors have worked to improve the relatively slow $O(n^{-1/10})$ convergence of CV bandwidth algorithms. Along the way to the best $O(n^{-1/2})$ rate, algorithms were proposed with such interesting rates as $O(n^{-4/13})$. All share some features: for example, all have U-statistic formulas similar to UCV and BCV (Jones and Kappenman, 1992). All strive to improve estimates of quantities such as $R(f'')$ and $R(f''')$. The improvements come from expected sources—the use of higher order kernels, for example. Some examples include Chiu (1991) and Sheather and Jones (1991).

For purposes of illustration, the discussion is limited to one particular $O(n^{-1/2})$ PI algorithm due to Hall et al. (1991). They discovered that simply improving the estimation of $R(f'')$ was not sufficient; rather, a more accurate approximation for the AMISE is required (see Problem 6.10):

$$\text{AMISE}(h) = \frac{R(K)}{nh} - \frac{R(f)}{n} + \frac{1}{4}h^4\mu_2^2 R(f'') - \frac{1}{24}h^6\mu_2\mu_4 R(f'''),$$

which is accurate to $O(n^{-7/5})$. The second term, which is constant, may be ignored. The two unknown roughness functionals are estimated not as in BCV, but rather with two auxiliary smoothing parameters, λ_1 and λ_2, so that

$$\widehat{\text{AMISE}}(h) = \frac{R(K)}{nh} + \frac{1}{4}h^4\mu_2^2 \hat{R}_{\lambda_1}(f'') - \frac{1}{24}h^6\mu_2\mu_4 \hat{R}_{\lambda_2}(f'''). \tag{6.79}$$

Since there is no simple formula linking the three smoothing parameters in this formula, the authors propose selecting λ_1 and λ_2 based on a robust version of the normal reference rule. Several practical observations may be made. First, since the selection of the two auxiliary bandwidths is a onetime choice, the total computational effort is much less than for the BCV or UCV approaches. Second, this new approximation to the AMISE diverges to $-\infty$ as $h \to \infty$; hence, the PI rule is also looking for a local minimizer less than the oversmoothed bandwidth. However, given the simple form of Equation (6.79), it is relatively easy to provide a closed-form, asymptotic approximation to its (local) minimizer (see Problem 6.32):

$$\hat{h}_{\text{PI}} = \left[\frac{\hat{J}_1}{n}\right]^{\frac{1}{5}} + \left[\frac{\hat{J}_1}{n}\right]^{\frac{3}{5}} \cdot \hat{J}_2; \quad \hat{J}_1 = \frac{R(K)}{\mu_2^2 \hat{R}_{\lambda_1}(f'')}, \quad \hat{J}_2 = \frac{\mu_4 \hat{R}_{\lambda_2}(f''')}{\mu_2 \hat{R}_{\lambda_1}(f'')}. \tag{6.80}$$

A portion of the details of a particular implementation given in the authors' paper is outlined in the next paragraph. The PI bandwidths computed in this fashion indeed have rapidly vanishing noise.

The estimates of $R(f'')$ and $R(f''')$ follow from the identities

$$\int f''(x)^2\,dx = +\int f^{iv}(x)f(x)\,dx$$

$$= +\mathrm{E}f^{iv}(X) \leftarrow \frac{1}{n(n-1)\lambda_1^5}\sum_{i=1}^{n}\sum_{j=1}^{n}K^{iv}\left(\frac{x_i-x_j}{\lambda_1}\right)$$

$$\int f'''(x)^2\,dx = -\int f^{vi}(x)f(x)\,dx$$

$$= -\mathrm{E}f^{vi}(X) \leftarrow \frac{1}{n(n-1)\lambda_2^7}\sum_{i=1}^{n}\sum_{j=1}^{n}\phi^{vi}\left(\frac{x_i-x_j}{\lambda_2}\right).$$

These two estimates are suggested by the UCV estimator. The authors chose a particular order-4 polynomial kernel for $K(x)$, which is supported on the interval $(-1,1)$, whose fourth derivative is equal to

$$K^{iv}(x) = \tfrac{135,135}{4,096}(1-x^2)(46,189x^8 - 80,036x^6 + 42,814x^4 - 7,236x^2 + 189).$$

Of course, for the normal kernel, $\phi^{vi}(x) = (x^6 - 15x^4 + 45x^2 - 15)\,\phi(x)$. Using normal reference rules and the interquartile range (IQR) as the measure of scale, the formulas for the auxiliary smoothing parameters are calculated to be

$$\hat{\lambda}_1 = 4.29\ \mathrm{IQR}\ n^{-1/11} \quad \text{and} \quad \hat{\lambda}_2 = 0.91\ \mathrm{IQR}\ n^{-1/9}.$$

To illustrate the performance of the PI bandwidth, 21 $N(0,1)$ simulations are displayed in Figure 6.18 for several sample sizes. The rapid convergence of the bandwidths to the h^* is apparent. For small samples, there may be no minimizer of the PI AMISE estimate given in Equation (6.79); examine some of the individual risk curves in Figure 6.18. The lack of a local minimum is a feature observed for small samples with the BCV criterion. It would appear that for small samples, the PI formula (6.80) is essentially returning a version of the normal reference rule rather than a true minimizer of the risk function. As the sample size increases, the "strength" of the PI estimate grows as the bowl shape of the AMISE estimate widens.

The PI AMISE function estimates for the Buffalo snowfall data ($n = 63$) and the Old Faithful eruption duration data ($n = 107$) are shown in Figure 6.19. Neither estimate has a local minimum, although the PI formula gives reasonable results. This illustrates the danger inherent in not plotting the risk function. It also suggests that the excellent small sample behavior of PI estimators involves subtle factors. In any case, the greater the number of reasonable bandwidth algorithms available to attack a dataset, the better.

For a very large dataset, such as the steel surface data, the PI, BCV, and UCV bandwidths are often identical. The PI risk curve is shown in Figure 6.19. The value of $\hat{h}_{PI} = 0.424$ for the Gaussian kernel. The conversion factor to the triweight kernel ($\sigma_K = 1/3$) is 3 or $\hat{h} = 1.27$, which is identical to the UCV and BCV predictions in

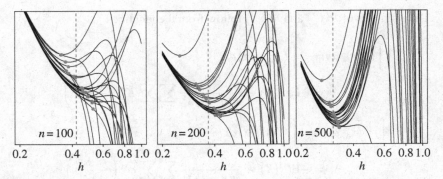

FIGURE 6.18 Twenty-one examples of the AMISE approximation of the plug-in rule with $N(0,1)$ data and a normal kernel. The PI bandwidth for each simulation is shown by the black dot on the risk curve. The vertical dotted line indicates the normal reference rule (with $\sigma = 1$). Note that the horizontal axis is the same for each sample size, but the vertical scale (not labeled) zooms in on the relevant area.

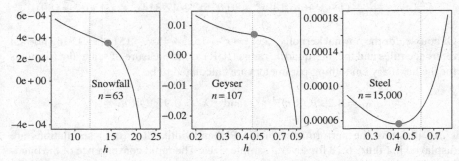

FIGURE 6.19 Plug-in cross-validation curves for the snowfall data ($n = 63$), the geyser dataset ($n = 107$) and the steel surface data ($n = 15,000$) for the normal kernel. The plug-in bandwidth obtained by formula (6.80) is indicated by the black dot, and the oversmoothed bandwidth by the dashed line.

Figure 6.15. The PI AMISE curve for the bimodal geyser dataset has no local minima. The PI bandwidth is $\hat{h}_{\mathrm{PI}} = 0.472$ (or $3 \times 0.472 = 1.42$ on the triweight scale). Thus the PI bandwidth matches the oversmoothed bandwidth but misses the local bandwidth found by the BCV and BMISE curves in Figure 6.17. For the smaller snowfall dataset, the PI bandwidth is greater than the oversmoothed bandwidth. Improved PI algorithms are under development, using generalized scale measures appropriate for multimodal data, and rapid changes and new successes can be expected. In any case, the experienced worker can expect to be able to judge the success or failure of any cross-validation bandwidth readily with modern interactive computing. Sheather (2004) indicates that the conservative choice of bandwidths for the auxiliary parameters tends to make the PI bandwidths oversmooth for complex densities.

6.5.1.6   Constrained Oversmoothing   Oversmoothing has been presented strictly as a means of bounding the unknown smoothing parameters by choosing a measure of scale, such as the sample range or standard deviation, for the data. It is easy to find cases where the oversmoothed bandwidths are much too wide. If the sampling density is in fact the oversmoothed density, then the bandwidth $\hat{h}_{OS}$ is not strictly an upper bound, as $\hat{h}_{OS}$ varies about h^*. In most instances, the two bandwidths will be within a few percentage points of each other.

The variational problems considered in oversmoothing can be generalized to provide much more relevant bandwidths in almost every situation. The basic idea is to add constraints in addition to the one measure of overall scale. The new proposal for constrained oversmoothing (CO) is to require that several of the percentiles in the oversmoothed density match the sample percentiles in the data. Specifically,

$$f_{CO}^* = \arg\min_f \int f''(x)^2\, dx \quad \text{s/t} \quad \int_{-\infty}^{\alpha_i} f(x)\, dx = F_n(\alpha_i), \quad i \in [1, k],$$

where $k \geq 2$. In other words, the cdf of the oversmoothed distribution should match the empirical cdf in several intermediate locations. This problem has already been solved in two instances when $k = 2$, once with the range constraint and again with the interquartile range constraint. The new suggestion is to choose to match the 10th, 30th, 50th, 70th, and 90th sample percentiles, for example. The resulting constrained oversmoothed density may or may not be close to the true density, but computing the roughness of the CO density provides a significantly improved *practical* estimate for use in the usual asymptotic bandwidth formula. With f_{CO}^* in hand, the constrained oversmoothed smoothing parameter is found as

$$\hat{h}_{CO} = \left[\frac{R(K)}{n\sigma_K^4 R[(f_{CO}^*)'']} \right]^{1/5}.$$

In Figure 6.20 for a sample of 1000 $N(0,1)$ points, several possible solutions to the variational problem are displayed along with the computed roughness. The location of the constraints is indicated by the dashed lines. The solution is a quartic spline. The relevant quantity is the fifth root of the ratio of that roughness to the true roughness, $R(\phi'') = 3/(8\sqrt{\pi}) = 0.2115$. The first CO solution is based on matching the 1, 50, and 99% sample percentiles. Now $(0.2115/0.085)^{1/5} = 1.20$, so that the constrained oversmoothed bandwidth is 1.2 times wider than h^*. Thus $h^* < h_{CO}$ as usual. However, by including more constraints, this inequality will not be strictly observed. For example, two solutions were found with 11 constraints. One used the 1, 10, 20, ..., 90, and 99% sample percentiles. Since the roughness of this solution is slightly *greater* than the true roughness, the CO bandwidth will be *smaller* than h^*. In fact, $h_{CO} = 0.94\, h^*$. A second problem with 11 constraints used a fixed width mesh as shown. The sample percentiles matched in this mesh were 1.0, 3.4, 7.8, 16.6, 30.6, 49.9, 68.9, 82.7, 92.5, 97.3, and 99.0%. For this solution, $h_{CO} = 0.82\, h^*$. Matching

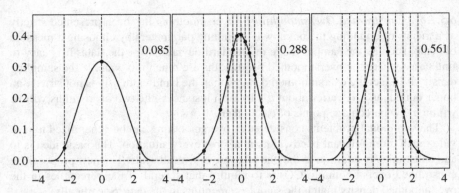

FIGURE 6.20 Constrained oversmoothed density solutions for a $N(0,1)$ sample of 1000 points. The true roughness is 0.212.

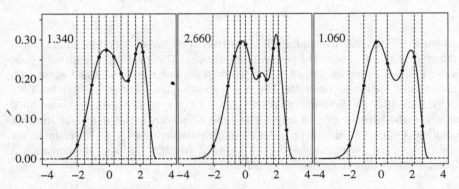

FIGURE 6.21 Constrained oversmoothed density solutions for a sample of 1000 points from the mixture density, $0.75N(0,1) + 0.25N(0, 1/9)$. The true roughness is 3.225.

5–10 percentiles would seem adequate in practice. For example, the percentile mesh with seven constraints (not shown) has a roughness of 0.336, which corresponds to $h_{CO} = 0.91\, h^*$.

This procedure was repeated for a sample of 1000 points from a mixture density, $\frac{3}{4}\phi(x|0,1) + \frac{1}{4}\phi(x|2, 1/9)$ (see Figure 6.21). The true roughness is 3.225. Solutions for a fixed bin width and 2 percentile meshes are shown. For each solution $h^* < h_{CO}$ as the roughness of each solution is less than the true roughness. The difference is greatest with the solution in the last frame, for which $h_{CO} = 1.25\, h^*$. The ordinary oversmoothed rule, which is based on the variance 55/36, leads to a lower bound of 0.0499 for the roughness and $h_{OS} = 2.30\, h^*$ from Equation (6.68). Thus the constrained oversmoothed procedure gives a conservative bandwidth, but not *so* conservative. The solution in the middle frame indicates how the constrained oversmoothed density solution may contain features not in the true density (the small bump at $x = 1$), but still be conservative ($h_{CO} = 1.04\, h^*$).

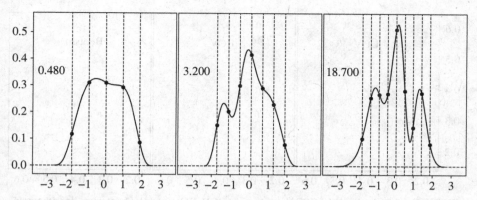

FIGURE 6.22 Constrained oversmoothed density solutions for Buffalo snowfall data.

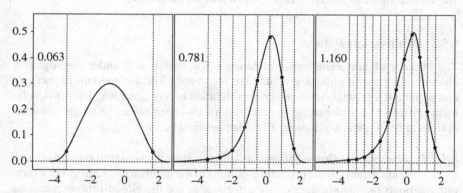

FIGURE 6.23 Constrained oversmoothed density solutions for steel surface data. The bin count data were jittered.

The application to real data is promising. (In all the examples, the original data were centered and standardized.) Several constrained oversmoothed solutions are shown in Figure 6.22 for the Buffalo snowfall data. In this case, all equally spaced meshes were selected. For small samples, this selection seems to work better than percentile meshes.

The final application is to a large dataset. Several constrained oversmoothed solutions are shown in Figure 6.23 for the steel surface data. Again, all equally spaced meshes were selected. Clearly, a roughness of about 0.8 is indicated—this leads to a bandwidth that is 60% of $(0.063/0.8)^{1/5}$, the usual oversmoothing shown in the first frame.

Clearly, the number and location of the constraints serve as surrogate smoothing parameters. However, if not too many are chosen, the solution should still serve as a more useful point of reference (upper bound). It is possible to imagine extending the problem to adaptive meshes.

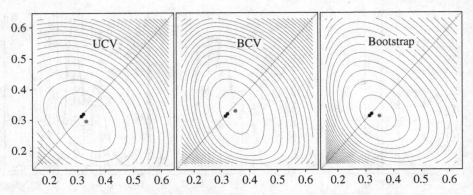

FIGURE 6.24   Estimated MISE(h_x, h_y) using UCV, BCV, and the bootstrap algorithms on 1500 $N(\mathbf{0}_2, I_2)$ points. The two dots on each diagonal are h^* and the oversmoothed bandwidths. The dot locating the minimizer of each criterion is below the diagonal.

6.5.2   Multivariate Data

The efficacy of univariate cross-validation algorithms is still under investigation, although the general options available are reasonably well understood. PI methods can be quite stable, while unbiased cross-validation is very general and easy to implement. Progress on multivariate generalizations has been made, although again the options are still under discussion. These are outlined below.

6.5.2.1   *Multivariate Cross-Validation*   In principle, it is straightforward to extend each of the algorithms in the preceding sections to the multivariate setting. For example, the bootstrap algorithm is easily extended, as are the closed-form expressions for BCV and UCV (Sain et al., 1994). Examples of UCV, BCV, and BMISE$_*(h_1, h_2)$ based on 1500 $N(\mathbf{0}_2, I_2)$ points are shown in Figure 6.24. All are reasonably close in this case.

Figure 6.25 shows the same criteria applied to the standardized log-lipid dataset. However, the BMISE$_*(h_1, h_2)$ estimate for the lipid values ($n = 320$) has its minimum beyond the normal reference rule and the oversmoothed bandwidths. The UCV and BCV bandwidths are similar, although the BCV is (surprisingly) a bit more aggressive. Both show the two and possibly third mode/bump, whereas the Bootstrap estimate is unimodal. In general, it may be expected that the performance of all multivariate CV algorithms is more asymptotic and hence slower to converge in practice.

Retaining a smoothing parameter for each coordinate direction is important. Even if the data have been standardized, it is unlikely that $h_i = h_j$ will be satisfactory, (see Nezames (1980) and Wand and Jones (1991)). Less clear is whether the gain using a full smoothing matrix, which allows for elliptical contours, is sufficient to account for the dramatic increase in the number of parameters that must estimated by cross-validation. In many or most instances, the gains are not realized in practice. Furthermore, a full covariance may degenerate into a singular form, which can have "infinite" likelihood. Different starting values may avoid this problem.

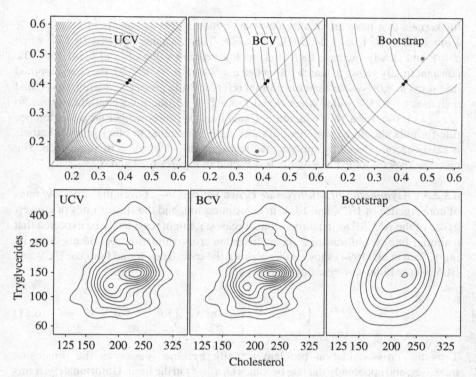

FIGURE 6.25 Same criterion as in Figure 6.24 for the standardized log lipid dataset ($n = 320$), together with the corresponding kernel estimates.

Density estimation is also quite difficult if the data cloud is highly skewed or falls (nearly) onto a lower dimensional manifold. This topic is taken up in detail in Chapter 7. However, marginal transformations using the Tukey ladder, for example, are always recommended. But as in the univariate setting, an absolutely optimal choice of bandwidth is not critical for exploratory purposes.

6.5.2.2 Multivariate Oversmoothing Bandwidths

The extension of the oversmoothing bandwidth to $\Re^d$ has been solved by Terrell (1990). The easiest multivariate density is spherically symmetric. Thus the general kernel formulation is required with the constraint that all the marginal bandwidths are equal. By symmetry, finding the form of the multivariate oversmoothed density along the x-axis, for example, will be sufficient. The variational problem turns out to be identical to the one-dimensional problem in Equation (6.67), but in polar coordinates: $2f^{iv} + \lambda_1 + \lambda_2 r^2 = 0$, which implies that $f(r) = a(c^2 - r^2)^3$. The general form of the solution has been given in Theorem 3 of Terrell (1990). The result is that

$$h_{OS} = \left[\frac{R(K)d}{nC_f} \right]^{1/(d+4)}, \quad \text{where} \quad C_f = \frac{16\Gamma\left(\frac{d+8}{2}\right)d(d+2)}{(d+8)^{(d+6)/2}\pi^{d/2}},$$

for kernels that have an identity covariance matrix. The constants in h_{OS} for $d \geq$ 1 are 1.14, 1.08, 1.05, 1.04, 1.03, and so on. The constants decrease to 1.02 when $d = 8$ and slowly increase thereafter, with a limiting value of $\sqrt{e/2} = 1.166$. The constant finally grows to match 1.14 when $d = 348$. For $1 \leq d \leq 10$, the oversmoothed rule starts at 8.0% wider than the normal reference rule given in Equation (6.43) and increases to 10.5% wider when $d = 10$. As $d \to \infty$, the ratio continues grows to $\sqrt{e/2} = 1.166$, for a 16.6% increase. Using the easy-to-remember normal reference rule formula should be sufficient in most cases. Rescaling to other product kernels is easily accomplished by applying the rescaling rule dimension by dimension. The bivariate oversmoothed bandwidth is depicted in Figures 6.24 and 6.25.

6.5.2.3 Asymptotics of Multivariate Cross-Validation
Given the very slow rates of convergence of UCV and BCV in one-dimension, and the slower rates of convergence of the AMISE as the dimension increases, it might reasonably be expected that a similar fate would transpire for multivariate cross-validation algorithms. Surprisingly, Sain et al. (1994) showed this was not the case. They found both the UCV and BCV bandwidths converged at the rates

$$n^{-d/(2d+8)} = \left(n^{-\frac{1}{10}}, n^{-\frac{1}{6}}, n^{-\frac{3}{14}}, n^{-\frac{1}{4}} \right) \quad \text{for} \quad 1 \leq d \leq 4. \qquad (6.81)$$

Thus the cross-validation problem actually becomes easier as the dimension increases, and apparently the rate becomes $O(n^{-1/2})$ in the limit. Unfortunately, terms which could be ignored when $d < 4$ come into play, and dominate when $d > 4$.

This fact was discovered by Duong (2004) in a truly impressive thesis of wide scope. He re-analyzed the asymptotics of some half-dozen cross-validation algorithms, and introduced a new variation in the multivariate case of the smoothed cross-validation algorithm of Hall et al. (1992). These results were reported in Duong and Hazelton (2005a,b). They show that the formula in Equation (6.81) is correct when $1 \leq d \leq 4$, but that the correct rates for $d > 4$ for UCV and BCV are

$$n^{-2/(d+4)} = \left(n^{-\frac{2}{9}}, n^{-\frac{1}{5}}, n^{-\frac{2}{11}}, n^{-\frac{1}{6}} \right) \quad \text{for} \quad 5 \leq d \leq 8. \qquad (6.82)$$

These rates, which are plotted in Figure 6.26, are decreasing and the cross-validation problem does become increasing difficult with dimension, in line with the general curse-of-dimensionality experienced by all nonparametric procedures. Fortunately, in the practical dimensions discussed in this book where $1 \leq d \leq 6$, there is good news. Also shown in Figure 6.26 are the rates for the PI bandwidth of Wand and Jones (1994). The faster rates occur for a diagonal smoothing parameter matrix, and PI-2 refers to the full smoothing matrix case. SCV-2 is described in Duong (2004). The formulae for these three criteria are $\min(8, d+4)/(2d+12)$, $4/(d+12)$, and $2/(d+6)$, respectively.

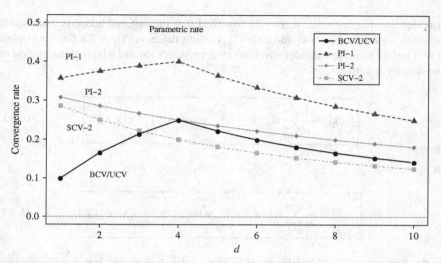

FIGURE 6.26 Magnitude of convergence rate exponents of several cross-validation algorithms. The best rate of $O(n^{-1/2})$ for parametric models would appear as $1/2$ on this graph.

In summary, the array of multivariate cross-validation algorithms available should suggest that all be tried and compared, as each has its failings in unpredictable ways. If all return similar bandwidths, so much the better. UCV remains the most flexible and general algorithm, and it is interesting to see it is quite competitive for $d > 2$. Its useful empirical performance in dimensions 1 and 2 has also been noted.

6.6 ADAPTIVE SMOOTHING

6.6.1 Variable Kernel Introduction

Consider the multivariate fixed kernel estimator

$$\hat{f}(\mathbf{x}) = \frac{1}{nh^d} \sum_{i=1}^{n} K\left(\frac{\mathbf{x} - \mathbf{x}_i}{h}\right) = \frac{1}{n} \sum_{i=1}^{n} K_h(\mathbf{x} - \mathbf{x}_i).$$

The most general adaptive estimator within this simple framework allows the bandwidth h to vary not only with the point of estimation but also with the particular realization from the unknown density f:

$$h \leftarrow h(\mathbf{x}, \mathbf{x}_i, \{\mathbf{x}_j\}) \approx h(\mathbf{x}, \mathbf{x}_i, f).$$

The second form indicates that asymptotically, the portion of the adaptive bandwidth formula dependent upon the whole sample can be represented as a function of the true density. Furthermore, it may be assumed that the optimal adaptive bandwidth function, $h(\mathbf{x}, \mathbf{x}_i)$, is smooth and a slowly varying function. Thus, for finite samples, it will be sufficient to consider two distinct approaches toward adaptive estimation of the density function:

$$\hat{f}_1(\mathbf{x}) = \frac{1}{n} \sum_{i=1}^{n} K_{h_{\mathbf{x}}}(\mathbf{x} - \mathbf{x}_i) \quad \text{where} \quad h_{\mathbf{x}} \equiv h(\mathbf{x}, \mathbf{x}, f) \tag{6.83}$$

or

$$\hat{f}_2(\mathbf{x}) = \frac{1}{n} \sum_{i=1}^{n} K_{h_i}(\mathbf{x} - \mathbf{x}_i) \quad \text{where} \quad h_i \equiv h(\mathbf{x}_i, \mathbf{x}_i, f). \tag{6.84}$$

In the first case, a fixed bandwidth is used for all n data points, but that fixed bandwidth changes for each estimation point $\mathbf{x}$. In the second case, a different bandwidth is chosen for each $\mathbf{x}_i$, and then applied to estimate the density globally. Each is justified asymptotically by the local smoothness assumption on $h(\mathbf{x}, \mathbf{x}_i, f)$, since asymptotically only those data points in a small neighborhood of $\mathbf{x}$ contribute to the density value there. Presumably, all the optimal bandwidths in that neighborhood are sufficiently close so that using just one value is adequate. The choice of $\hat{f}_1$ or $\hat{f}_2$ depends on the practical difficulties in specifying the adaptive bandwidth function. For small samples, one may expect some difference in performance between the two estimators. Jones (1990) has given a useful graphical demonstration of the differences between the two adaptive estimators.

Examples of $\hat{f}_1$ include the k-NN estimator of Loftsgaarden and Quesenberry (1965) (see Section 6.4.1) with $h_{\mathbf{x}}$ equal to the distance to the kth nearest sample point:

$$h_{\mathbf{x}} = d_k(\mathbf{x}, \{\mathbf{x}_i\}) \approx \left(\frac{k}{n V_d f(\mathbf{x})} \right)^{1/d}, \tag{6.85}$$

where the stochastic distance is replaced by a simple histogram-like formula. The second form was introduced by Breiman et al. (1977), who suggested choosing

$$h_i = h \times d_k(\mathbf{x}_i, \{\mathbf{x}_1, \mathbf{x}_2, \dots, \mathbf{x}_n\}) \approx h \times \left(\frac{k}{n V_d f(\mathbf{x})} \right)^{1/d}. \tag{6.86}$$

The similarity of these two particular proposals is evident. However, the focus on the use of the k-NN distance is simply a matter of convenience.

When estimated optimally point by point, the first form provides the asymptotically best possible estimate, at least from the MISE point of view. However, the estimator is not by construction a density function. For example, the k-NN estimator is easily seen to integrate to ∞ (see Problem 6.34). The second estimator, on the other hand, is by construction a *bona fide* density estimator for nonnegative kernels.

In practice, adaptive estimators in the very general forms given in Equations (6.83) and (6.84) can be very difficult to specify. The "adaptive smoothing function" $h_{\mathbf{x}}$ for $\hat{f}_1$ is ∞-dimensional. The specification for $\hat{f}_2$ is somewhat easier, since the "adaptive smoothing vector" $\{h_i\}$ is only n-dimensional. As usual, the "correct" choices of these quantities rely on further knowledge of unknown derivatives of the density function.

In practice, adaptive estimators are made feasible by significantly reducing the dimension of the adaptive smoothing function. One simple example is to incorporate the distance function $d_k(\cdot,\cdot)$ as in Equations (6.85) and (6.86). Abramson (1982a) proposed a variation on the Breiman et al.'s formula (6.86):

$$h_i = h/\sqrt{f(\mathbf{x}_i)} \quad \text{for } \mathbf{x}_i \in \Re^d.$$

In a companion paper, Abramson (1982b) proves that using a nonadaptive pilot estimate for f is adequate. Observe that these two proposals agree when $d = 2$, where Breiman et al. (1977) discovered empirically that their formula worked well.

In the univariate setting, the simple idea of applying a fixed kernel estimator to transformed data and then backtransforming falls into the second category as well. If u is a smooth monotone function selected from a transformation family such as Box and Cox (1964), then when the fixed kernel estimate of $w = u(x)$ is retransformed back to the original scale, the effect is to implicitly specify the value of h_i, at least asymptotically. The transformation approach has a demonstrated ability to handle skewed data well, and symmetric kurtotic data to a lesser extent (see Wand et al. (1991)). The transformation technique does not work as well with multimodal data.

In each case, the potential instability of the adaptive estimator has been significantly reduced by "stiffening" the smoothing function or vector. This may be seen explicitly by counting the number of smoothing parameters s that must be specified: $s = 1$ for k-NN (k); $s = 2$ for Breiman et al. (h,k); $s = 2$ for Abramson (h,h_{pilot}); and $s = 2$, 3, or 4 for the transformation approach (h plus 1–3 parameters defining the transformation family).

Theoretical and practical aspects of the adaptive problem are investigated further. There is much more known about the former than the latter.

6.6.2 Univariate Adaptive Smoothing

6.6.2.1 Bounds on Improvement   Consider a pointwise adaptive estimator with a pth-order kernel:

$$\hat{f}_1(x) = \frac{1}{nh(x)} \sum_{i=1}^{n} K\left(\frac{x - x_i}{h(x)}\right) \quad \text{letting } h(x) \equiv h_{\mathbf{x}}.$$

The pointwise AMSE properties of this estimator can be recognized in Theorem 6.3 for a pth-order kernel:

$$\text{AV}(x) = \frac{R(K)f(x)}{nh(x)} \quad \text{and} \quad \text{ASB}(x) = \frac{\mu_p^2 f^{(p)}(x)^2}{(p!)^2} h(x)^{2p}$$

from which the optimal pointwise $\text{AMSE}(x)$ may be obtained:

$$h^*(x) = \left[\frac{(p!)^2 R(K)f(x)}{2p\,\mu_p^2 f^{(p)}(x)^2} \right]^{\frac{1}{2p+1}} n^{-\frac{1}{2p+1}}$$

$$\text{AMSE}^*(x) = \frac{2p+1}{2p} \left[\frac{2p\,\mu_p^2 R(K)^{2p} f(x)^{2p} f^{(p)}(x)^2}{(p!)^2} \right]^{\frac{1}{2p+1}} n^{-\frac{2p}{2p+1}}. \tag{6.87}$$

Recall that the MISE accumulates pointwise errors. Thus accumulating the minimal pointwise errors obtained by using $h^*(x)$ gives the asymptotic lower bound to the adaptive AMISE:

$$\text{AAMISE}^* = \int_{-\infty}^{\infty} \text{AMSE}^*(x)\,dx$$

$$= \frac{2p+1}{2p} \left[\frac{2p\,\mu_p^2 R(K)^{2p}}{(p!)^2} \right]^{\frac{1}{2p+1}} \int \left[f(x)^{2p} f^{(p)}(x)^2 \right]^{\frac{1}{2p+1}} dx \times n^{-\frac{2p}{2p+1}}. \tag{6.88}$$

Comparing Equations (6.23) and (6.88), it follows that the bound on the improvement of an adaptive kernel estimator is

$$\frac{\text{AAMISE}^*}{\text{AMISE}^*} = \frac{\int \left[f(x)^{2p} f^{(p)}(x)^2 \right]^{\frac{1}{2p+1}} dx}{\left[\int f^{(p)}(x)^2\,dx \right]^{\frac{1}{2p+1}}}. \tag{6.89}$$

An application of Jensen's inequality to the quantity

$$\mathrm{E}[f^{(p)}(X)^2/f(X)]^{1/(2p+1)}$$

shows that the ratio in (6.89) is always ≤ 1 (see Problem 6.35). In Table 6.4, this lower bound ratio is computed numerically for the normal and Cauchy densities. Observe that the adaptivity potential decreases for higher order kernels if the data are normal, but the opposite holds for Cauchy data. The table gives further evidence of the relative ease when estimating the normal density. Rosenblatt derived (6.88) in the positive kernel case $p = 2$.

TABLE 6.4 Ratio of AAMISE* to AMISE*
for Two Common Densities as a Function of
the Kernel Order

Kernel order	Density	
p	Normal(%)	Cauchy(%)
1	89.3	84.0
2	91.5	76.7
4	94.2	72.0
6	95.6	70.0
8	96.5	68.9

For the case of a positive kernel $p = 2$, the asymptotically optimal adaptive mesh is in fact equally spaced when

$$\frac{f''(x)^2}{f(x)} = c \quad \Rightarrow \quad f(x) = \frac{c}{144}(x-a)^4, \tag{6.90}$$

where a is an arbitrary constant (see Section 3.2.8.3). Thus the null space for kernel estimators occurs in intervals where the density is a pure quartic function. Piecing together pure segments of the form (6.90), while ensuring that f and f' are continuous implies that f is monotone; thus there does not exist an entire null adaptive density in C^1 or C^2 unless boundary kernels are introduced.

6.6.2.2 Nearest-Neighbor Estimators

Using the asymptotic value for the adaptive smoothing parameter from Equation (6.85) with a positive kernel ($p = 2$) and $d = 1$,

$$h(x) = \frac{k}{2nf(x)}, \tag{6.91}$$

the adaptive asymptotic integrated squared bias is given by

$$\text{AAISB}(k) = \int_x \text{AISE}(x)dx = \int_x \frac{1}{4}h(x)^4 f''(x)^2 dx = \frac{k^4}{64n^4}\int_x \frac{f''(x)^2}{f(x)^4}dx.$$

Surprisingly, the latter integral is easily seen to diverge for such simple densities as the normal. This divergence does not imply that the bias is ∞ for finite samples, but it does indicate that no choice of k can be expected to provide a satisfactory variance/bias trade-off.

The explanation is quite simple: asymptotically, optimal adaptive smoothing depends not only on the density level as in Equation (6.91) but also on the curvature as in Equation (6.87). Thus the simple rule given by Equation (6.91) does not represent an improvement relative to nonadaptive estimation. This phenomenon of performing worse will be observed again and again with many simple ad hoc adaptive procedures, at least asymptotically. Some do provide significant gain for small samples or certain densities. Other approaches, such as transformation, include the fixed bandwidth setting as a special case, and hence need not perform significantly worse asymptotically.

6.6.2.3 *Sample-Point Adaptive Estimators*   Consider the second form for an adaptive estimator with different smoothing parameters at the sample points:

$$\hat{f}_2(x) = \frac{1}{n}\sum_{i=1}^{n}\frac{1}{h(x_i)}K\left(\frac{x-x_i}{h(x_i)}\right) \quad \text{letting } h(x_i) \equiv h_i. \tag{6.92}$$

Terrell and Scott (1992, appendix) proved under certain conditions that

$$\text{AV}\{\hat{f}_2(x)\} = f(x)R(K)/[nh(x)]$$

and

$$\text{ASB}\{\hat{f}_2(x)\} = \{(p!)^{-1}[h(x)^p f(x)]^{(p)}\}^2,$$

where the superscript $^{(p)}$ indicates a pth-order derivative. Abramson (1982a) proposed what is obvious from this expression for the bias, namely, that when $p = 2$, the choice

$$h(x) = h/\sqrt{f(x)} \tag{6.93}$$

implies that the second-order bias term in ASB vanishes! The bias is actually

$$\frac{1}{4!}\left[h(x)^4 f(x)\right]^{(iv)} = \frac{1}{24}h^4\left[\frac{1}{f(x)}\right]^{(iv)},$$

as shown by Silverman (1986). This fourth-order bias is usually reserved for negative $p = 4$th order kernels and apparently contradicts Farrell's (1972) classical result about the best bias rates with positive kernels.

In fact, Terrell and Scott (1992) have provided a simple example that illustrates the actual behavior with normal data ($f = \phi$):

$$\hat{f}_2(x) = \frac{1}{n}\sum_{i=1}^{n}\frac{\sqrt{\phi(x_i)}}{h}K\left(\frac{(x-x_i)\sqrt{\phi(x_i)}}{h}\right),$$

from which it follows that

$$\text{E}\hat{f}_2(x) = \int \frac{\sqrt{\phi(t)}}{h}K\left(\frac{(x-t)\sqrt{\phi(t)}}{h}\right)\phi(t)\,dt, \tag{6.94}$$

with a similar expression for the variance. The exact adaptive MISE is difficult to obtain, but may be computed numerically for specific choices of h and n. The authors showed that the exact adaptive MISE was half that of the best fixed bandwidth estimator when $n < 200$; however, the fixed bandwidth estimator was superior for $n > 20,000$. This finding suggests the procedure does not have $O(n^{-8/9})$ MISE.

In fact, it is easy to demonstrate the source of the difference with Silverman (1986) and Hall and Marron (1988). Consider an asymptotic formula for the MSE of the estimator at $x = 0$, without loss of generality. The subtle point is made most clearly by choosing the boxcar kernel $K = U(-1, 1)$ so that Equation (6.94) becomes

$$\mathrm{E}\hat{f}(0) = \frac{1}{2h} \int \phi(t)^{3/2} \left\{ I_{[-1,1]} \left(\frac{t\sqrt{\phi(t)}}{h} \right) \right\} dt. \tag{6.95}$$

Usually, the limits of integration would extend from $-h$ to h. However, a closer examination shows that the argument of the kernel is not monotone increasing and, in fact, approaches zero as $|t| \to \infty$. Thus the integral in Equation (6.95) covers three intervals, call them $(-\infty, -b), (-a, a)$, and (b, ∞), where a and b are solutions to the equation

$$\frac{t\sqrt{\phi(t)}}{h} = 1. \tag{6.96}$$

Define $c = (2\pi)^{1/4}h$. Then Equation (6.96) takes two forms that give sufficient approximations to the interval endpoints a and b:

$$t\,e^{-t^2/4} = c \quad \Rightarrow \quad a \approx c + c^3/4 + 5c^5/32$$

$$\log t - t^2/4 = \log c \quad \Rightarrow \quad b \approx \left(-4\log c + 4\log\sqrt{-4\log c} \right)^{1/2}.$$

Now taking a Taylor's series of the integrand in Equation (6.95) and integrating over $(-a, a)$ gives

$$\mathrm{E}\hat{f}(0) = \phi(0) + (2\pi)^{1/2}h^4/40 + O(h^6),$$

which gives the predicted $O(h^4)$ bias. However, the contribution toward the bias from the remaining two intervals totals

$$2 \cdot \frac{1}{2h} \int\limits_{-\infty}^{-b} \phi(x)^{3/2} dx = \left(\frac{2}{9\pi} \right)^{1/4} \frac{1}{h} \Phi\left(-b\sqrt{3/2} \right),$$

which, using the approximation for b and the tail approximation $\Phi(x) \approx -\phi(x)/x$ for $x \ll 0$, equals

$$h^2 / \left(24[\log\{(2\pi)^{1/4}h\}]^2 \right).$$

Thus the tails exert an undue influence on the estimate in the middle and destroy the apparent gain in bias. With a smoother kernel, the same effect is observed but not so clearly. Abramson recognized this practical problem and suggested putting an upper bound on h_i by "clipping" the pilot estimator in Equation (6.93) away from zero. Other authors have missed that suggestion. However, the asymptotic inefficiency does not negate the good small-sample properties observed by Abramson (1982a), Silverman (1986), and Worton (1989).

This same analysis may be applied to the original proposal of Breiman et al. (1977). The contribution from the tails to the bias turns out to be $O(h/\log h)$. These authors had noted that despite excellent empirical bivariate performance, the univariate performance was poor. This slow bias rate helps to explain that observation.

6.6.2.4 Data Sharpening

The bias of a positive kernel estimate, $\hat{f}(x)$, is controlled by the second derivative there (see Equation (6.16)). Thus modes will generally be underestimated and antimodes will always be overestimated. (Even if higher-order negative kernels are employed, this same phenomena will be observed.) Recall that the gradient method minimizes a function by moving in the direction of the *negative* gradient. Samiuddin and El-Sayyad (1990) proposed adjusting the data towards the nearest peak by following the *positive* gradient. Specifically, in the univariate setting where $X_i \sim f(x)$, they propose using a kernel estimate on the adjusted data

$$\tilde{x}_i = x_i + \frac{h^2}{2} \frac{f'(x_i)}{f(x_i)} \qquad \text{for } i = 1,\ldots,n \tag{6.97}$$

$$= x_i - \frac{h^2}{2} \cdot \frac{x_i - \mu}{\sigma^2} \qquad \text{for } N(\mu,\sigma^2) \text{ data,} \tag{6.98}$$

where we assume the kernel is scaled so that $\sigma_K = 1$. Note that for a normal density, Equation (6.98) may be rewritten as $(\tilde{x}_i - \mu) = (x_i - \mu)\left[1 - h^2/(2\sigma^2)\right]$; therefore, the data are shrunk toward the mean (and mode) by the factor $\frac{1}{2}(2 - h^2/\sigma^2)$. Versions of Equation (6.97) that are data-based and also of a more general form have been considered by Choi and Hall (1999) and Hall and Minnotte (2002).

In order to investigate the effectiveness of this proposal, we focus on the expectation of the kernel estimator using the modified data:

$$\hat{f}(x) = \frac{1}{n}\sum_{i=1}^{n} K_h(x - \tilde{x}_i) = \frac{1}{n}\sum_{i=1}^{n} \frac{1}{h} K\left(\frac{x - x_i - \frac{1}{2}h^2 f'(x_i)/f(x_i)}{h}\right), \tag{6.99}$$

where the true density is used in the adjustment (6.97). Assume that the kernel K is symmetric, $K(-w) = K(w)$, and that $\sigma_K = 1$. Using the change of variables $w = (x - y)/h$ and a Taylor's Series in the kernel, the expectation is given by

$$E\hat{f}(x) = \int \frac{1}{h} K\left(\frac{x - y - \frac{1}{2}h^2 f'(y)/f(y)}{h}\right) f(y)\,dy$$

$$= \int K\left(w - \frac{h}{2}\frac{f'(x - hw)}{f(x - hw)}\right) f(x - hw)\,dy$$

$$= \int f(x - hw)\left[K(w) - \frac{h}{2}\frac{f'(x - hw)}{f(x - hw)}K'(w) + \frac{h^2}{8}\frac{f'(x - hw)^2}{f(x - hw)^2}K''(w)\right]dw$$

$$= \int K(w)f(x - hw) - \frac{h}{2}\int K'(w)f'(x - hw) + \frac{h^2}{8}\int K''(w)\frac{f'(x - hw)^2}{f(x - hw)}$$

$$= \left[f(x) + \frac{1}{2}h^2 f''(x) + O(h^4)\right] - \frac{h}{2}\left[hf''(x) + O(h^3)\right] + \frac{h^2}{8}\left[O(h^2)\right]$$

$$= f(x) + \frac{1}{2}h^2 f''(x) - \frac{1}{2}h^2 f''(x) + O(h^4) = f(x) + O(h^4).$$

The result is that the bias of the Samiuddin and El-Sayyad proposal is of higher order $O(h^4)$, even with a positive kernel. In the third to last expression, the first integral is the familiar bias expansion. The second integral may be approximated by applying a Taylor's Series to $f'(x - hw)$:

$$\int K'(w)\left[\sum_{\ell=0}(-hw)^{\ell}f^{(1+\ell)}(x)/\ell!\right]dw = 0 - hf''(x) + 0 + O(h^3),$$

since $\int K' = 0$, $\int wK' = -1$, $\int w^2 K' = 0$, and $\int w^3 K' = -3\sigma_K^2$. There is little to be gained by deriving the explicit bias expression, which is quite complicated (see Hall and Minnotte (2002)). The leading variance terms turn out to be the same as for the ordinary kernel estimator. While this derivation adjusts the data using $f(x)$, which is unknown, the data-based algorithm explicated by Hall and Minnotte (2002) achieves any higher order desired.

As we have seen so many times before, the search for an algorithm that reduces bias by local adapting or adjustment results in a higher order bias result instead. Hall and Minnotte (2002) and others report promising examples with real data and some simulations. However, the application of the data adjustment (6.97) globally has an obvious deficiency, as can be seen by observing its action on the mixture density in Equation (3.78). From Equation (6.98), for standard normal data, the adjustment is $-h^2 x_i/2$, while for $N(3, 1/3^2)$ data, the adjustment is $-9h^2(x_i - 3)/2$, which is actually 27 times greater in standard units. Intuitively, the adjustment for the narrower mixture data should be one-third as wide since the standard deviation is 1/3. One possible fix would be to incorporate the magnitude of the second derivative at the local mode. Examining the published case studies, many had the same curvature at the modes and the effect was missed. Nevertheless, this idea holds good promise with a multistage estimation implementation to find the location of the modes and antimodes and local scale. It may also be advisable to use different smoothing parameters in Equations (6.97) and (6.99) and determine both smoothing parameters by unbiased cross-validation. In general, each modal region may require its own smoothing parameter as well.

6.6.3   Multivariate Adaptive Procedures

Multivariate adaptive procedures contain some interesting and unique features. These results are available in more detail in Terrell and Scott (1992).

6.6.3.1   Pointwise Adapting   Let $\nabla^2 f(\mathbf{x})$, which is the matrix of second partial derivatives of f at $\mathbf{x}$, be denoted by $S_{\mathbf{x}}$. From Equation (6.48), the pointwise asymptotic bias is

$$\mathrm{AB}(\mathbf{x}) = \frac{1}{2} h^2 \mathrm{tr}\{A^T S_{\mathbf{x}} A\} = \frac{1}{2} h^2 \mathrm{tr}\{A A^T S_{\mathbf{x}}\}. \tag{6.100}$$

In the univariate setting, the bias is controlled entirely by the scale of the kernel, while in the multivariate setting, the shape of the kernel is also available to control the bias. The importance of the shape in minimizing $\mathrm{tr}\{A^T S_{\mathbf{x}} A\}$ depends on the properties of the matrix $S_{\mathbf{x}}$.

Case I $S_{\mathbf{x}}$ is positive or negative definite. As H (and hence A) is full rank by assumption, then $A^T S_{\mathbf{x}} A$ is also positive or negative definite. Because the sum and product of the eigenvalues of a definite matrix equal its trace and determinant, respectively, the matrix with minimum (absolute) trace and determinant equal to 1 has all of its eigenvalues equal. Thus the matrix A should be chosen to satisfy

$$A A^T S_{\mathbf{x}} = |S_{\mathbf{x}}|^{1/d} I_d.$$

Observe that with this choice, the matrix on the right-hand side has the same determinant as $S_{\mathbf{x}}$, and all the eigenvalues of $A A^T S_{\mathbf{x}}$ equal $|S_{\mathbf{x}}|^{1/d}$. Therefore, the best $\mathrm{tr}\{A^T S_{\mathbf{x}} A\} = d |S_{\mathbf{x}}|^{1/d}$. The pointwise asymptotic MSE of $\hat{f}(\mathbf{x})$ follows from Equations (6.49) and (6.100) and may be optimized to yield

$$h^*(\mathbf{x}) = \left[\frac{f(\mathbf{x}) R(K)}{n d |S_{\mathbf{x}}|^{2/d}} \right]^{1/(d+4)}$$

so that

$$\mathrm{AMSE}^*(\mathbf{x}) = \left(\frac{d+4}{4d} \right)^{\frac{2(d+2)}{d+4}} \left[f(\mathbf{x}) R(K) \sqrt{|S_{\mathbf{x}}|} \right]^{4/(d+4)} n^{-4/(d+4)}.$$

Case II   The density is *saddle-shaped* at $\mathbf{x}$; that is, the matrix $S_{\mathbf{x}}$ has both positive and negative eigenvalues. The density is curved upward in some directions and downward in others. In this case, it is possible to construct the matrix A so that sum of the eigenvalues of $AA^T S_x$ equals 0 (see Terrell and Scott (1992)). Thus the order h^2 bias terms vanish with higher order terms dominating. Therefore, regions where the density is saddle-shaped asymptotically contribute nothing to the AAMISE compared to regions where the density is definite.

How common are saddle-shaped regions? Consider the multivariate normal density $N(\mathbf{0}_d, I_d)$. Then the gradient and Hessian of $f(\mathbf{x})$ are

$$\nabla f(\mathbf{x}) = -f(\mathbf{x})\mathbf{x} \quad \text{and} \quad \nabla^2 f(\mathbf{x}) = S_x = f(\mathbf{x})(\mathbf{x}\mathbf{x}^T - I_d).$$

There are $d-1$ eigenvalues of the matrix $(\mathbf{x}\mathbf{x}^T - I_d)$ equal to -1 and one equal to $\mathbf{x}^T\mathbf{x} - 1$. Therefore, the multivariate normal density is negative definite when $\|\mathbf{x}\| < 1$ and saddle-shaped *everywhere* outside the unit sphere. Given that the fraction of probability mass inside the unit sphere decreases as the dimension increases, the potential practical significance of this finding seems promising. Finally, observe that exactly on the unit sphere where $\|\mathbf{x}\| = 1$, one of the eigenvalues vanishes. This leads to the final case.

Case III S_x is semidefinite with at least one zero eigenvalue. The density is flat in certain directions. There is no contribution to the bias in those directions and hence the bias contribution is again of lower order than in Case I. The problem at that point can be reduced to a lower dimension by projection if desired.

6.6.3.2 *Global Adapting*   The problem of selecting the best global adaptive fixed kernel estimator was formulated by Deheuvels (1977a, b), who characterized the solution in differential equation form. The global criterion to be optimized is

$$\text{AAMISE}(h, A) = \frac{R(K)}{nh^d} + \frac{1}{4}h^4 \int_{R^d} \text{tr}\{A^T S_x A\}\, d\mathbf{x}.$$

Minimizing over h and A separately gives

$$\text{AAMISE}^* = \left[\min_A \int_{R^d} \text{tr}^2\{A^T S_x A\}\, d\mathbf{x}\right]^{d/(d+4)} \left[\frac{(d+4)R(K)}{4nd}\right]^{4/(d+4)}.$$

For $N(\mathbf{0}_d, I_d)$ data, the advantage of the adaptive scheme compared to the fixed kernel is great (see Table 6.5 (from Terrell and Scott, 1992)). The minimization was done numerically by Terrell. The advantage comes from the large saddle-shaped portion of the density.

The final global comparison is between the fixed kernel estimator and the k-NN multivariate density estimator. The latter estimator will be treated as an adaptive

TABLE 6.5   Relative Efficiency of Transformed Adaptive Density Estimate Compared to Fixed Kernel for the Multivariate Normal Density

Dimension d	1	2	3	4	5	6
Adapt/fixed efficiency	1	45.5%	30.2%	18.6%	10.6%	5.7%

TABLE 6.6   Relative Efficiency of k-NN Estimator to Fixed Kernel for the Multivariate Normal Density

Dimension d	1	2	3	4	5	15	100
k-NN/fixed efficiency	0	0	0.48	0.87	1.15	1.55	1.49

kernel estimator, with the kernel being a uniform density over the unit sphere and the bandwidth being taken as the asymptotic form

$$\bullet h \leftarrow [k/(nf(\mathbf{x})V_d)]^{1/d}.$$

The AMISE of the fixed kernel estimator using this same kernel follows from the general kernel results earlier (see Terrell and Scott, 1992). It is shown that

$$\frac{\text{AMISE}_h^*}{\text{AAMISE}_k^*} = \left(\frac{d-2}{d}\right)^{\frac{d(d+2)}{2(d+4)}} \left(\frac{4(d^2-4)}{d^2-6d+16}\right)^{\frac{d}{d+4}} \to \frac{4}{e}$$

as $d \to \infty$ (see Table 6.6). Thus the nearest-neighbor estimator, which overadapts to the tails in the univariate and bivariate cases, is seen to perform better than the fixed kernel estimator when $d \geq 5$, at least for normal data. This superiority is reassuring since the algorithm has a proven track record in high-dimensional applications such as clustering.

6.6.4   Practical Adaptive Algorithms

We conclude this topic with a discussion of two algorithms that use the unbiasedness property of UCV in a fundamental manner. PI approaches cannot compete with UCV because of the unwieldly asymptotics encountered.

6.6.4.1   Zero-Bias Bandwidths for Tail Estimation   One common truism about nonparametric density estimates is that their quality degrades in the tails where data are sparse. The estimates are generally "lumpy" around whatever data points may occur. However, there is an interesting phenomena for fixed bandwidth estimators that can provide remarkably accurate and useful estimates in the tails. The visual presentation can be much improved as well.

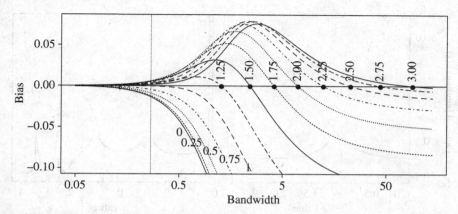

FIGURE 6.27 Bias of normal kernel density estimator for a standard normal sample of size $n = 1000$ as function of bandwidth.

Let us focus on the general behavior of the bias of a positive kernel estimator as the bandwidth h ranges from 0 to ∞. Recall the introduction to data-sharpening in Section 6.6.2.4, where it was observed that peaks are underestimated and valleys are overestimated. This is the direct result of the fact that the bias of a kernel estimate is proportional to $h^2 f''(x)$. In Figure 6.27, the exact bias (Fryer, 1976) of a normal kernel estimator for a normal sample is essentially 0 for small bandwidths, then becomes negative or positive depending whether $0 < x < 1$ or $x > 1$, respectively. (Since the $N(0,1)$ density is symmetric, the bias curves for negative values of x are identical; the density is concave for $-1 < x < 1$ and convex for $|x| > 1$.) At the other extreme, as $h \to \infty$ the density estimate $\hat{f}(x) \to 0$ for all x; hence, the bias$(x) \to -f(x)$, which is negative everywhere for a $N(0,1)$ density (see Figure 6.27).

Recall from Equation (6.13) that the expected value of a kernel estimate does not directly depend on the sample size n, except possibly through h. Since the expectation is a continuous function of x, we may conclude that when $x > 1$, there must be a fixed a fixed bandwidth h_x where $E\hat{f}_{h_x}(x) = f(x)$ exactly. This bandwidth is highlighted in Figure 6.27. We shall refer to this bandwidth as the *zero-bias bandwidth* and denote it by $h_0(x)$; see Sain (1994, 2003), Sain and Scott (2002), and also Hazelton (1998) who independently investigated what he called the bias-annihilating bandwidth.

The existence of the zero-bias bandwidth can result in more than one local minima in the pointwise MSE(x) and more than one "optimal" bandwidth (see Figure 6.7). For n sufficiently large, there will be a local minimum in the MSE(x) near $h_0(x)$, which we will denote by $h_0^*(x)$. The smaller optimal bandwidth corresponds to the usual asymptotic theory, that is, $h_x^* \propto n^{-1/5} \to 0$ as $n \to \infty$. However, if we choose to use the zero-bias bandwidth $h_0(x)$, then the squared bias is 0 and the MSE(x) is simply the variance $R(K)f(x)/(nh_0(x))$ to good approximation (see Equations (6.11)) and (6.15). This parametric rate $O(n^{-1})$ is not achievable in practice. Both Hazelton (1998) and Sain (2003) show the rate is actually the familiar $O(n^{-4/5})$ with real data.

In Figure 6.28, the values of the bandwidths that (locally) minimize the MSE(x) are shown for the standard normal density and for the two-component mixture density

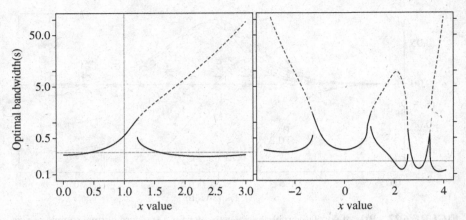

FIGURE 6.28 (Left) Asymptotically optimal bandwidths and zero-bias bandwidths for normal sample of size $n = 1000$. There are two optimal bandwidths when $x > 1.218$. The dashed line shows bandwidths close to zero-bias ones. (Right) Same for normal mixture but $n = 500$.

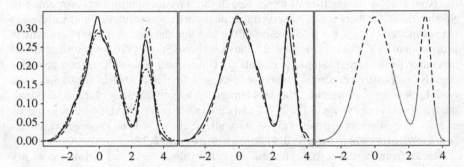

FIGURE 6.29 Several density estimates (solid lines) of a sample of size $n = 500$ from the mixture density in Equation (3.78), which is shown as a dashed line in each frame. (Left) A fixed kernel estimate with $h^* = 0.176$ and the normal reference rule $h = .473$ (dotted line). (Middle) Kernel estimate using $h^*(x)$, which is the smallest bandwidth in the right frame of Figure 6.28. (Right) Kernel estimate using $h_0^*(x)$ where it exists.

in Equation (3.78). In the right frame, there is even a region near $x \approx 3.7$ where there are three such bandwidths.

In Figure 6.29, we compare the various available kernel estimates for a sample of size $n = 500$ from the two-component mixture density. The left frame uses the asymptotically optimal fixed bandwidth $h^* = 0.176$, which undersmooths on the left and oversmooths on the right. The middle frame shows the use of $h^*(x)$ pointwise. This is much improved over the fixed kernel estimate, except for some minor anomolies near the inflection points. Finally, in the right frame, we show the zero-bias estimates where they exist. To the naked eye, there is no error in this estimate (even with a sample of only 500 points). Of course, without knowledge of the exact zero-bias bandwidths, such a result should not be expected. Note that for the fixed bandwidth estimator, the normal reference rule, UCV, BCV, and Sheather-Jones bandwidths are 0.473, 0.130, 0.222, and 0.218, respectively. The $h = 0.130$ UCV bandwidth (not shown) is well-calibrated for the narrow right component, leaving the left component quite undersmoothed.

Finally, to implement a data-based procedure for zero-bias estimation, the UCV formulation in Equation (3.52) can simply be modified by including an indicator function for data in an assumed known interval (a, b),

$$\int_{-\infty}^{\infty} \left[\hat{f}(x) - g(x) \right]^2 I(x \in (a, b)) = \int_a^b \hat{f}(x)^2 - 2\mathrm{E}\left[\hat{f}(X) I(X \in (a, b)) \right] + c,$$

leading to

$$\mathrm{UCV}_{(a,b)}(h) = \int_a^b \hat{f}(x)^2 dx - \frac{2}{n} \sum_{i=1}^n \hat{f}_{-i}(x_i) I(x_i \in (a, b)). \qquad (6.101)$$

The simplest implementation is to assume that the function $h_0(x)$ can be approximated by a constant over the interval (a, b). Obviously, other functional forms can be chosen, from polynomials to splines.

Returning to the mixture example, consider the interval $(a, b) = (1.35, 2.51)$ in the convex region around the anti-mode. The left frame in Figure 6.30 displays the data-based $\mathrm{UCV}(h)$ function in Equation (6.101) assuming a constant bandwidth over (a, b). Two local minima are observed, as hoped for. The larger bandwidth, $\hat{h}_0^* = 5.71$, is in the zero-bias region, and might be used at the midpoint of (a, b) rather than over the entire interval. Note, however, that the predicted error is not smaller than that at the asymptotically optimal bandwidth, $\hat{h}^* = 0.149$. The middle frame displays the $\mathrm{UCV}(h)$ bandwidth fits, where the logarithm of bandwidth h is taken to be a polynomial of order 0, 1, or 2 on (a, b). Depending on the starting values, there are

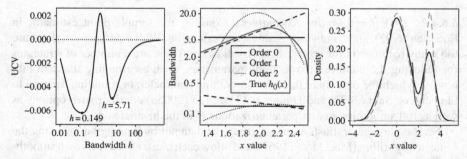

FIGURE 6.30   (Left) The UCV function over the interval $(a, b) = (1.35, 2.51)$ for the normal mixture dataset. (Middle) Log-polynomial fits to $h_0(x)$ that minimize UCV over (a, b). (Right) The zero-bias estimate over (a, b) together with the true mixture density and normal reference rule kernel estimate.

two solutions, corresponding to the two bandwidth types. For reference, the exact zero-bias MSE(x) minimizing bandwidth, $h_0^*(x)$, is shown as the thick solid curved line. Notice that none of the polynomial fits has sufficient degrees of freedom to approximate that curve. The order-1 (log-linear) fit is good for $x < 2.1$, and the order-2 (log-quadratic) fit good for $x > 2.3$. The right frame shows the zero-bias density estimate using quadratic fit, which has the smallest UCV score (20% lower than a constant bandwidth). The overestimation of the fixed kernel estimate over (a,b) is corrected, and the zero-bias estimate is almost perfect for $x > 2$. (The log-linear fit, not shown, is also almost perfect for $x < 2$.) A spline model for $h_0^*(x)$ is indicated.

We conclude by demonstrating a simple approximation for the zero-bias bandwidth due to Sain (2003). Since the bandwidth h is very large, we take a Taylor series of the *kernel* about 0 in Equation (6.12) (rather than the change of variables) and obtain

$$
\begin{aligned}
E\hat{f}(x) &= \int \frac{1}{h}\left[K(0) + \left(\frac{x-t}{h}\right)K'(0) + \left(\frac{x-t}{h}\right)^2 K''(0) + \cdots\right]f(t)\,dt \\
&= \frac{K(0)}{h} + 0 + \frac{K''(0)}{2h^3}\int (x - \mu_f + \mu_f - t)^2 f(t)\,dt + \cdots \\
&= \frac{K(0)}{h} + \frac{K''(0)}{2h^3}\left[(x - \mu_f)^2 + \sigma_f^2\right] + \cdots,
\end{aligned}
$$

since $K'(0) = 0$ for symmetric kernels. If $E\hat{f}(x) = f(x)$, that is, is unbiased, then $f(x) \approx K(0)/h_0(x)$ or $h_0(x) \approx K(0)/f(x)$, which is quite accurate in the tails. Observe that finding the zero-bias bandwidth function is related to estimation of the inverse density.

To summarize, zero-bias bandwidths hold interesting potential for estimating the density nonparametrically in areas typically deemed too difficult for quality estimation. The unbiasedness property of UCV permits investigation of these unusually large bandwidths.

6.6.4.2 UCV for Adaptive Estimators

Consider the sample-point estimator in Equation (6.92), and its obvious multivariate extension. With n parameters, there are too many parameters for practical cross-validation. There are a number of strategies for reducing the number of smoothing parameters, such as grouping the data into bins, in which the data share the same smoothing parameter, or rounding the data to bin centers. Sain (2002) and Duong and Hazelton (2005a) consider such options as well as full smoothing matrix parametrizations H in the bivariate case.

For the purpose of illustration, we group the data not by binning but by using the k-means algorithm (MacQueen, 1967), and allow each cluster to have its own smoothing parameter. The usual UCV data-based criterion is optimized numerically over the vector of smoothing parameters, for example, using R function *nlminb*. Figure 6.31 displays the fixed and adaptive kernel estimates. Seven clusters were selected using

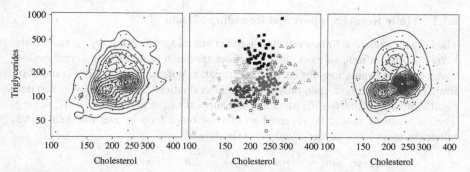

FIGURE 6.31 (Left) Twelve contours of the UCV-calibrated ($\hat{h} = 0.276$) bivariate Gaussian fixed-kernel estimate of the standardized log cholesterol and triglyceride data. (Middle) Seven clusters from k-means. (Right) The adaptive kernel estimator. The seven bandwidths range from 0.174 to 2.36. The mode is 54% greater than in the left frame. The 19 contour levels are the same as in the left frame plus seven more at higher levels.

k-means and are shown in the middle frame. To reduce the number of parameters, the data were standardized and a spherical gaussian kernel was selected that has only one smoothing parameter (instead of 2). The fixed bandwidth is 0.174 while the adaptive bandwidths are 0.17, 0.25, 0.47, 1.02, 1.48, 1.50, and 2.36. The adaptive estimate enhances the two primary modes as well as the smaller third bump, and reduces noise in the tails.

However, other choices of k lead to quite different estimates, some with fewer, some with more modes. Artifacts are also a potential problem with over-parameterized estimates. This one was selected as more faithful to the basic structure in the fixed kernel estimator. When experimenting with a fully parameterized elliptical kernel, Sain (2002) and Duong and Hazelton (2005a) produced estimates that could have even more extreme anomalies. Users of normal mixture models and the expectation-maximization (EM) algorithm are familiar with the potential for singularities when choosing a fully parameterized covariance matrix (when the fitted covariance matrix is nearly singular). UCV is also susceptible to such singularities. A well-written mixture package, Mclust, was applied to these data (see Fraley and Raftery (2002)). The BIC criterion evaluated mixtures with k in the range of 1–10; $k = 1$ was indicated by BIC. The single bivariate normal fit misses the bimodal feature altogether.

6.7 ASPECTS OF COMPUTATION

We conclude our survey of kernel methods by revisiting several options for computing a kernel estimator on an equally spaced mesh $\{a = t_0, t_1, \ldots, t_m = b\}$ based on a sample of size n. How can this be made more efficient, as alternatives to the histogram or ASH?

6.7.1 Finite Kernel Support and Rounding of Data

The kernel estimate is to be evaluated at the m bin midpoints, $\{m_1, m_2, \ldots m_m\}$. If the kernel, K, has infinite support, then the direct approach requires $n \cdot m$ kernel evaluations, n kernel evaluations for each of m estimation points. However, if a kernel with finite support is chosen, then only $f \cdot m$ kernel evaluations are required for each data point, where $f = 2h/(b-a)$, assuming the support of the kernel is $(-1, 1)$, w.l.o.g. The calculations should be reversed so that the outer loop is over the data points, rather the inner loop, as this pseudo R code shows:

$$\delta = (b-a)/m; \quad m_k = \text{seq}(a + \delta/2, b - \delta/2, \delta); \quad y = \text{rep}(0, m)$$
$$\text{for}(i \text{ in } 1:n)\{ k_0 = 2 + \lfloor (x_i - h - m_1)/\delta \rfloor; k_1 = 1 + \lfloor (x_i + h - m_1)/\delta \rfloor$$
$$\text{for}(k \text{ in } \max(1, k_0) : \min(m, k_1))\{ y_k = y_k + K_h(x_i - m_k) \}\}; y = y/n.$$

However, the work is still proportional to the sample size n. Prebinning or rounding the data to the same mesh reduces the work from $f \cdot m \cdot n$ to $f \cdot m^2$. The portion of the code that needs to modified is

$$\text{for}(\ell \text{ in } 1:m)\{ k_0 = \ell + 1 - \lfloor h/\delta \rfloor; k_1 = \ell + \lfloor h/\delta \rfloor$$
$$\text{for}(k \text{ in } \max(1, k_0) : \min(m, k_1))\{ y_k = y_k + K_h(m_\ell - m_k) \}\}.$$

The use of rounded data in a kernel density estimator attracted much attention with regards to loss of accuracy (see Scott (1981), Scott and Sheather (1985), and Silverman (1982); see also Jones (1989)). Härdle and Scott (1988) coined the phrase *weighted averaging of rounded points* or the WARPing method to describe the general approach and applied the WARPing method to other multivariate algorithms. With modern computers, m can be quite large and δ quite small so that the approximation error introduced is negligible. In multiple dimensions, however, keeping an entire grid in core must be considered when $d \geq 3$.

6.7.2 Convolution and Fourier Transforms

In Section 6.1.2, we explored one motivation for the fixed kernel estimator via convolution. This engineering approach also leads to efficient computational algorithms using Fourier methods for performing the convolution. We start with three definitions:

$$\text{Convolution} \qquad (f * g)(x) = \int_{-\infty}^{\infty} f(t)\, g(x-t)\, dt \qquad (6.102)$$

$$\text{Fourier Xfm} \qquad \tilde{f}(\xi) = \int_{-\infty}^{\infty} f(x)\, e^{-2\pi i x \xi}\, dx \qquad (6.103)$$

$$\text{Inverse Fourier Xfm} \qquad f(x) = \int_{-\infty}^{\infty} \tilde{f}(\xi)\, e^{2\pi i \xi x}\, d\xi. \qquad (6.104)$$

For example, representing the data $\{x_1, \ldots, x_n\}$ via the epdf, $f_n(x)$, in (2.2),

$$\tilde{f}_n(\xi) = \int \left[\frac{1}{n} \sum_{j=1}^{n} \delta(x - x_j) \right] e^{-2\pi i x \xi} \, dx$$

$$= \frac{1}{n} \sum_{j=1}^{n} \int \delta(x - x_j) e^{-2\pi i x \xi} \, dx = \frac{1}{n} \sum_{j=1}^{n} e^{-2\pi i x_j \xi}. \tag{6.105}$$

An alternative to performing the direct convolution operation (6.102), which can be very slow, is to perform three Fourier Transforms to obtain $c(x) = (f * g)(x)$ via:

$$\tilde{c}(\xi) = \tilde{f}(\xi) \cdot \tilde{g}(\xi) \quad \Longrightarrow \quad c(x) = \widetilde{\tilde{c}(\xi)} = \widetilde{\tilde{f}(\xi) \cdot \tilde{g}(\xi)}.$$

6.7.2.1 *Application to Kernel Density Estimators*   Applied to a kernel estimator, Equation (6.102) becomes

$$\tilde{\hat{f}}(x) = \int \hat{f}(x) e^{-2\pi i x \xi} \, dx \tag{6.106}$$

$$= \int \left[\frac{1}{n} \sum_{j=1}^{n} \frac{1}{h} K \left(\frac{x - x_j}{h} \right) \right] e^{-2\pi i x \xi} \, dx$$

$$= \frac{1}{n} \sum_{j=1}^{n} \left[\int K(s) e^{-2\pi i (x_j + h s_j) \xi} \, ds_j \right]$$

$$= \frac{1}{n} \sum_{j=1}^{n} \left[e^{-2\pi i x_j \xi} \int K(s) e^{-2\pi i (h s_j) \xi} \, ds_j \right]$$

$$= \frac{1}{n} \sum_{j=1}^{n} e^{-2\pi i x_j \xi} \tilde{K}(h s_j) \tag{6.107}$$

$$= \tilde{f}_n \cdot \tilde{K}(h s),$$

using the change of variables $s_j = (x - x_j)/h$ and noting that $\tilde{K}(h s_j)$ is the same for all j.

Now $\tilde{K}$ is known for many kernels, including the normal kernel:

$$\tilde{\phi}_h(\xi) = e^{-2 h^2 \pi^2 \xi^2}.$$

Therefore, (6.107) becomes

$$\tilde{\hat{f}}(\xi) = \frac{1}{n} \sum_{j=1}^{n} e^{-2\pi i x_j \xi} e^{-2 h^2 \pi^2 \xi^2}. \tag{6.108}$$

We may check the correctness of (6.108) by computing its inverse Fourier Transformation using Mathematica (2012):

$$\tilde{\hat{f}}(x) = \hat{f}(x) = \int \tilde{\hat{f}}(\xi)\, e^{2\pi i \xi x}\, d\xi \tag{6.109}$$

$$= \frac{1}{n} \sum_{j=1}^{n} \int e^{-2\pi i x_j \xi}\, e^{-2h^2 \pi^2 \xi^2}\, e^{2\pi i \xi x}\, d\xi \tag{6.110}$$

$$= \frac{1}{n} \sum_{j=1}^{n} \frac{1}{\sqrt{2\pi} h}\, e^{-(x-x_j)^2 / 2h^2}, \tag{6.111}$$

which is the Gaussian kernel density estimate exactly.

6.7.2.2 FFTs In practice, the discrete Fourier Transform and the FFT are employed to compute a very good approximation to $\hat{f}(x)$. The FFT is applied to a finite interval, say $(-\frac{1}{2}, \frac{1}{2})$ w.l.o.g., and the resulting estimate is periodic. To avoid end effects, the data are rescaled to a subinterval such as $(-0.3, 0.3)$. Silverman (1986) recommends a buffer of at least $3h$ on each side. In the time series context, padding with zeroes in spectral density estimation is a well-known requirement to avoid aliasing, that is, frequencies overlapping (Blackman and Tukey, 1958).

Binning of the data is critical to the approach, that is,

$$\hat{f}(x) = \frac{1}{n} \sum_{i=1}^{n} K_h(x - x_i) \approx \frac{1}{n} \sum_{k=1}^{m} \nu_k K_h(x - m_k),$$

and if we compute $\hat{f}(x)$ at the same bin midpoints

$$\hat{f}(m_\ell) = \frac{1}{n} \sum_{k=1}^{m} \nu_k K_h(m_\ell - m_k) = \sum_{k=1}^{m} \nu_k \cdot \frac{1}{n} K_h\big(\delta(\ell - k)\big), \tag{6.112}$$

which is precisely a discrete convolution. In R, this may be implemented via the *convolve* function, with arguments given by the bin counts and the kernel values at $t = m_\ell$ divided by n. So that the final values are properly aligned, the kernel values should be circularly shifted by half.

Multivariate extension of the FFT approach is discussed by Wand (1994) and Wand and Jones (1995). These authors and Silverman (1986) advocate replacing simple binning by linear binning, although with modern computing and very fine meshes, the impact may be rather minor.

6.7.2.3 Discussion The default density function in R now uses the FFT algorithm in the univariate case. In situations where a boundary kernel is called for, then other algorithms will need to be employed. The use of the FFT in more than one dimension is also available, but the mesh grows exponentially so that three dimensions is a practical limit.

The ASH approach goes beyond three dimensions by computing slices or conditional densities. It is also possible to look at subsets of the domain. In any case, there are many more opportunities to try kernels of ASH estimators in the multivariate setting, and the reader is encouraged to try these on their data. In addition to the ASH and *ashn* software available in R and from the author, the *ks* package is recommended (Duong, 2014). The package uses the *rgl* animation toolbox.

6.8 SUMMARY

This chapter has covered kernel methods, which is often the sole topic of a book. The kernel approach was motivated in different ways that might appeal to a statistician, numerical analyst, engineer, or mathematician. Issues related to the choice of kernel—its order, boundary properties, and design—were surveyed. Extensions to multivariate data and the use of the product kernel reviewed. Cross-validation ideas discussed in earlier chapters were extended to kernels. Finally, the important topic of locally adaptive methods was introduced. Adaptive methods hold much promise, but usually introduce many new parameters that are difficult to estimate, and frequently introduce artifacts of the sample (rather than the underlying density). An interesting survey is provided by Jones and Signorini (1997), who cover these adaptive algorithms and others in the context of higher order bias strategies. They observe there are many technical details to overcome. We have examined but a few. A conservative strategy that may be recommended is to stick with a fixed bandwidth kernel estimate and be aware of and sensitive to its limitations.

PROBLEMS

6.1 Compute the IV of estimator (6.2) and compare to the histogram result. What is the kernel if $\hat{f}(x) = [F_n(x+h) - F_n(x)]/h$ is used?

6.2 Compute the bias and variance directly for estimator (6.3). Demonstrate that the equivalent kernel $K = U(-0.5, 0.5)$.

6.3 Check that the ASH results are obtained from Theorem 6.1 when the isosceles triangle kernel is specified.

6.4 Examine the graphs of Wahba's equivalent kernel (6.9) for several choices of the two parameters p and λ.

6.5 Following the discussion of the rootgram for the histogram, show that the square root of the kernel estimate is variance stabilizing compared to (6.15). Show that the variance of the root-kernel estimate is $R(K)/(4nh)$.

6.6 Verify Equations (6.21). Find the optimal bandwidths for estimating the first and second derivatives of f. Evaluate these for the $N(\mu, \sigma^2)$ density.

6.7 Consider kernel $K_4(t)$ in Table 6.1. What if you make this kernel smoother by increasing the power on the first factor, such as $c(1-t^2)^k(a+bt^2)$, where k is 2 or 3, and the constants c, a, and b are chosen so the kernel is order-4 and integrates to 1. How do these alternative kernels behave theoretically? Visually?

6.8 Show that two kernel estimates with sample sizes in the ratio given in Equation (6.26) have the same AMISE.

6.9 Show that the IV of the fixed kernel estimate equals (exactly)

$$\mathrm{IV}(h) = \frac{1}{n} \int E K_h(x - X)^2 dy - \frac{1}{n} \int [E K_h(x - X)]^2 \, dx.$$

Show that the first term in the IV is *exactly* $R(K)/(nh)$. Show that the next terms are

$$-\frac{R(f)}{n} + \frac{h^2 \mu_2 R(f')}{n} - \frac{h^4}{n} \left(\frac{\mu_2^2}{4} + \frac{\mu_4}{24} \right) R(f'') + \cdots,$$

where μ_k is the kth moment of the kernel.

6.10 Show that the ISB of a fixed kernel estimate is

$$\mathrm{ISB}(h) = \int [E K_h(x - X) - f(x)]^2 \, dx.$$

Suppose that the kernel K is symmetric around zero so that $\int w^i K(w) dw = 0$ for i odd. Show that the first few bias terms equal

$$h^4 \frac{\mu_2^2}{4} R(f'') - h^6 \frac{\mu_2 \mu_4}{24} R(f''') + h^8 \left(\frac{\mu_4^2}{476} + \frac{\mu_2 \mu_6}{720} \right) R(f^{vi}).$$

Hint: Watch the cross-product terms and note that $\int f'' f^{iv} = -\int (f''')^2$, for example.

6.11 Verify the bias expressions for the higher order finite difference estimators in Equation (6.32). Devise your own higher order boxcar kernels using different spacings than integer multiples of h.

6.12 Compare the AMSE(x) values for four combinations of pointwise kernel estimates—at 0 and 1 with a second-order and fourth-order kernel.

6.13 Empirically compare the order-4 kernel method to the Terrell–Scott fourth-order ratio estimator for simulated normal data as well as the snowfall data.

6.14 Derive the Terrell–Scott fourth-order kernel ratio estimator. Extend the procedure to a sixth-order estimator.

6.15 Compute the equivalent kernel of the parametric estimator $\phi(x|0, s^2)$.

6.16 Compute the "theoretical" kth moment of $\hat{f}$, treating the kernel estimator as a "true" density.

6.17 Find the indifferent frequency polygon kernel mentioned in Table 6.2.

6.18 Consider the class of shifted-Beta kernels, $c_k(1 - t^2)^k_+$. Find their variance and rescale so that each has variance 1. Show that these rescaled kernels converge to a standard normal kernel as $k \to \infty$.

6.19 Derive the product kernel AMISE results from the general multivariate kernel formulas.

6.20 Verify the equivalent-bandwidth formula for higher order kernels in Equation (6.31). Check that the last factor is approximately equal to 1 for several kernels.

6.21 Using a Taylor's series on $\Delta F_n(x, kh)$, verify that the finite difference estimates in Equation (6.32) are of higher order. Derive some additional estimates based on the spacings $h, 2h, 4h, 8h, \ldots$.

6.22 Check by direct integration that the "variance" of the kernel in Equation (6.36) is 0.

6.23 Find boundary modification kernels based on the Epanechnikov kernel. Compare with kernels supported on $[c, 1]$. Investigate the increase in $R(K_c)$. How much wider (on the right) should the kernel be so that the roughness is the same as for the biweight kernel? Does such an "equivalent roughness" always exist?

6.24 Verify Equation (6.40).

6.25 Recall estimator (6.45). Show that it is functionally equivalent to choose K to be $N(\mathbf{0}_d, \Sigma)$ with $H = I_d$, or to choose K to be $N(\mathbf{0}_d, I_d)$ with $H = \Sigma^{1/2}$. Thus the linear transformation may be applied to either the data or the kernel as a matter of preference.

6.26 Finite support kernels need not have only a finite number of derivatives. For example, consider

$$K(t) \propto e^{-1/(1-t^2)} I_{(-1,1)}(t).$$

Show that the normalizing constant is $\sqrt{\pi}/e \times$ Hypergeometric $U\left[\frac{1}{2}, 0, 1\right]$. Plot the kernel. *Hint:* Use the change of variable $t = (1 - x^2)^{-1} - 1$ for $x \in (0, 1)$.

6.27 Show that $c(x - a)^4/144$ is the solution to the null adaptive density differential Equation (6.90).

6.28 Verify the equivalent kernel for the parametric estimator $N(\bar{x}, 1)$. Plot $K(x, t)$. Compute the equivalent kernel for the parametric estimator $N(0, s^2)$ and plot it.

6.29 Recall that $\sigma_f^2 = s_x^2 + h^2 \sigma_K^2$. From this result, argue that if $s_x > \sigma_f$, then $\hat{h}_{\text{ISE}} > \hat{h}_{\text{MISE}}$ is the likely result. Is the converse true?

6.30 Try the simple orthogonal series estimator on some Beta data with $0 \leq m \leq 6$. With a larger sample, is there sufficient control on the estimate with m alone?

6.31 Using the simple estimate for the Fourier coefficients in Equation (6.6), find unbiased estimates of the two unknown terms in Equation (6.62).

6.32   Show that if a correction factor of the form $(1 + bn^{-\delta})$ is applied to h^* in (6.79), then the best choice is $\delta = 1/5$.

6.33   Derive Equation (6.85). *Hint*: The fraction of points, k/n, in the ball of radius h centered on $\mathbf{x}$ is approximately equal to $f(\mathbf{x})$ times the volume of the ball.

6.34   Show that the k-NN estimator using the d_k distance function has infinite integral. *Hint*: In the univariate case, look at the estimator for $x > x_{(n)}$, the largest order statistic.

6.35   Use Jensen's inequality to show that the ratio in Equation (6.89) is ≤ 1. *Hint*: The integral in brackets in the denominator is $\mathrm{E}[f^{(p)}(X)^2/f(X)]$.

6.36   (Research). Rather than abandoning UCV in favor of more efficient asymptotic estimators, consider adjusting the data before computing the UCV curve. The noise in UCV is due in part to the fact that the terms $(x_i - x_j)/h$ are very noisy for $|x_i - x_j| < h$. Imagine placing small springs between the data points with the result that the interpoint distances become very smoothly changing. The resulting UCV seems much better behaved. A locally adaptive method replaces $x_i \leftarrow (x_{i-1} + x_{i+1})/2$, assuming that the data have been sorted. Try this modification on simulated data.

6.37   Prove that the average bias of a kernel estimate is *exactly* 0. *Hint*: The pointwise bias is $\int K(w)f(x - hw)dw - f(x)$.

6.38   Wand and Jones (1995) use two matrix identities, which may be found in Neudecker and Magnus (1988), to derive Equation (6.51). Suppose B and C are symmetric $d \times d$ matrices. Then $\mathrm{hvec}(B)$ vectorizes the lower triangular portion, while $\mathrm{vec}(B)$ vectorizes the entire matrix. These two vectors are connected by the $d^2 \times \frac{d(d+1)}{2}$ *duplication matrix*, D_d, whose entries are all 0's and 1's; that is, $D_d\,\mathrm{vech}(B) = \mathrm{vec}(B)$. The identities and relationships are

$$\mathrm{vec}(B) = D_d\,\mathrm{vech}(B)$$
$$D_d^T\,\mathrm{vec}(B) = \mathrm{vech}[\,2B - \mathrm{Diag}(B)\,] \qquad\qquad \text{and}$$
$$\mathrm{tr}\,(B^T C) = \mathrm{vec}(B)^T\,\mathrm{vec}(C) = \mathrm{vec}(C)^T\,\mathrm{vec}(B).$$

Complete the following argument. Let $B = AA^T$ in Equation (6.50). Write the integrand as

$$\mathrm{tr}\big[B\,\nabla^2 f(\mathbf{x})\big]^2 = (\mathrm{vec}\,B)^T[\mathrm{vec}\,\nabla^2 f(\mathbf{x})] \cdot [\mathrm{vec}\,\nabla^2 f(\mathbf{x})]^T(\mathrm{vec}\,B)$$
$$= \mathrm{vech}(B)^T D_d^T[\mathrm{vec}\,\nabla^2 f(\mathbf{x})] \cdot [\mathrm{vec}\,\nabla^2 f(\mathbf{x})]^T D_d\,\mathrm{vech}(B).$$

The derivation may be completed by using the second identity on $D_d^T[\mathrm{vec}\,\nabla^2 f(\mathbf{x})]$.

7

THE CURSE OF DIMENSIONALITY AND DIMENSION REDUCTION

7.1 INTRODUCTION

The practical focus of most of this book is on density estimation in "several dimensions" rather than in very high dimensions. While this focus may seem misleading at first glance, it is indicative of a different point of view toward counting dimensions. Multivariate data in $\Re^d$ are almost never d-dimensional. That is, the *underlying structure* of data in $\Re^d$ is almost always of dimension lower than d. Thus, in general, the full space may be usefully partitioned into subspaces of signal and noise. Of course, this partition is not precise, but the goal is to eliminate a significant number of dimensions so as to encourage a parsimonious representation of the underlying structure.

For example, consider a parametric multivariate linear regression problem in $\Re^{d+1}$. The data $\{(\mathbf{x}_i, y_i) : 1 \le i \le n\}$ are modeled by

$$y_i = \sum_{j=1}^{d} a_j x_{ij} + \epsilon_i = \mathbf{a}^T \mathbf{x}_i + \epsilon_i.$$

While d may be very large, the relevant structure of the solution data space is precisely two-dimensional: $\{(w_i, y_i) : 1 \le i \le n\}$ where $w_i \equiv \mathbf{a}^T \mathbf{x}_i$. While this may seem unfamiliar, try to imagine looking at a billion random projections of the data matrix $\mathbf{X}$ to one dimension of the form $\mathbf{X}\alpha$, and choosing the projection that correlates the most with the response vector $\mathbf{y}$. Then the α so selected will be nearly proportional to the regression coefficient $\hat{\beta}$. In fact, the correlation will be identical to the

Multivariate Density Estimation, First Edition. David W. Scott.
© 2015 John Wiley & Sons, Inc. Published 2015 by John Wiley & Sons, Inc.

multiple correlation coefficient. Thus multiple linear regression may be thought of as a dimension reduction technique from $\Re^{d+1} \to \Re^2$.

Often the motivation in regression extends beyond the prediction of y and focuses on the structure of that prediction as it is reflected in the relative magnitudes and signs of the coefficients a_j. However, when some of the predictor variables in $\mathbf{x}$ are highly correlated, many choices of weights on those variables can give identical predictions. Collinearity in the $\mathbf{x}$ data (reflected by a nearly singular covariance matrix) adversely affects the interpretability of the coefficients $\mathbf{a}$. One successful approach is to find a lower dimensional approximation to the covariance matrix. Factor analysis is a statistical technique for culling out structure in the correlation matrix with only several "factors" being retained (Johnson and Wichern, 1982). Additional dimensions or factors may provide more information on the margin, but their use requires more careful analysis.

The dimension reduction afforded by these models results in a tremendous increase in the signal-to-noise ratio. The full information is contained in the raw multivariate data themselves, which are sufficient statistics. In more than a few dimensions, noise limits visual interpretability of raw data. Exploratory graphical methods described in Chapter 1 such as brushing work best when the structure in the data is very low-dimensional, two at most. If the structure is too complicated, then the sequence of points highlighted during brushing can only hint at that structure's existence. More often than not, any subtle structure is lost in apparently random behavior.

With ordinary multivariate data, the conclusions are similar. If the structure in 100-dimensional data falls on a 20-dimensional nonlinear manifold, then without prior information, the task of detection is futile. With prior information, appropriate modeling can be brought to bear, but parsimonious representations of nonlinear 20-dimensional manifolds are few. Fortunately, it appears that in practical situations, the dimension of the structure seldom exceeds four or five. It has been suggested that speech recognition, surely a complex example, requires a feature space of five dimensions. That is only one expert's opinion, but even if a nonparametric surface in $\Re^d$ could be well estimated, the ability to "fully explore" surfaces in dimensions beyond 5 is limited (see Tukey and Tukey, 1981; Huber, 1985).

Projection of high-dimensional data onto subspaces is unavoidable. The choice remains whether to work with techniques in the full dimension, or to first project and then work in the subspace. The clear preference expressed here is for the latter. The two tasks need not be totally decoupled, but the separation aids in the final interpretation.

The precise cutoff dimension is a subject of intense research. Can a kernel estimate of 20-dimensional data be constructed to aid projection? Almost no one believes so. Even a few decades ago, many held the belief that bivariate nonparametric estimation required prohibitively large samples to be sufficiently accurate. In higher dimensions, the limitation focuses on an understanding of the *curse of dimensionality*. The term was first applied by Bellman (1961) to combinatorial optimization over many dimensions, where the computational effort was observed to grow exponentially in d. In statistics, the phrase reflects the sparsity of data in multiple dimensions. Indeed, it was shown in Section 1.5.3.1 that given a uniform sample over the hypercube $[-1, 1]^d$, almost no sample elements will be found in the inscribed hypersphere. Even the "large" histogram bin $[0, 1]^d$ in the first quadrant contains only the fraction 2^{-d} of the data. In 10 dimensions, this fraction is only $1/10,000$. The bin $[-0.5, 1.0]^{10}$, which covers three-quarters of each edge of the hypercube, contains only 5.6% of the data. The conclusion is that in order to include sufficient data, the smoothing parameters must be so large that no local behavior of the function can be reasonably approximated without astronomical sample sizes. Epanechnikov (1969) sought to quantify those sample sizes. His and several other arguments are presented below.

In density estimation, two quite different kinds of structure can have contours that are similar in appearance. The first is data with spherical symmetry, for which many examples have already been presented. The second is of more pedagogical than practical value. Consider a density that is spherically symmetric about the origin but with a "hole" in it. Now the hole should not be a discontinuous feature, but rather a dip in the density in the middle (visualize a volcano). There are two notable features about the contours of this density. First, the mode is not a single point; but rather, it is a connected set of points falling on a circle (sphere) in two (three) dimensions. Thus the mode is not a zero-dimensional object, but rather a $(d-1)$-dimensional object. Second, when "falling off the mode," contours occur in *nested pairs*; a bivariate example is illustrated in Figure 7.1. Contours in the bimodal setting also occur in pairs, but are not nested. In fact, without any labeling of contours, the nested contours of the "volcano" might be mistaken for the ordinary normal density. As the result of a finite sample size, the sample mode is not a circle, but is rather a set of several points located around the circle. A trivariate example is presented in Section 9.3.4.

Assessing the dimension of structure is not always easy. For example, if the regression model includes a nonlinear term of the kth variable,

$$y_i = \mathbf{a}^T \mathbf{x}_i + g(x_{ik}) + \epsilon_i, \tag{7.1}$$

then it seems reasonable to describe the structure of the model as three-dimensional. Suppose g was a nonparametric estimator in (7.1). Even though g is usually thought of as being infinite-dimensional, three-dimensional is still an appropriate description.

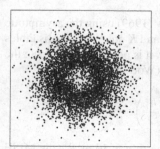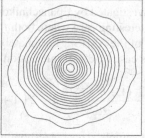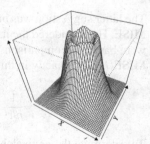

FIGURE 7.1 Scatter diagram and ASH representations of bivariate data from a density with a hole ($n = 5000$). Observe how some contours occur in nested pairs.

These *semiparametric regression* or *partial spline* models (Speckman, 1987) are reviewed by Wahba (1990).

The illustration of the histogram in $\Re^{10}$ misses the additional complication of rank deficient data. It was shown in Equation (6.42) that the asymptotic mean integrated squared error (AMISE) of a kernel estimate in two dimensions blows up as the correlation $\rho \to \pm 1$. Real data in 10 dimensions are never independent nor spread uniformly on a cube. Serious density estimation is never performed with an inefficient histogram or uniform kernel estimator. What is in fact feasible? This issue is explored in Section 7.2.

7.2 CURSE OF DIMENSIONALITY

The curse of dimensionality describes the apparent paradox of "neighborhoods" in higher dimensions—if the neighborhoods are "local," then they are almost surely "empty," whereas if a neighborhood is not "empty," then it is not "local." This description is simply another way of saying that the usual variance—bias trade-off cannot be accomplished very well in higher dimensions without very large samples. If the bandwidth is large enough to include enough data to hold down the variance, the bias is intolerable due to the large neighborhood, and vice versa. Stated in the usual MISE terms, the effects of the curse of dimensionality are investigated for some simple examples.

7.2.1 Equivalent Sample Sizes

In order to demonstrate the progressive deterioration of kernel density estimation as the dimension d increases, it is sufficient to follow the increase of the sample size, $n_d(\epsilon)$, required to attain an equivalent amount of accuracy ϵ. The multivariate MISE has the same units as $(\sigma_1 \sigma_2 \cdots \sigma_d)^{-1}$, which is not dimensionless. Thus direct comparison of MISE across different dimensions is meaningless. Two suggestions for a dimensionless measure of accuracy are

$$(1) \quad [\sigma_1 \sigma_2 \cdots \sigma_d \times \text{MISE}]^{1/2} \quad \text{and} \quad (2) \quad [\text{MISE}/R(f)]^{1/2}. \tag{7.2}$$

The latter suggestion was investigated by Epanechnikov (1969) using the asymptotic MISE. He considered the interesting case where both f and K are $N(0, I_d)$. His tables were based on comparisons of the AMISE. Fortunately, it is possible to compute the $\text{MISE}(h, n, d)$ exactly in this case by direct integration (Worton, 1989):

$$(4\pi)^{\frac{d}{2}} \text{MISE} = \frac{1}{nh^d} - \frac{(1+h^2)^{-\frac{d}{2}}}{n} + 1 - 2\left(1 + \frac{h^2}{2}\right)^{-\frac{d}{2}} + (1+h^2)^{-\frac{d}{2}}.$$

To determine the equivalent sample sizes corresponding to $n = 50$ in $\Re^1$, the MISE expression in Equation (7.2) was minimized numerically to find the best bandwidth $h^* = 0.520$ and the corresponding $\text{MISE}^* = 0.00869$. Now $R(f) = 2^{-d} \pi^{-d/2}$ and

$\sigma_1\sigma_2\cdots\sigma_d = 1$; hence, the Epanechnikov criterion in 1-D is 0.176. In $\Re^2$, the sample size n is found by numerically searching over the corresponding optimal bandwidth so that the criteria match. For example, with the Epanechnikov criterion, $n = 258$ in $\Re^2$ (with $h^* = 0.420$) and $n = 1,126$ in $\Re^3$ (with $h^* = 0.373$). However, the first criterion in (7.2) yields quite a different sequence: $n = 29$ in $\Re^2$ and $n = 6$ in $\Re^3$. Clearly, not all choices of dimensionless MISE-based criteria are equivalent. Since the estimation problem does not appear to get easier as d increases, the first proposal is ignored.

Epanechnikov's equivalent sample sizes corresponding to 50 points in $\Re^1$ are displayed in Figure 7.2. The graph confirms that the growth in sample size is at least exponential, since the plot is on a log–log scale. Certainly, the sample sizes for dimensions below 6 are manageable with the averaged shifted histogram (ASH) technology.

Epanechnikov's particular choice of a dimensionless MISE criterion lacks theoretical foundation. The following pointwise criterion (relative root MSE) may also be computed exactly (Fryer, 1976; Silverman, 1986) and used for extrapolation:

$$\text{RRMSE}(0) = \sqrt{\text{MSE}\{\hat{f}(0)\}/f(0)}.$$

This criterion is the most pessimistic (see Figure 7.2), which is expected as the origin is the most difficult point of the density to estimate. While this criterion is dimensionless, it is only indirectly related to the noise-to-signal ratio.

Scott and Wand (1991) advocate a related criterion (root coefficient variation):

$$\text{RCV}(0) = \sqrt{\text{Var}\{\hat{f}(0)\}/\text{E}\hat{f}(0)}.$$

This criterion measures the amount of noise superimposed on the average value of the estimate at the multivariate origin. As a rule of thumb, if $\text{RCV} < 1/3$, then the estimate is unlikely to be totally overwhelmed by noise. In fact, $\text{RCV}(0) = 12.7\%$

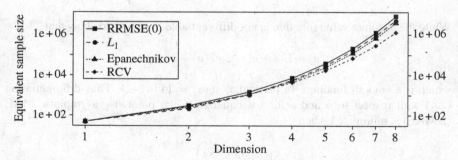

FIGURE 7.2   Equivalent sample sizes for several criteria that have the same value as in one dimension with 50 sample points. The density and kernel are both $N(0, I_d)$. The criterion values for RRMSE(0), AMIAE, Epanechnikov, and RCV(0) are 0.145, 0.218, 0.176, and 0.127, respectively.

in Figure 7.2. The equivalent sample size in eight dimensions is 33% of the next smallest criterion. Even so, a million points (which the eight-dimensional equivalent requires) is not a routine sample size. If $\text{RCV}(0) = 30\%$ is acceptable, then "only" 20,900 points are required in $\Re^8$. This prediction is by far the most optimistic.

The pointwise criteria are well-founded but not global. The most attractive criterion for the purpose at hand is the L_1 error. It was noted in Section 2.3.2.2 that the L_1 criterion was a naturally dimensionless criterion suitable for comparisons across dimensions. Fortunately, the univariate asymptotic L_1 formula of Hall and Wand (1988a) can be extended into a multivariate setting. A derivation is given in the next section. Observe that in Figure 7.2, the L_1 equivalent sample sizes are smallest in two and three dimensions, but are second from the top beyond three dimensions. This result should be expected, as the L_1 criterion places more emphasis on errors in the tails. In higher dimensions, virtually all of the probability mass is in the tails. For a much larger error value (corresponding to $n = 4$ in $\Re^1$) given in Scott and Wand (1991), the L_1 equivalent sample sizes were smallest.

7.2.2 Multivariate L_1 Kernel Error

Hall (1984) has shown that the multivariate kernel estimate for fixed $\mathbf{x}$ is asymptotically normal, so that

$$\hat{f}(\mathbf{x}) - f(\mathbf{x}) \approx b_d(\mathbf{x})h^2 + Z_1\sqrt{\frac{\sigma_d^2(\mathbf{x})}{nh^d}} = \sqrt{\frac{\sigma_d^2(\mathbf{x})}{nh^d}}\left[Z_1 + b_d(\mathbf{x})\sqrt{\frac{nh^{d+4}}{\sigma_d^2(\mathbf{x})}}\right], \quad (7.3)$$

where Z_1 is univariate $N(0, 1)$; the pointwise bias and variance quantities were given in Section 6.3.1. Define the function ψ by $\psi(u) = \text{E}|Z + u|$. Then the mean absolute error (MAE) may be approximated by

$$\text{MAE}(\mathbf{x}) = \text{E}|\hat{f}(\mathbf{x}) - f(\mathbf{x})| \approx \sqrt{\frac{\sigma_d^2(\mathbf{x})}{nh^d}}\,\psi\left\{b_d(\mathbf{x})\sqrt{\frac{nh^{d+4}}{\sigma_d^2(\mathbf{x})}}\right\}. \quad (7.4)$$

While the absolute value function is not differentiable, it is easy to show that

$$\psi(u) = \text{E}|Z + u| = 2u\Phi(u) + 2\phi(u) - u,$$

which is a smooth function. In particular, $\psi'(u) = 2\Phi(u) - 1$. Thus differentiating (7.4) with respect to h and setting it equal to 0, the pointwise asymptotic MAE (AMAE) is minimized when

$$h_1^*(\mathbf{x}) = \left[\frac{\alpha_d^2 \sigma_d^2(\mathbf{x})}{nb_d^2(\mathbf{x})}\right]^{1/(d+4)},$$

where α_d is the unique positive solution to $4\alpha_d\left[\Phi(\alpha_d) - \frac{1}{2}\right] = \phi(\alpha_d)\,d$. For example, $\alpha_1 = 0.48$ and $\alpha_{10} = 1.22$.

Using the same notation as above, the smoothing parameter which minimizes MSE(**x**) is

$$h_2^*(\mathbf{x}) = \left[\frac{\sigma_d^2(\mathbf{x})d}{4nb_d(\mathbf{x})} \right]^{1/(d+4)} \quad \Rightarrow \quad \frac{h_1^*(\mathbf{x})}{h_2^*(\mathbf{x})} = \left[\frac{4\alpha_d^2}{d} \right]^{1/(d+4)}.$$

Thus, by observation

Theorem 7.1:   For any kernel, for all qualifying density functions, and for all $\mathbf{x} \in \Re^d$, the asymptotically optimal L_1 and L_2 pointwise bandwidths satisfy

$$0.9635 \le \frac{h_1^*(\mathbf{x})}{h_2^*(\mathbf{x})} \le 1. \tag{7.5}$$

Remarkably, this ratio of asymptotically optimal bandwidths does not depend on the particular choice of kernel, the underlying density function, or the point of estimation (see Figure 7.3). Pointwise, the choice of criterion does not appear to be critical.

A global multivariate L_1 approximation may be obtained by integrating the pointwise result above. The asymptotic mean integrated absolute error (AMIAE) follows directly from Equation (7.4):

$$\text{AMIAE} = \text{E} \int_{\Re^d} |\hat{f}(\mathbf{x}) - f(\mathbf{x})| \approx \sqrt{\frac{1}{nh^d}} \int_{\Re^d} \sigma_d(\mathbf{x}) \, \psi \left\{ b_d(\mathbf{x}) \sqrt{\frac{nh^{d+4}}{\sigma_d^2(\mathbf{x})}} \right\} d\mathbf{x}.$$

Proceeding as before, the optimal width is $h_1^* = [v_d^2/n]^{-1/(d+4)}$, where v_d is the unique positive solution to

$$\int_{\Re^d} \sigma_d(\mathbf{x}) \left(4v_d r_d(\mathbf{x}) \left[\Phi\{v_d r_d(\mathbf{x})\} - \tfrac{1}{2} \right] - d\,\phi\{v_d r_d(\mathbf{x})\} \right) d\mathbf{x} = 0,$$

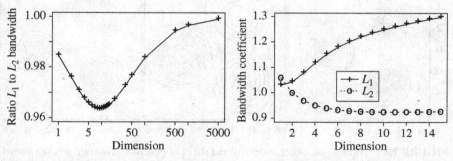

FIGURE 7.3 Ratio of pointwise optimal L_1 and L_2 bandwidths for all situations. The right frame displays the coefficients of $n^{-1/(d+4)}$ for the global L_1 and L_2 bandwidths for normal data.

letting $r_d(\mathbf{x}) = b_d(\mathbf{x})/\sigma_d(\mathbf{x})$ (Scott and Wand, 1991). By spherical symmetry, this integral may be computed as a univariate integral. The coefficients of $n^{-1/(d+4)}$ for the optimal L_1 and L_2 bandwidths for the case of multivariate normal data are graphed in Figure 7.3. Thus the difference in the emphasis placed on the tails does lead to somewhat larger bandwidths for the absolute error criterion.

7.2.3 Examples and Discussion

The theoretical arguments discussed earlier generally suggest that kernel estimation beyond five dimensions is fruitless. However, the empirical evidence is actually less pessimistic. Friedman et al. (1984) , discuss a particular simulation example in $\Re^{10}$. The (x_1,x_2)-space contains the signal, which is an equal mixture of three shifted bivariate $N(0,I_2)$ densities with means $(\mu,0)$, $(-\mu,3)$, and $(-\mu,-3)$ with $\mu = 3^{3/2}/2$. The marginal variances are both 7. Next, eight pure normal noise dimensions are added, each with variance 7. The authors investigated the task of estimating the following slice of $f(\mathbf{x})$:

$$f(x_1,x_2,0,0,0,0,0,0,0,0).$$

The authors easily accomplished this task using their projection pursuit density estimation algorithm with only 225 data points.

Given the results of the preceding section, what could be expected of a kernel estimate with only 225 points? Figure 7.4 displays a product kernel estimate of such a sample using a triweight kernel. The same bandwidth $h = 4.0$ was used in each dimension (as suggested by an AMISE computation). Surprisingly, the trimodal structure is quite clearly represented, even though the bandwidth is hardly local. The optimal product kernel bandwidths are 2.0 in the first two dimensions and 5.25 in the remaining eight. The structure is even clearer, of course, although the equal bandwidth choice is a fairer test.

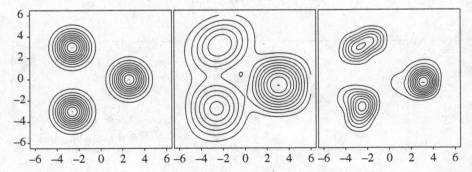

FIGURE 7.4 Bivariate slice of true density and slices of two 10-D triweight product kernel estimates. The nine contour levels are equally spaced up to the mode. The bandwidths for the middle frame were $h_i = 4.0$; for the right frame $h_1 = h_2 = 2.0$ and $h_3 = \cdots = h_{10} = 5.25$ (see Scott and Wand (1991)).

The kernel estimate is not as good as the figure suggests. The middle estimate is quite biased at the peaks, only a third of the true modal value. Thus the global error measure is certainly large. However, the important structure in the data is revealed by a nonadaptive kernel estimate with a very modest sample size.

What is to be made of such conflicting theoretical and empirical evidence? The answer lies in the recognition that it may not be unreasonable to accept less global accuracy in 10 dimensions than in 2 dimensions. In some situations, a high-dimensional kernel estimate may be useful, taking into account the known limitations. In other situations, the bias must be examined closely. But one situation where high-dimensional kernel estimation does seem to work is in helping to estimate which subspace on which to focus (see Section 7.3.3).

A rather different kind of evidence comes from a simple hill-climbing algorithm investigated by Boswell (1983). He generated samples of size 100 from $N(0, I_d)$ in up to 100 dimensions. Starting at random points, he used a gradient algorithm to find the nearest local peak. Somewhat to his surprise, the algorithm invariably went to the sample mode near the origin and did not find local maxima. The exact role played by the smoothing parameter choice could be looked at more closely, but the result is impressive in any case.

Neither of these two normal examples explore the other aspect of the curse of dimensionality that results from rank deficiencies in the data. Indeed, these two examples are somewhat unrealistic in that the variables are uncorrelated. The presence of even moderate skewness in several dimensions can adversely affect kernel density estimates. Consider the PRIM4 dataset ($n = 500$), one of several sets originating from the Stanford Linear Accelerator Center during development of the original PRIM system (related datasets include PRIM7 and PRIM9). Each of the raw variables is very strongly skewed (see Figure 7.5). By successive application of Tukey's transformation ladder (Equation (3.38)) to each variable, the skewness was reduced. The four transformations were

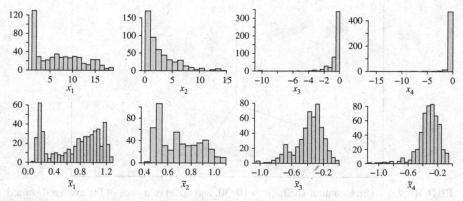

FIGURE 7.5 Histograms of the original four variables in PRIM4 before (top row) and after (bottom row) transformation; $n = 500$.

$$\log_{10}(x_1), \quad \sqrt{\log_{10}(1+x_2)}, \quad -\sqrt{\log_{10}(1-x_3)}, \quad -\sqrt{\log_{10}(1-x_4)}.$$

Note that when a marginal variable is bimodal, eliminating the skewness of that variable may *not* be the most desirable result. Rather, the intent is to reduce or possibly eliminate the skewness of each component mixture within that variable. Histograms of the transformed variables are displayed in Figure 7.5.

What is the practical effect on estimation of the multivariate density surface? The same slice $\hat{f}(x_1, x_2, x_3 | x_4)$ of the ASH estimator was computed using the data before and after the transformation. Those slices are shown in Figures 7.6 and 7.7. The original data were crowded into a corner and along a face of the surrounding four-dimensional hyper-rectangle. After transformation, the data were clustered around the center of the hyper-rectangle. The advantage of the transformation has been to simultaneously reduce the variance and bias in the kernel estimate. The two primary clusters are more clearly represented, as are the two lesser "bumps" that are suggested.

The role of transformations is of great practical importance. Each marginal variable should be examined and transformed if skewed. The PRIM4 example used simple transformation ideas. It is possible to try to find optimal transformations using a cross-validation criterion. The transformation families need to cope not only with skewness but also with other problems. This task has been accomplished in the univariate setting by Wand et al. (1991). They advocate backtransforming to the original

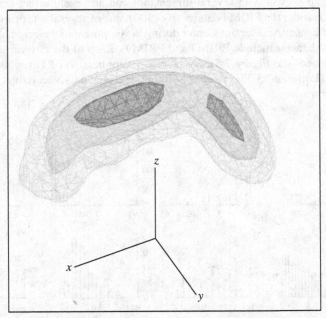

FIGURE 7.6   Three contour shells ($\alpha = 10$, 30, and 60%) of a slice of the averaged shifted histogram of the four-dimensional PRIM4 dataset with 500 points. These variables are heavily skewed, and the resulting density estimation problem more difficult.

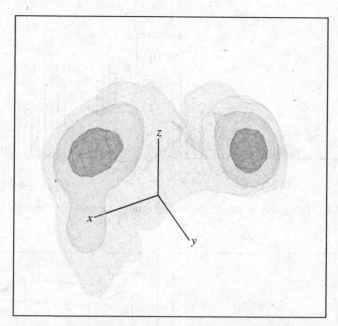

FIGURE 7.7   Three contour shell levels as in Figure 7.6 based on an ASH of the transformed PRIM4 data. The transformation was chosen to reduce skewness in each marginal variable. Such marginal transformations can greatly improve the quality of density estimation in multiple dimensions.

scale after applying the kernel estimate in the transformed space. That may or may not be desirable. For example, backtransformation can introduce artifacts such as extra bumps (see Duan (1991) and Scott (1991c)). For the multivariate setting, visualization is facilitated when the density mass is away from the edges of the support. Backtransformation returns the mass to the edges, although in a smoother manner. It is a trade-off between the ease of examining the contours and the ease of using the scales of the original variables.

The earthquake data preceding the 1982 eruption of Mount St. Helens (Weaver et al., 1983) also illustrates the value of transformation in separating tightly located clusters. A histogram of the depth of the epicenters of the 510 earthquakes is shown in Figure 7.8, together with the transformed variable $-\log_{10}(-z)$. In this case, the recognition of two of the three clusters of earthquake epicenters was almost missed in a histogram of the (untransformed) depth variable. However, the beauty of these data is best captured in the trivariate and quadravariate ASHs displayed in Figure 7.9. As this figure is based on 2 months of data preceding the eruption, it is of interest to note that the earlier eruptions are the deep eruptions and the eruptions near the surface concentrated in the last weeks. It would be informative to correlate the contour surfaces with geological structures.

Thus the presence of rank deficiencies in the multivariate data, rather than the fact of high dimensions *per se*, is the more important component of the curse of

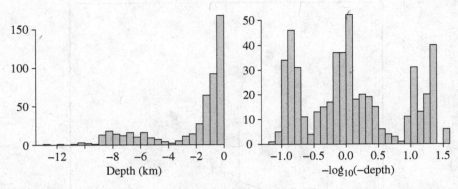

FIGURE 7.8   Histograms of the depths of 510 earthquake epicenters and the transformed depths.

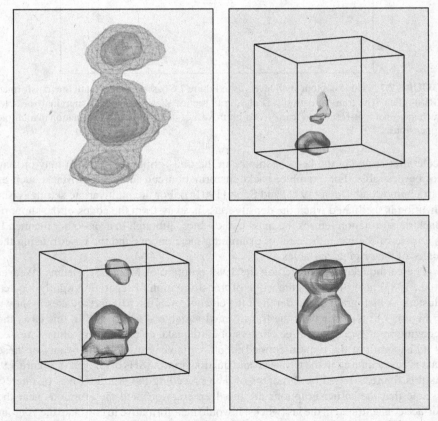

FIGURE 7.9   (Top left) ASH of the location of 510 earthquake epicenters and the transformed depths. The next three frames show the space-time ASH at approximately 1-week intervals leading up to the eruption (all at the same α level.)

dimensionality. The recognition of rank deficiencies and corresponding projection algorithms are the focus of the next section.

7.3   DIMENSION REDUCTION

The strategy advocated here is a two-stage approach: first reduce the dimension of the data, and then perform density estimation and use other graphical techniques in the smaller-dimensional subspace. This section deals with dimension reduction technology. Three steps are outlined for accomplishing the task of reducing the dimension from $\Re^d \rightarrow \Re^{d'}$, where $d' \in [1,5]$. First, marginal variables are transformed to enhance the shape of the data cloud and move points away from "edges" in the surrounding hyper-rectangle. Second, a linear criterion is used to strip away obviously redundant variables. Finally, a nonlinear criterion is applied that retains most of the remaining structure. These steps are discussed later.

7.3.1   Principal Components

If the data cloud takes on the form of a hyper-ellipse, then it may be completely summarized by the mean vector and the covariance matrix. Such is the case with a multivariate normal density, $\mathbf{X} \sim N(\mu, \Sigma_d)$. Thus the shape and the "dimension" of the data cloud may be determined by examining the eigenvalues of the positive definite covariance matrix Σ_d, assuming Σ_d is of rank d.

The covariance matrix may be computed in the following compact form:

$$\Sigma \equiv \Sigma_d = \mathrm{E}\{(\mathbf{X} - \mu)(\mathbf{X} - \mu)^T\},$$

dropping the subscript on Σ_d. Suppose that the eigenvalues of Σ are denoted by $\lambda_1 \geq \lambda_2 \geq \ldots \geq \lambda_d > 0$, with corresponding eigenvectors $\mathbf{a}_1, \mathbf{a}_2, \ldots, \mathbf{a}_d$. The eigenvectors are orthonormal; that is, $\mathbf{a}_i^T \mathbf{a}_j = \delta_{ij}$. Then $\Sigma \mathbf{a}_i = \lambda_i \mathbf{a}_i, 1 \leq i \leq d$, which may be summarized by the matrix equation known as the *spectral representation*:

$$\Sigma \mathbf{A} = \mathbf{A}\Lambda, \qquad (\Rightarrow \quad \Sigma = \mathbf{A}\Lambda\mathbf{A}^T) \tag{7.6}$$

where

$$\mathbf{A} = [\mathbf{a}_1 \mathbf{a}_2 \cdots \mathbf{a}_d] \quad \text{and} \quad \Lambda = \begin{pmatrix} \lambda_1 & 0 & & 0 \\ 0 & \lambda_2 & & 0 \\ & & \ddots & \\ 0 & 0 & & \lambda_d \end{pmatrix}.$$

The *sphering transformation* of a multivariate random variable is defined by

$$\boxed{\text{Sphering transformation:} \quad \mathbf{Z} = \Sigma^{-1/2}(\mathbf{X} - \mu),} \tag{7.7}$$

where $\Sigma^{-1/2} = \mathbf{A}\Lambda^{-1/2}\mathbf{A}^T$. The random variable $\mathbf{Z}$ has mean $\mu_{\mathbf{Z}} = \mathbf{0}_d$. Its covariance matrix is easily seen to be the identity matrix I_d, recalling that Σ and hence $\Sigma^{-1/2}$ are symmetric matrices:

$$\Sigma_{\mathbf{Z}} = \mathrm{E}\{\mathbf{Z}\mathbf{Z}^T\} = \mathrm{E}\{\Sigma^{-1/2}(\mathbf{X}-\mu)(\mathbf{X}-\mu)^T\Sigma^{-1/2}\} = \Sigma^{-1/2}\Sigma\Sigma^{-1/2} = I_d.$$

The sphering transformation destroys all first- and second-order information in the density or data. This transformation is useful when the covariance structure is not desired.

A related transformation is the *principal components transformation*:

> Principal components transformation: $\mathbf{Y} = \mathbf{A}^T(\mathbf{X} - \mu)$.   (7.8)

This transformation again centers the density, $\mu_{\mathbf{Y}} = \mathbf{0}_d$, and produces uncorrelated variables, but without wiping out the variance information:

$$\Sigma_{\mathbf{Y}} = \mathrm{E}\{\mathbf{Y}\mathbf{Y}^T\} = \mathrm{E}\{\mathbf{A}^T(\mathbf{X}-\mu)(\mathbf{X}-\mu)^T\mathbf{A}\} = \mathbf{A}^T\Sigma_{\mathbf{X}}\mathbf{A}$$
$$= \mathbf{A}^T(\mathbf{A}\Lambda\mathbf{A}^T)\mathbf{A} = \Lambda.$$

Thus the principal components are *uncorrelated*, and the variance of the ith principal component, $Y_i = \mathbf{a}_i^T(\mathbf{X} - \mu)$, is the ith eigenvalue λ_i. It is shown in any multivariate analysis textbook (e.g., Johnson and Wichern, 1982) that there is no other linear combination with larger variance than Y_1, and that Y_2 is the linear combination (uncorrelated with Y_1) with the second largest variance, and so on.

The results obtained hold for data as well as for the density function if Σ is replaced with the maximum likelihood estimate $\hat{\Sigma}$, which is positive definite with eigenvalues $\hat{\lambda}_1 > \hat{\lambda}_2 > \cdots > \hat{\lambda}_d > 0$, and μ is replaced with $\bar{\mathbf{x}}$. The dimension of the data may be reduced by choosing the smallest $d' < d$ such that

$$\frac{\sum_{i=1}^{d'} \hat{\lambda}_i}{\sum_{i=1}^{d} \hat{\lambda}_i} = \frac{\sum_{i=1}^{d'} \hat{\lambda}_i}{\mathrm{tr}\{\hat{\Sigma}\}} > 90\%,   (7.9)$$

for example. The reduced data are said to have retained at least 90% of the variance ("information") in the original, higher dimensional data. The formula for transforming data to the principal components subspace in $\Re^{d'}$ is given by

> PC data transformation: $\mathbf{Y}_{n\times d'} = (\mathbf{X}_{n\times d} - \mathbf{1}_{n\times 1}\bar{\mathbf{x}}_{1\times d}^T) \cdot [\mathbf{a}_1\mathbf{a}_2\cdots\mathbf{a}_{d'}]_{d\times d'}.$

The matrix $\mathbf{1} \cdot \bar{\mathbf{x}}^T$ is simply a device used to remove the sample mean from each of the n data vectors, $\mathbf{x}_i$. In practical situations, d' is typically in the range $1 \leq d' \leq 5$.

In the current setting with non-normal data, the principal components transformation could be used as follows. Histograms of each of the original d are examined in order to determine the need for marginal transformations. The new marginal variables are then standardized to have mean 0 and variance 1. After this preprocessing, the covariance matrix $\hat{\Sigma}$ is replaced by the correlation matrix $\hat{\mathbf{R}}$. As $r_{ii} = 1$, then

$\text{tr}\{\hat{\mathbf{R}}\} = d$ exactly. The goal is to get rid of dimensions that contain no independent linear information, so that $d' < d$ is chosen differently than in (7.9):

$$\frac{\sum_{i=1}^{d'} \hat{\lambda}_i}{\sum_{i=1}^{d} \hat{\lambda}_i} = \frac{\sum_{i=1}^{d'} \hat{\lambda}_i}{d} > 95\%,$$

or perhaps 99%. The goal is to avoid losing prematurely possible nonlinear information in the data. With this choice, $1 \leq d' \leq 10$ is expected in practice.

7.3.2 Projection Pursuit

With the data "cleaned up" and stripped of obviously redundant information, a final step is proposed that will reduce the data to the final "working dimension" by $\Re^{d'} \to \Re^{d''}$. To simplify notation, these dimensions are relabeled $\Re^d \to \Re^{d'}$. If the data fall into clearly defined clumps, then the data may be partitioned at this point, and the further transformations applied separately to each clump or cluster. For the remainder of the discussion, the data will be analyzed as a single unit.

In general, the data should be "sphered" at this point. As a result, all possible projections will have moments $(\mathbf{0}_{d'}, I_{d'})$. With this step, no second-order information can be included in the final projection algorithm. Once the small variance dimensions have been removed, the variance criterion is no longer of interest as a measure of structure in the data. Principal components search for projections with maximum variance, while nonparametric procedures find projections with minimum variance more difficult (recall IMSE is inversely proportional to scale) and therefore more interesting. The need to perform separately principal components and projection pursuit is now evident.

The idea of a projection criterion other than variance was discussed by Kruskal (1969, 1972) and Switzer (1980). Kruskal discussed an index that would detect clusters. The first successful implementation was reported by Friedman and Tukey (1974), who coined the phrase *projection pursuit* (PP). PP is the numerical optimization of a criterion in search of the most interesting low-dimensional linear projection of a high-dimensional data cloud.

Consider a projection from $\Re^d \to \Re^{d'}$. Denote the projection matrix by $P_{d \times d'}$. Let the density functions of the random variables $\mathbf{X} \in \Re^d$ and $\mathbf{Y} = P^T \mathbf{X} \in \Re^{d'}$ be denoted by $f(\mathbf{x})$ and $g(\mathbf{y}|P)$, respectively. Epanechnikov attempted to find the *smoothest density* in $\Re^1$ by minimizing the dimensionless quantity $\sigma_K R(K)$. The multivariate version of the same quantity may be used as a measure of the *least-smooth density*, which is the most informative in $\Re^{d'}$, in the optimization problem

$$\max_{P} \quad \sigma(\mathbf{y}|P) R[g(\mathbf{y}|P)],$$

where $\sigma(\mathbf{y}|P)$ is the product of the d' marginal standard deviations. A sample version of this problem, letting $\mathbf{X}_{n \times d}$ denote the sample data, would be

$$\max_{P} \quad \hat{\sigma}(\mathbf{X}P) R\{\hat{f}_h(\mathbf{y}|\mathbf{X}P)\}, \tag{7.10}$$

where the kernel estimate is based on the data matrix $\mathbf{Y} = \mathbf{X}P$ and the smoothing parameter h. The optimization assumes that h is fixed, because the kernel estimate becomes infinitely rough as $h \to 0$. However, the optimal projection matrix P should be relatively insensitive to the choice of h. The numerical optimization of Problem (7.10) is nontrivial. The problem is a constrained nonlinear optimization problem, the constraints being $P^T P = I_{d'}$. Further, the criterion is not strictly concave; it exhibits many local maxima. Taken together, these observations emphasize the necessity of a prior principal components transformation to the original data.

The original univariate Friedman and Tukey (1974) PP index from $\Re^d \to \Re^1$ was

$$\hat{\sigma}_\alpha(\mathbf{Y}) \sum_{i,j} (h - |y_i - y_j|)_+ , \qquad (7.11)$$

where $\hat{\sigma}_\alpha$ denoted the α-trimmed standard deviation (i.e., the standard deviation after $\alpha/2$ of the data are deleted on each end). The criterion is large whenever many points are clustered in a neighborhood of size h. Criterion (7.11) does not obviously resemble criterion (7.10). However, Huber (1985) noted that with the Uniform kernel $U(-0.5, 0.5)$ (see Problem 7.1),

$$R\{\hat{f}_h(y|\mathbf{Y})\} = \frac{1}{n^2 h^2} \sum_{i,j} (h - |y_i - y_j|)_+ ,$$

so that the two proposals are in fact proportional, except for the modification to $\hat{\sigma}$. The choice of both a discontinuous kernel and a discontinuous measure of scale made the numerical optimization of criterion (7.11) unnecessarily difficult. The use of smoother kernels seems worth further investigation. Smoother kernels could also be used in related PP indices, such as $\sigma^3 R(f')$ and $\sigma^5 R(f'')$. If the data have been sphered, the appearance of σ in these formulas is unnecessary.

Huber (1985) also proposed using information criteria as candidates for PP indices. These were formulated in such a manner that any linear transformation of the data left the index unchanged. Two criteria proposed were standardized Fisher and negative Shannon entropy, which are defined to be

$$\sigma^2(y) \int \frac{f'(y)^2}{f(y)} dy - 1 \quad \text{and} \quad \int f \log f + \log \left[\sigma(y) \sqrt{2\pi e} \right] ,$$

respectively. These indices are especially interesting because they are each minimized when $f = \phi$. Thus maximizing them provides the *least* normal projection. This interpretation is significant, as Diaconis and Freedman (1984) proved that most random one-dimensional projections of multivariate data are approximately normal, even if the underlying density is multimodal. Projection indices based on higher-order moments were considered by Jones and Sibson (1987).

Jee (1985, 1987) compared the optimal projections for theoretical densities that were mixtures of multivariate normal pdfs. Observe that the linear projection from $\Re^d \to \Re^{d'}$ of a mixture of k normals in $\Re^d$ is also a mixture of k normals in $\Re^{d'}$ with component mean and covariance matrices given by $\mu_i P$ and $P^T \Sigma_i P$, respectively. Jee also considered two other least-normal indices. Assume that the random variables have been sphered, so that $(\mu, \sigma) = (0, 1)$ for all possible transformations. Then the L_1 and Hellinger indices are, respectively,

$$\int \left| f(y) - \phi(y) \right| dy \quad \text{and} \quad \int \left[\sqrt{f(y)} - \sqrt{\phi(y)} \right]^2 dy.$$

Jee constructed several bivariate mixtures of two normal densities, each with $\Sigma = I_2$. All four indices were maximized and each returned the identical "best" projection to $\Re^1$: namely, the angle where $f(y)$ displayed the two-component normal densities with maximal separation. As expected, the four indices were minimized at the angle where the two bivariate densities were superimposed so that $f(y) = \phi(y)$. Note that there are no data associated with these calculations—the four indices are calculated exactly as a function of θ over the interval $(0°, 180°)$ degrees, where θ is the angle that the vector P makes with the positive x-axis.

A second PP example from $\Re^2 \rightarrow \Re^1$ provided different results. Three bivariate normal densities with identity covariance matrices were placed at three of four corners of a square $[-4, 4]^2$. The density is shown at three angles of rotation in Figure 7.10. The last orientation with $\theta = 135°$ clearly displays the trimodal structure after projection. However, only Fisher information takes on its maximum at $\theta = 135°$. The other three criteria all take on their maxima at $\theta = 0°, 90°$, and $180°$. These angles all correspond to a bimodal projection density with two of the densities being superimposed. The angle $\theta = 45°$ is a local maximizer for all three; it superimposes the diagonal pair of densities. The resulting univariate density is bimodal but not as well separated. From this result, it would appear that only Fisher information can successfully identify multiple clusters. Jee confirmed this finding for the projection $\Re^3 \rightarrow \Re^2$. Placing four trivariate normals at corners of a tetrahedron, Fisher information was maximized with a projection that was a mixture of four bivariate normals, while bivariate density with optimal Shannon information was only trimodal, with two of the original four trivariate densities superimposed.

Jee discussed sample-based estimates of these information criteria estimated from histograms. In simulations from the earlier examples, he found 200 points usually inadequate, 800 points stable, and 3200 points very reliable. He also applied the criterion based on Fisher information (trace of the standardized Fisher information matrix)

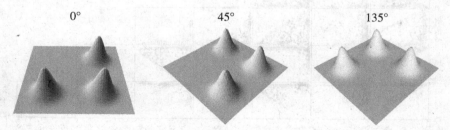

$\quad$ 0° $\qquad\qquad\qquad\qquad$ 45° $\qquad\qquad\qquad\qquad$ 135°

FIGURE 7.10   Three possible projection angles ($\theta = 0°$, 45°, and 135°) for a bivariate mixture of three normal densities.

to the PRIM7 data (see Figure 7.11) presented by Friedman and Tukey (1974) in the setting $\Re^7 \to \Re^2$. The numerical optimization was started and restarted from many random initial guesses. Scatter diagrams of the best and second best projections are displayed in Figure 7.12. The criterion values were 9.6 and 7.8, respectively. The second scatter diagram is quite similar to that found by Friedman and Tukey (1974). The special structure in the first seems a remarkable find in this dataset. Observe the higher density at the corners of the triangle.

Much work on projection pursuit remains. Many workers skip the PP step and simply rely on the first two, three, or four principal components. The interesting variables often may be found by looking among all (but not necessarily the first few) of the pairwise scatter plots of the principal components. For example, the third and fourth principal components of the PRIM7 are somewhat close to the optimal Fisher information solution.

7.3.3 Informative Components Analysis

There are noniterative alternatives to projection pursuit besides principal components that are directly related to density estimation. For example, the results in Section 6.6.3.2 concerning optimal adaptive kernel density estimation in terms of a smoothing parameter and rotation matrix (h, A) could lead to possible algorithms. Deheuvels (1977a) characterized the solution (h^*, A^*) by a system of nonlinear differential equations. A similar approach for the classical multivariate histogram results

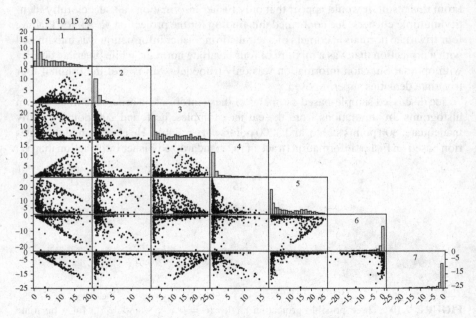

FIGURE 7.11   Pairs of the seven variables in the PRIM7 data.

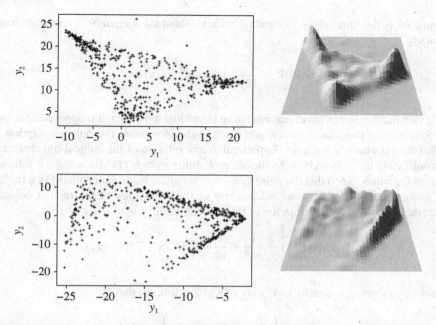

FIGURE 7.12 Bivariate projections of the PRIM7 data using the global and best local maxima of Fisher information. The ASH estimates for each are shown beside.

in some practical and quite simple projection algorithms. These results are due to Terrell (1985).

Assume that the data have been sphered. As there is no *a priori* knowledge of the scale of the remaining structure in the data, a histogram with square bins will be employed. The pointwise bias of the histogram over bin $B_0 \equiv (-h/2, h/2)^d$ given in Equation (3.58) may be written in vector form as

$$\text{Bias}\{\hat{f}(\mathbf{x})\} = -\mathbf{x}^T \nabla f(\mathbf{0}),$$

where $\nabla f(\mathbf{0})$ is the gradient vector. Now the matrix of integrals $\int_{B_0} \mathbf{x}\mathbf{x}^T d\mathbf{x} = \frac{1}{12} h^{d+2} I_d$, as $\int_{B_0} x_i x_j d\mathbf{x} = 0$ and $\int_{B_0} x_i^2 d\mathbf{x} = h^{d+2}/12$. Therefore, the asymptotic integrated squared bias (AISB) over bin B_0 equals

$$\int_{B_0} \text{Bias}^2\{\hat{f}(\mathbf{x})\} d\mathbf{x} = \int_{B_0} \nabla f(\mathbf{0})^T \mathbf{x}\mathbf{x}^T \nabla f(\mathbf{0}) d\mathbf{x} = h^d \times \left[\frac{h^2}{12} \nabla f(\mathbf{0})^T \nabla f(\mathbf{0}) \right],$$

which is the vector form of Equation (3.59) with $h_i = h \ \forall i$. A similar expression holds in every bin B_k with $\mathbf{0}$ replaced by the bin center ξ_k. Summing over all bins, h^d is the area of the integrating unit, so standard Riemannian approximation yields

$$\text{AISB} = \frac{1}{12}h^2 \int_{\Re^d} \nabla f(\mathbf{x})^T \nabla f(\mathbf{x}) \, d\mathbf{x}.$$

Reviewing the derivation of accumulating individual bin errors, it is apparent that any orientation of the cubical bins would result in the same bias. The only assumption is that the bins have volume h^d. In particular, any rotation of the cubical bin structure would yield the same AISB. As the bin probability $p_k \approx h^d f(\mathbf{x})$ for $\mathbf{x} \in B_k$, it follows from Equation (3.55) that the pointwise variance of the histogram equals $f(\mathbf{x})/(nh^d)$. Obviously, this result is also independent of the orientation of the regular cubical mesh. Therefore, for any rotation of the cubical mesh,

$$\text{AMISE}(h) = \frac{1}{nh^d} + \frac{1}{12}h^2 \int_{\Re^d} \nabla f(\mathbf{x})^T \nabla f(\mathbf{x}) \, d\mathbf{x}.$$

Now by the matrix identity $\text{tr}(\mathbf{y}^T\mathbf{y}) = \text{tr}(\mathbf{y}\,\mathbf{y}^T)$, it follows that

$$\text{ISB} = \frac{1}{12}h^2 \text{tr}\left\{ \int_{\Re^d} \nabla f(\mathbf{x}) \nabla f(\mathbf{x})^T d\mathbf{x} \right\} \equiv \frac{1}{12}h^2 \text{tr}\{\Upsilon_f\}.$$

The matrix Υ_f was shown by Terrell to be positive definite. Thus the spectral representation (7.6) of Υ_f is

$$\Upsilon_f = B\Lambda B^T \quad \Rightarrow \quad \text{tr}\{\Upsilon_f\} = \sum_{i=1}^{d} \lambda_i.$$

Therefore, the AMISE may be written as

$$\text{AMISE}(h) = \frac{1}{nh^d} + \frac{1}{12}h^2(\lambda_1 + \lambda_2 + \cdots + \lambda_{d'} + \cdots + \lambda_d).$$

The eigenvalues λ_i of Υ_f reflect the contribution to the overall bias from the particular direction $\mathbf{b}_i$, because the eigenvectors are orthogonal. Thus the subspace spanned by the first d' eigenvectors is the "most interesting" since that is where the largest contribution to the bias occurs. Terrell calls these directions the "informative components." The choice of d' in informative components is entirely analogous to the choice of d' in principal components. In practice, as the multivariate histogram is constructed with square bins, even along the "singular" dimensions, the eigenvalues in those directions will be overestimated due to the variance inflation of the histogram. The most important directions should still be correctly ranked as the inflation applies equally to all dimensions, but the smallest eigenvalues will not be close to zero as in principal components.

The practical estimation of the matrix Υ_f is facilitated by the observation that

$$(\Upsilon_f)_{ij} = \int_{\Re^d} \frac{\partial f(\mathbf{x})}{\partial x_i} \cdot \frac{\partial f(\mathbf{x})}{\partial x_j} \, d\mathbf{x} = -\mathrm{E}\, \frac{\partial^2 f(\mathbf{x})}{\partial x_i \partial x_j}$$

upon integration by parts. $(\Upsilon_f)_{ij}$ may be estimated by a leave-one-out estimator

$$-\hat{\mathrm{E}}\, \frac{\partial^2 f(\mathbf{x})}{\partial x_i\, \partial x_j} = -\frac{1}{n}\sum_{k=1}^{n} \frac{\partial^2 f_{-k}(\mathbf{x}_k)}{\partial x_i\, \partial x_j}.$$

The informative components analysis (ICA) estimator was used on a sample from Jee's bivariate example shown in Figure 7.10. In Figure 7.13, the raw data with 200 points in each cluster are shown as well as the data on the estimated ICA coordinates. The eigenvalues of 0.377 and 0.283 correctly order the interesting projections. Note that y_1 is trimodal while y_2 is bimodal. Similar reproducible results were obtained with 50 points in each cluster.

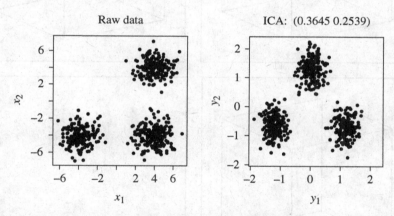

FIGURE 7.13   ICA of simulated bivariate trimodal data with 200 points in each cluster.

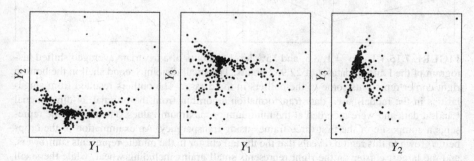

FIGURE 7.14   Pairwise scatter diagrams of first three ICA components for the PRIM7 data.

The ICA estimator was also used on the PRIM7 example shown in Figure 7.12. The pairwise scatter diagrams of the first three of seven ICA coordinates are shown in Figure 7.14. The seven eigenvalues of $\hat{\Upsilon}_f$ were (relative to the largest) 1.0, 0.996, 0.856, 0.831, 0.761, 0.308, and 0.260. Spinning the 3-D scatterplot reveals a data cloud in the shape of a dragonfly with its wings spread open. The ICA coordinates are promising, but more work is required in choosing the bandwidth and coping with outliers.

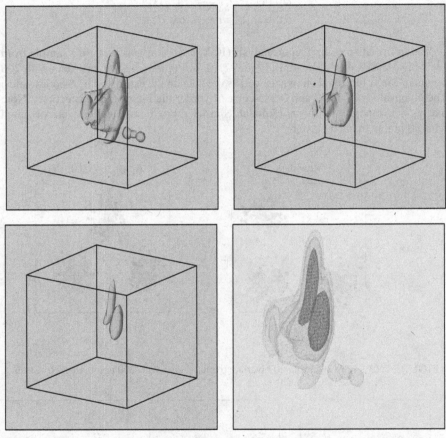

FIGURE 7.15   The $\alpha = 1$, 3.5, and 15% level contours of a trivariate averaged shifted histogram of the Landsat dataset of 22,932 points. The small disjoint second shell in the bottom right corner represents some of the outliers in the dataset. The outliers resulted from singularities in the model-based data transformation algorithm from the original 24-dimensional Landsat data and were recorded at the minimum or maximum values. The final frame represents a composite of the first three frames using transparency. An examination of the crops being grown in this region reveals that the the tall cluster in the middle represents sunflowers, and the largest cluster on the right represents small grains including wheat, while the small small cluster on the far left represents sugar beets (see Sections 1.4.3).

7.3.4   Model-Based Nonlinear Projection

Statisticians are often justly criticized for using techniques that ignore known scientific models when analyzing data. Dimension-reduction transformations based on prior knowledge are common in applied sciences. Often the transformations reflect an expert's opinion based on years of experience. Objective comparisons of model-based projections and "blind" statistical methodology are rare.

The Landsat data presented earlier fall into the model-based category. The raw four-dimensional data from the satellite sensors (four channels) were projected by principal components into a single dimension know as the "greenness" of a pixel. Each pixel was imaged every 3 weeks, except when clouds obscured the region. As the region was known to be largely agricultural, the scientists' idea was to plot the greenness as a function of time and then fit a shape-invariant growth model to those data (Badhwar et al., 1982). The model chosen had three parameters. Thus the data in

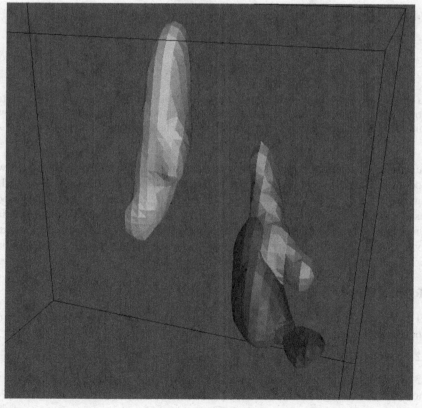

FIGURE 7.16   Median trivariate contours for sunflower ($n = 3694$, left cluster), spring wheat ($n = 3811$, lower right cluster), and barley ($n = 892$, middle right cluster). The median contour contains 50% of the labelled data.

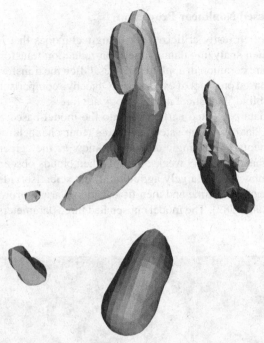

FIGURE 7.17 Median contours shown in Figure 7.16 with spring oats ($n = 459$, overlaps spring wheat), peanuts ($n = 304$, lower center cluster), soybeans ($n = 731$, overlaps lower sunflower), and sugar beets ($n = 506$, several disjoint clusters in upper and lower left).

the feature space for each pixel are the three parameters of the fitted greenness growth model on a pixel-by-pixel basis. The trivariate ASH of these data is displayed in Figure 7.15. An evaluation of this model-based model compared to a statistical model has not yet been performed. Finally, in Figure 7.16 overlays the median trivariate contours of the three most common crops in the dataset, while Figure 7.17 shows seven crops. These views are more revealing for the discrimination task. Note how spring oats is nestled between spring wheat and barley.

PROBLEMS

7.1 Show that the Friedman–Tukey criterion (7.11) is exactly equal to $\hat{\sigma}_\alpha R(\hat{f}_h)$, where $\hat{f}_h$ is a kernel estimate with the $U(-0.5, 0.5)$ kernel.

7.2 If $Z \sim N(0,1)$, show that $E|Z + u| = 2u\Phi(y) + 2\phi(u) - u$.

7.3 Perform an informative components analysis on the *Iris* data. Examine how the choice of smoothing parameter affects your answer.

7.4 Repeat Problem 3 for the PRIM4 dataset found in the S system, both with and without marginal transformations used in Figure 7.5.

8

NONPARAMETRIC REGRESSION AND ADDITIVE MODELS

The most commonly used statistical technique, parametric or nonparametric, is regression. The basic assumption in nonparametric regression is the existence of a smooth function $r(\cdot)$ relating the response y and the predictor x:

$$y_i = r(x_i) + \epsilon_i, \quad \text{for } 1 \le i \le n, \quad \text{where} \quad \epsilon_i \sim g(\mu, \sigma_\epsilon^2(x)). \qquad (8.1)$$

Often g is assumed to be normal and $\sigma_\epsilon^2(x) = \sigma_\epsilon^2$, a constant.

In this chapter, the relationship of the underlying density estimator to the regression estimator will be explored. Many regression estimators discussed in this chapter are *linear smoothers*, that is, linear combinations of the observed responses. Issues of robustness are relevant, particularly in higher dimensions. Nonlinear nonparametric smoothers will be discussed. For high-dimensional regression problems, additive models of the form $r(y_i) = \sum_{j=1}^d r_j(x_{ij})$ have received much attention; (see Stone (1985), Wahba (1990), and Hastie and Tibshirani (1990)). An implementation based on average shifted histogram (ASH) estimates will be presented. Other promising techniques, including the little known modal regression method and nonparametric L_1 regression procedure, will be introduced. Several other books provide comprehensive treatments of various aspects of this broad field (see Eubank (1988), Müller (1988), Härdle (1990), and Simonoff (1996)).

Multivariate Density Estimation, First Edition. David W. Scott.
© 2015 John Wiley & Sons, Inc. Published 2015 by John Wiley & Sons, Inc.

8.1   NONPARAMETRIC KERNEL REGRESSION

8.1.1   The Nadaraya–Watson Estimator

There are two distinct cases of (8.1) that should be considered, depending on the probabilistic structure in the data $\{(x_i, y_i) : 1 \leq i \leq n\}$. The first case occurs when the $\{x_i\}$ data come from a *fixed design*; that is, the data x_i are not random but chosen by the experimenter. The second case is the *random design*, which occurs when the data come from a joint probability density function $f(x, y)$. The emphasis in this chapter will be on the latter case, in keeping with the focus of this book.

The theoretical regression function is defined to be

$$r(x) = \mathrm{E}(Y|X = x) = \int y f(y|x) dy = \frac{\int y f(x, y) dy}{\int f(x, y) dy}. \tag{8.2}$$

Consider the construction of a nonparametric regression estimator obtained by computing (8.2) using the bivariate product kernel estimator

$$\hat{f}(x, y) = \frac{1}{n h_x h_y} \sum_{i=1}^{n} K\left(\frac{x - x_i}{h_x}\right) K\left(\frac{y - y_i}{h_y}\right)$$

$$= \frac{1}{n} \sum_{i=1}^{n} K_{h_x}(x - x_i) K_{h_y}(y - y_i)$$

in place of the unknown bivariate density $f(x, y)$. The denominator in (8.2) contains the marginal density function, which here becomes

$$\int \hat{f}(x, y) \, dy = \frac{1}{n} \sum_{i=1}^{n} K_{h_x}(x - x_i) \int K_{h_y}(y - y_i) \, dy = \frac{1}{n} \sum_{i=1}^{n} K_{h_x}(x - x_i)$$

as $\int K_h(t) dt = 1$. The restriction for an order-2 kernel that $\int t K_h(t) dt = 0$ implies that $\int y K_h(y - y_i) dy = y_i$; hence, the numerator becomes

$$\int y \hat{f}(x, y) dy = \frac{1}{n} \sum_{i=1}^{n} y_i K_{h_x}(x - x_i).$$

Therefore, the nonparametric kernel regression estimator corresponding to (8.2) is

$$\hat{r}(x) = \frac{\frac{1}{n} \sum_{i=1}^{n} y_i K_{h_x}(x - x_i)}{\frac{1}{n} \sum_{i=1}^{n} K_{h_x}(x - x_i)} = \sum_{i=1}^{n} w_{h_x}(x, x_i) y_i, \tag{8.3}$$

where

$$w_{h_x}(x, x_i) = \frac{K_{h_x}(x - x_i)}{\sum_{j=1}^n K_{h_x}(x - x_j)}. \tag{8.4}$$

This estimator was proposed by Nadaraya (1964) and Watson (1964).

There are two observations to be made. First, the kernel regression estimator (8.3) is *linear in the observations* $\{y_i\}$ and is, therefore, a linear smoother. This feature is shared by many other nonparametric regression estimators. Second, the kernel regression estimate is independent of the particular choice of the smoothing parameter h_y. This feature is not totally unexpected, given the earlier result in Problem 6.16 that $\int x \hat{f}_h(x) dx = \bar{x}$ for any choice of h. However, certain deficiencies in the model such as the nonrobustness of kernel regression follow from this feature.

Finally, suppose that the data are not random but come from a fixed design $\{x_i = i/n, i = 1, \ldots, n\}$, and that the regression function is periodic on [0, 1]. If the regression estimate is to be computed at the design points, then the denominator in Equation (8.3) is constant, and the weight vector $w_h(x_j, x_i)$ need only be computed once. As evaluating the kernel repeatedly can entail much of the computational burden, the equally spaced fixed design is particularly advantageous. In fact, it is possible to mimic this efficient computation load even with a random design by "rounding" the design points $\{x_i\}$ to an equally spaced fixed mesh. This proposal is reminiscent of the ASH algorithm and has been investigated by Härdle and Scott (1988) (see Section 8.4.1).

8.1.2   Local Least-Squares Polynomial Estimators

The use of higher order polynomials in *parametric regression* to approximate a larger class of possible regression curves deserves some consideration. Here, the degree of the polynomial plays a role corresponding to the smoothing parameter in nonparametric regression. However, parametric polynomial fits can exhibit rapid oscillations as the order of the polynomial grows. Thus the use of higher and higher order polynomials for "nonparametric" regression does not seem practical. If the true regression curve is smooth, then a low-order polynomial fit should be adequate *locally* (see Stone (1977)). This remark is nothing but a restatement of Taylor's theorem. Two well-known nonparametric estimators emerge from this line of thinking.

8.1.2.1   *Local Constant Fitting*   Globally, the *best constant regression fit* (polynomial of degree 0) to a scatter diagram is $\hat{r}(x) = \bar{y}$, which is the estimate that minimizes the least-squares criterion:

$$\bar{y} = \arg\min_a \sum_{i=1}^n (a - y_i)^2; \tag{8.5}$$

here the $\arg\min_a$ notation indicates that the constant $a = \bar{y}$ is the choice (argument) that minimizes the criterion. Consider a *local constant fit* at x to the data. Here, "local" may mean including only those data (x_i, y_i) for which $x_i \in (x - h, x + h)$ in the sum in (8.5), or it may mean including only the k design points nearest to x. It is convenient to introduce the kernel function $K_h(x - x_i)$ to indicate precisely which terms are included, and the weights (if any) on those terms. Thus the best local constant fit is

$$\hat{r}(x) = \arg\min_{a} \sum_{i=1}^{n} \left[K_h(x - x_i) \times (a - y_i)^2 \right]. \tag{8.6}$$

Minimizing the right-hand side of (8.6) with respect to a leads to

$$\hat{r}(x) = \frac{\sum_{i=1}^{n} y_i K_h(x - x_i)}{\sum_{j=1}^{n} K_h(x - x_j)}, \tag{8.7}$$

which is precisely the Nadaraya–Watson kernel estimator (8.3). This pointwise result can be extended to the entire regression function by noting that

$$\hat{r}(\cdot) = \arg\min_{a(\cdot)} \int \sum_{i=1}^{n} K_h(x - x_i) \times [a(x) - y_i]^2 \, dx,$$

as the integrand is minimized by $\hat{a}(x) = \hat{r}(x)$ in (8.7) for each x.

8.1.2.2   Local Polynomial Fitting

The local constant fits in the preceding section are zero-order polynomials. The use of local linear or quadratic polynomial fits is intuitively appealing. The use of higher order polynomials locally results in a different order of bias, as with higher order kernels. This result has been shown for the fixed design by Härdle (1990) and for the random design by Fan (1992). The local fitting of linear polynomials has been especially advocated by Cleveland (1979) with his LOWESS procedure distributed in the S language (Becker et al., 1988). Cleveland shows that by careful algorithm organization, the work in fitting the many local polynomials may be minimized by adding and deleting design points one at a time to the sum of squares. Wang (1990) has considered local polynomial fitting but with an absolute error criterion applied to the residuals rather than a squared-error criterion.

8.1.3   Pointwise Mean Squared Error

The mean squared error (MSE) properties of the Nadaraya–Watson estimator are rather complicated, because the estimator is the ratio of two correlated random variables. The bias and variance of the Nadaraya–Watson estimator in Equation (8.3) may be obtained by using approximations to the standard errors of functions of random variables (Stuart and Ord, 1987, Section 10.5).

   If the numerator and denominator in (8.3) each converge to a (positive) constant, then the asymptotic expectation of the ratio is the ratio of the asymptotic expectations of the numerator and denominator, to first order. The properties of the kernel estimator in the denominator have been well-studied (see Equations (6.16) and (6.17)), with the result that

$$\mathrm{E}\hat{f}(x) \approx f(x) + h^2\sigma_K^2 f''(x)/2 \quad \text{and} \quad \mathrm{Var}\,\hat{f}(x) \approx \frac{R(K)f(x)}{nh}. \tag{8.8}$$

Next, consider the expectation of the numerator:

$$\mathrm{E}\left\{\frac{1}{n}\sum_{i=1}^{n} y_i K_h(x - x_i)\right\} = \int\int v\frac{1}{h}K\left(\frac{x-u}{h}\right)f_i(u,v)du\,dv$$

$$= \int\int vK(s)f(x - hs, v)ds\,dv, \tag{8.9}$$

after the change of variable $s = (x - u)/h$. Now the conditional density satisfies $f(v|x - hs)f(x - hs) = f(x - hs, v)$. Therefore, the integral over v in (8.9) equals (ignoring $K(s)$)

$$f(x - hs)\int vf(v|x - hs)\,dv = f(x - hs)\,r(x - hs),$$

as the integral is the conditional mean and $r(\cdot)$ is the true regression function defined in (8.1). Continuing with (8.9), we have

$$= \int K(s)f(x - hs)\,r(x - hs)\,ds$$

$$= f(x)r(x) + h^2\sigma_K^2\left[f'(x)r'(x) + f''(x)r(x)/2 + f(x)r''(x)/2\right] \tag{8.10}$$

after expanding $f(x - hs)$ and $r(x - hs)$ in Taylor's series up to order h^2, assuming that K is a second-order kernel. Thus the expectation of the Nadaraya–Watson estimator with a random design is the ratio of the expectations in (8.10) and (8.8), with the density f factored out of each:

$$\mathrm{E}\hat{r}(x) \approx \frac{f(x)\cdot\left[r(x) + h^2\sigma_K^2\{f'r'/f + f''r/(2f) + r''/2\}\right]}{f(x)\cdot\left[1 + h^2\sigma_K^2 f''/(2f)\right]}$$

$$\approx r(x) + \frac{1}{2}h^2\sigma_K^2\left\{r''(x) + 2r'(x)\frac{f'(x)}{f(x)}\right\} \tag{8.11}$$

using any symbolic manipulation program, or the approximation $(1 + h^2c)^{-1} \approx (1 - h^2c)$ for $h \approx 0$ in the factor in the denominator and multiplying through.

The bias in the fixed design case is $h^2\sigma_K^2 r''(x)/2$, which appears in Equation (8.11), (see Problem 8.1). The term $2r'(x)f'(x)/f(x)$ in (8.11) is small if there are many data points in the window (i.e., $f(x)$ large). If more design points are in the interval $(x, x+h)$ than in $(x - h, x)$ (i.e., $f'(x) > 0$), then the local average will be biased as the average will include more responses over $(x, x+h)$ than over $(x - h, x)$. The bias will be positive if $r'(x) > 0$ and negative otherwise. A similar interpretation exists when $f'(x) < 0$. The interaction of the various possibilities for the signs of $f'(x)$ and $r'(x)$

is summarized in that term. Although the expression for the bias with a fixed design matches the random design when $f'(x) = 0$, observe that these two settings are not identical. The random design has zero probability of being equally spaced even when $f = U(0,1)$.

The variance of $\hat{r}(x)$ may be computed using the following approximation for the variance of the ratio of two random variables (Stuart and Ord, 1987):

$$\text{var}\left[\frac{U}{V}\right] = \left[\frac{\text{E}\,U}{\text{E}\,V}\right]^2 \left[\frac{\text{Var}\,U}{(\text{E}\,U)^2} + \frac{\text{Var}\,V}{(\text{E}\,V)^2} - \frac{2\,\text{Cov}(U,V)}{(\text{E}\,U)(\text{E}\,V)}\right]. \qquad (8.12)$$

Calculations similar to those following Equation (8.9) show that

$$\text{Var}\left\{\frac{1}{n}\sum_{i=1}^{n} K_h(x-x_i)\,y_i\right\} = \frac{1}{n}\text{E}\{K_h(x-x_i)\,y_i\}^2 - O(n^{-1})$$

$$\approx \frac{R(K)f(x)}{nh}\left[\sigma_\epsilon^2 + r(x)^2\right], \qquad (8.13)$$

using the facts that $\int v^2 f(v|x-hs) = [\sigma_\epsilon^2(x-hs) + r(x-hs)^2]$ and $\sigma_\epsilon^2(x) = \sigma_\epsilon^2$ for all x. The variance of $\hat{f}(x)$ is given in Equation (8.8). Finally,

$$\text{Cov}\left\{\frac{1}{n}\sum_{i=1}^{n} K_h(x-x_i)\,y_i, \frac{1}{n}\sum_{i=1}^{n} K_h(x-x_i)\right\}$$

$$= \frac{1}{n}\text{E}\{K_h(x-x_i)^2\,y_i\} - O(n^{-1}) \approx \frac{R(K)f(x)\,r(x)}{nh}. \qquad (8.14)$$

Carefully substituting Equations (8.8), (8.10), (8.13), and (8.14) into (8.12) yields

$$\text{Var}\,\{\hat{r}(x)\} \approx \frac{R(K)\sigma_\epsilon^2}{nhf(x)}.$$

In addition to the familiar factor $R(K)/(nh)$, the variance of $\hat{r}(x)$ includes factors relating to the noise variance σ_ϵ^2 and the (relative) amount of data through $f(x)$.

These results may be collected into the following theorem (Rosenblatt, 1969). A multivariate version is given by Mack and Müller (1989).

Theorem 8.1:   *The asymptotic MSE(x) of the Nadaraya–Watson estimator is*

$$\text{AMSE}\{\hat{r}(x)\} = \frac{R(K)\sigma_\epsilon^2}{nhf(x)} + \frac{1}{4}h^4\sigma_K^4\left\{r''(x) + 2\,r'(x)\frac{f'(x)}{f(x)}\right\}^2. \qquad (8.15)$$

The fact that the Nadaraya–Watson estimator follows directly from the product kernel estimator in its functional form deserves closer examination. For the bivariate density case, $h_x^* = O(n^{-1/6})$; however, it happens that $h_x^* = O(n^{-1/5})$ for this particular estimator, which is the rate for univariate density estimation. Thus, the extra smoothing provided by integration of the bivariate density function obviates the need for a wider smoothing parameter.

Several authors have considered weights other than those of Nadaraya–Watson and obtained significantly simpler bias expressions. The scheme of integral or convolution weights devised by Gasser and Müller (1979) has certain advantages (Gasser and Engel, 1990). In addition to a bias expression similar to the fixed design case, the shape of the bias is favorable when $r(x)$ is linear, but the variance is increased. Fan (1992) has shown that the local polynomial fitting technique simultaneously achieves good bias and variance properties even with the random design.

8.1.4 Bandwidth Selection

The selection of smoothing parameters by cross-validation in nonparametric regression is easier than in density estimation, if only because the regression curve goes *through the data point cloud*. Two general classes of algorithms have emerged based on simple modifications to the naive average predictive squared error, which is defined by

$$G(h) = \frac{1}{n} \sum_{i=1}^{n} \left[y_i - \hat{r}(x_i) \right]^2.$$

Letting $\hat{r}_{-i}(\cdot)$ denote the Nadaraya–Watson estimator of the $n - 1$ points with the point (x_i, y_i) omitted, Allen (1974), Stone (1974), Clark (1975), and Wahba and Wold (1975) proposed choosing

$$\hat{h}_{\text{CV}} = \arg\min_h \text{CV}(h) \quad \text{where} \quad \text{CV}(h) = \frac{1}{n} \sum_{i=1}^{n} \left[y_i - \hat{r}_{-i}(x_i) \right]^2.$$

Leaving out the ith data point eliminates the degenerate solution $h = 0$, which corresponds to interpolation of the data points, and the estimate $\text{CV}(h) = 0$, a severe underestimate of the true predictive error. Alternatively, underestimation of the predictive error can be avoided by multiplication of a simple factor:

$$\hat{h} = \arg\min_h \left(1 + \frac{2K(0)}{nh} \right) \times G(h). \tag{8.16}$$

It may be shown that this choice provides a consistent estimator of the predictive error. This particular formula in (8.16) is due to Shibata (1981). Rice (1984b) uses the factor $[1 - 2K(0)/(nh)]^{-1}$. Härdle et al. (1988) show how five cross-validation algorithms all have the same Taylor's series as criterion (8.16).

8.1.5 Adaptive Smoothing

A quick review of Theorem 8.1 leads to the conclusion that construction of asymptotically optimal adaptive regression curves is difficult, certainly more difficult than the in density estimation case. The use of the Gasser–Müller weights simplifies matters somewhat since the term involving r' does not appear in the asymptotic mean squared error (AMSE) expansion. Müller (1988) describes several algorithms for adaptive estimation, as do Staniswalis (1989) and Schucany (1995).

One of the more innovative adaptive smoothers is Friedman's (1984) *super-smoother*. The algorithm constructs oversmoothed, average, and undersmoothed estimates, referred to as the "woofer," "midrange," and "tweeter," respectively. With nine passes through the data and the use of local cross-validation, an adaptive estimate is pieced together.

8.2   GENERAL LINEAR NONPARAMETRIC ESTIMATION

The kernel regression estimate is a linear combination of the observed responses of the form $\mathbf{w}^T\mathbf{y}$, which is a *linear smoother* by definition. In vector form, the bivariate regression data $\{(x_i, y_i)\}$ will be denoted by

$$\mathbf{x} = (x_1, \ldots, x_n)^T \quad \text{and} \quad \mathbf{y} = (y_1, \ldots, y_n)^T.$$

The weights for other linear estimators are derived in the next three sections.

8.2.1   Local Polynomial Regression

The study of local polynomial regression is only slightly more complicated than the global case. Therefore, the global case is considered first.

Any least-squares polynomial regression is easily shown to be "linear" in the sense described above, even when the polynomial terms are quadratic or cubic. Consider the case of straight-line regression with an intercept term, which may be incorporated into the model by appending a column of one's to the design points:

$$X = \begin{bmatrix} 1 & 1 & \cdots & 1 \\ x_1 & x_2 & \cdots & x_n \end{bmatrix}^T \quad \text{and} \quad \beta = (\beta_0, \beta_1)^T.$$

Then the regression problem may be written as

$$\mathbf{y} = X\beta + \epsilon \quad \Rightarrow \quad \hat{\beta} = (X^TX)^{-1}X^T\mathbf{y}, \tag{8.17}$$

which is the well-known least-squares estimate of β. Therefore, the best predictor of y at x is

$$\hat{y}(x) = \hat{r}(x) = \mathbf{w}_x^T\mathbf{y}, \quad \text{where} \quad \mathbf{w}_x^T\mathbf{y} = (1\ x)(X^TX)^{-1}X^T\mathbf{y}. \tag{8.18}$$

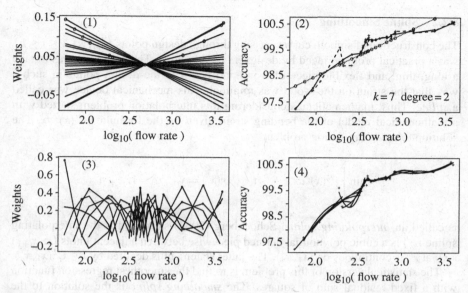

FIGURE 8.1 Linear smoothers applied to the gas flow dataset at 74.6 psia. In the first frame, the 33 weight vectors are plotted for a parametric straight-line fit, with the circle showing the weight on the data point itself. In the second frame, the raw data ("+") are shown connected by small dashes; the 33 estimates at $x = x_i$ from frame 1 are shown (the "o" symbols on the straight line connected by big dashes); the 33 estimates "o" connected by a solid line using a local quadratic model with $k = 7$ (three points on either side of x_i except at the boundaries). In the third frame, the 33 weight vectors for the local quadratic fits are shown with the circle showing the weight on x_i itself. The fourth frame shows the 33 local quadratic fits superimposed on the data.

Not only is the predictor a linear combination of the responses $\{y_i\}$, but the weights themselves fall on a straight line when plotted in the (x, y) plane. This can be observed empirically in Figure 8.1, but follows from Equation (8.18) by examining the way x enters into the definition of the weight vector $\mathbf{w}_x$ (see Problem 8.3).

This demonstration is really quite general, as the design matrix X can represent polynomial regression of any order, with multiple predictor variables and with arbitrary powers of those variables. Thus the general linear model in the parametric setting is a linear estimator according to the nonparametric definition as well.

For *local polynomial nonparametric regression* algorithms, such as LOWESS, the demonstration above is easily extended. Rather than a single linear model of the form (8.17) that holds for all x, a local linear model is formed *for each point x*. Only design points x_i "close" to x are included in the local linear model. The number of points included in the local linear fit plays the role of the smoothing parameter (although the degree of the polynomial also plays a role). The number of points included in the local fit can be designated directly as k, or indirectly as a fraction f of the sample size. For example, LOWESS chooses the $f \cdot n$ points closest to x; that is, $k = fn$.

8.2.2 Spline Smoothing

The construction of smooth curves passing through design points $\{(x_i, y_i), 1 \leq i \leq n\}$ was a practical problem faced by designers of ship hulls. Their solution was to take a long, thin, and flexible piece of wood and peg it to the design points in such a way that the strain on the wood was minimal. This mechanical device was called a *spline*. Thus, mathematicians considering this interpolation problem started with a mathematical model of the bending strain given by the formula $\int r''(x)^2 dx$. The solution to the interpolation problem

$$\min_r \int_{x_1}^{x_n} r''(x)^2 dx \quad \text{s/t} \quad r(x_i) = y_i, \quad i = 1, \dots, n$$

is called an *interpolating spline*. Schoenberg (1964) showed that the interpolating spline $r(\cdot)$ is a cubic polynomial defined piecewise between adjacent knots (x_i, x_{i+1}) so that r'' is continuous on (x_1, x_n). The latter statement is denoted by $r \in C^2(x_1, x_n)$.

The statistical version of this problem is to find the smoothest regression function with a fixed residual sum of squares. The *smoothing spline* is the solution to the variational problem

$$\hat{r}_\lambda(\cdot) = \arg\min_r S_\lambda(r) \equiv \sum_{i=1}^n \left[y_i - r(x_i)\right]^2 + \lambda \int_{x_1}^{x_n} r''(x)^2 \, dx. \tag{8.19}$$

Reinsch (1967) demonstrated that the solution to (8.19) is also a cubic spline. The $\lim\{\hat{r}_\lambda\}$ as $\lambda \to 0$ is the interpolating spline, while the $\lim\{\hat{r}_\lambda\}$ as $\lambda \to \infty$ is the least-squares linear fit to the data (observing that the bending strain of a line is 0). The parameter λ plays the role of the smoothing parameter. Given λ, the amount of computation required to find $\hat{r}_\lambda$ is $O(n)$. Packages such as IMSL (1991) provide software for this purpose.

Several properties of the smoothing spline may be derived without an explicit solution. For any function $q \in C^2(x_1, x_n)$ and constant α,

$$S_\lambda(\hat{r} + \alpha q) \geq S_\lambda(\hat{r}) \qquad \forall q \in C^2(x_1, x_n).$$

Therefore, $T(\alpha) \equiv S_\lambda(\hat{r} + \alpha q)$ has a local minimum at $\alpha = 0$ for all q. By the Euler–Lagrange equation, the derivative of T should vanish there. Now

$$T'(\alpha) = \frac{\partial}{\partial \alpha} \left\{ \sum_{i=1}^n \left[y_i - (\hat{r} + \alpha q)(x_i)\right]^2 + \lambda \int_{x_1}^{x_n} \left[(\hat{r} + \alpha q)''(x)\right]^2 dx \right\}$$

$$= \frac{\partial}{\partial \alpha} \left\{ \sum_{i=1}^n \left[y_i - \hat{r}(x_i) - \alpha q(x_i)\right]^2 + \lambda \int_{x_1}^{x_n} \left[\hat{r}''^2 + 2\alpha \hat{r}'' q'' + \alpha^2 q''^2\right] dx \right\}.$$

Performing the differentiation, the equation $T'(\alpha) = 0$ when $\alpha = 0$ becomes

$$-2\sum_{i=1}^{n}\left[y_i - \hat{r}(x_i)\right]q(x_i) + 2\lambda\int_{x_1}^{x_n}\hat{r}''(x)\,q''(x)dx = 0. \qquad (8.20)$$

It is now a short argument to show that $\hat{r}_\lambda$ is a linear smoother. Suppose that there are two response vectors, $\mathbf{y}^{(1)}$ and $\mathbf{y}^{(2)}$, each with the same design vector $\mathbf{x}$. Let the respective smoothing splines be denoted by $\hat{r}_\lambda^{(1)}$ and $\hat{r}_\lambda^{(2)}$. Then, if the smoothing spline is a linear smoother, the smoothing spline for the sum of the two data response vectors should be the sum of the individual smoothing splines; that is,

$$(\mathbf{x}, \mathbf{y}^{(1)} + \mathbf{y}^{(2)}) \quad \Rightarrow \quad \hat{r}^{(1+2)} = \hat{r}^{(1)} + \hat{r}^{(2)}. \qquad (8.21)$$

As $\{y_i^{(1)}, y_i^{(2)}, r_\lambda^{(1)}, r_\lambda^{(2)}\}$ enter linearly into Equation (8.20), the Euler–Lagrange equation holds for the proposed solution in Equation (8.21) by additivity (see Problem 8.2).

As a consequence of the linear smoothing property of the smoothing spline, the solution for the original dataset $(\mathbf{x}, \mathbf{y})$ can be thought of as the sum of the solutions to n individual problems, as

$$\mathbf{y} = \sum_{i=1}^{n} y_i \mathbf{e}_i, \quad \text{where} \quad \mathbf{e}_i = (0, \ldots, 0, 1, 0, \ldots, 0)^T.$$

If $\hat{r}_\lambda^{(i)}$ is the smoothing spline for the data $(\mathbf{x}, \mathbf{e}_i)$, then the spline solution for the original data $(\mathbf{x}, \mathbf{y})$ is given by

$$\hat{r}_\lambda = \sum_{i=1}^{n} y_i\,\hat{r}_\lambda^{(i)}.$$

If the design is periodic (also called circular) and equally spaced, then all of the n component smoothing splines $\hat{r}_\lambda^{(i)}$ are identical in shape. Silverman (1984) demonstrated that the shape converged to the kernel

$$K_s(t) = \frac{1}{2}e^{-|t|/\sqrt{2}}\sin\left(\frac{\pi}{4} + \frac{|t|}{\sqrt{2}}\right) \quad |t| < \infty. \qquad (8.22)$$

$K_s(t)$ may be shown to be an order-4 kernel. For data from a random design, Silverman proved that the smoothing spline is an adaptive rather than fixed bandwidth estimator, with weights

$$w_\lambda(x, x_i) \approx \frac{1}{f(x)h(x_i)}K_s\left(\frac{x - x_i}{h(x_i)}\right), \quad \text{where} \quad h(x_i) \approx \left(\frac{\lambda}{nf(x)}\right)^{1/4}.$$

A fourth-order kernel should have bandwidth $h(x_i) = O(n^{-1/9})$, which occurs if $\lambda = O(n^{5/9})$ is chosen. Unfortunately, the spline smoother adapts only to the design density $f(x)$ and not to the curvature of the unknown curve $r(x)$ as required by Theorem 8.1. While the estimator may not provide optimal adaptive estimation, it provides visually appealing estimates in practice. Reinsch's (1967) original proposal included a weight function in (8.19), which could be designed to provide improved adaptive properties.

8.2.3 Equivalent Kernels

A wide range of nonparametric regression algorithms, including LOWESS and smoothing splines, have been shown to be linear smoothers. To compare these estimates to the original kernel estimator in (8.3), it is natural to compare the weight vectors $\mathbf{w}_x$ multiplying the response vector $\mathbf{y}$. Following the definition of the kernel weights in (8.4), the weight vectors for other linear smoothers will be called the *equivalent kernel weights* of that estimator. For a particular dataset, these weights may be computed exactly for theoretical purposes. In practice, it is not necessary to obtain the weights explicitly to compute the estimate.

Consider the gas flow accuracy dataset of $n = 33$ readings taken when the pressure was set at 74.6 psia in the gas line. The accuracy of the meter is plotted as a function of the logarithm of the rate of the gas flow rate in the pipeline. It is clear from the upper right panel in Figure 8.1 that the relationship is not linear. The first panel shows the collection of equivalent kernel weights for the parametric straight-line fit, and the estimate is shown in the upper right frame. For example, consider the leftmost data point (1.76, 97.25). The 33 weights on the 33 responses used to compute the regression estimate were calculated according to (8.18). Those weights fall on a straight line plotted in the upper left frame of Figure 8.1, connecting the two extreme points $(x, w) = (1.76, 0.145)$ and $(3.59, -0.081)$. To identify the 33 lines, the weight (x_i, w_i) is shown as a small circle on the line representing the ith weight vector. Notice that the weights are not local (i.e., the weights do not vanish outside a neighborhood of x_i).

A local quadratic polynomial fit was also applied to these data. The fit was based on $k = 7$ design points, $\{x_{i-3}, \ldots, x_i, \ldots, x_{i+3}\}$. At the boundaries, the seven points closest to the boundary are used for the fit. The 33 local quadratic fits are shown in the bottom right frame of Figure 8.1, and the 33 equivalent weight vectors are shown in the bottom left frame. The small bandwidth allows the estimate to follow the small dip in the middle of the curve. The equivalent weight vectors are 0 where they are not explicitly shown. By construction, the weights are local. While the equivalent kernels are fairly rough and take on negative values, a portion of the roughness results from their ability to correctly adapt at the boundaries (Fan, 1992). Compare the boundary kernels here to those shown in Figure 6.10.

The Nadaraya–Watson kernel weights are shown in Figure 8.2 for the biweight kernel with a bandwidth of $h = 0.25$, which provides a smoother estimator than the local polynomial fit. However, a boundary kernel was not employed, as is evident in the estimate at the left boundary and in the picture of the kernel weights.

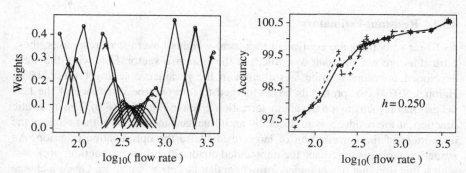

FIGURE 8.2 Biweight Nadaraya–Watson kernel weights and estimate for the gas flow dataset.

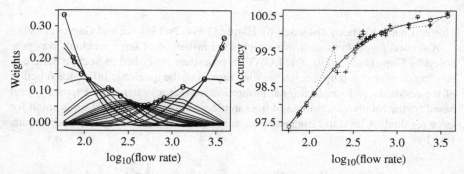

FIGURE 8.3 Smoothing spline equivalent kernel weights and estimate for gas flow dataset.

The smoothing spline fit is shown in Figure 8.3. Observe that the kernel weights are not as local for this small sample because the estimate is much smoother (less local). The boundary behavior is reasonable. For a discussion of the dual of equivalent bandwidths for linear smoothers, see Hastie and Tibshirani (1990).

8.3 ROBUSTNESS

In recent years, more effort was required for data entry and verification than for a complete regression and analysis of variance. While changes in computing have been dramatic, the problems of handling bad or influential data points have grown even faster than the growth of the size of datasets. In such cases it is increasingly difficult to identify hundreds of influential points, particularly in higher dimensions. Thus the practical importance of robust procedures that "automatically" handle outliers is ever growing (Rousseeuw and Leroy, 1987).

8.3.1 Resistant Estimators

As linear smoothers are equivalent to a local weighted average of responses, these estimators are not resistant to outliers in the response vector. Following the well-developed literature in robust estimation in the parametric setting (Huber, 1964; Hampel, 1974) two proposals have emerged to improve the resistance of the kernel estimator, focusing on the characterization given in Equation (8.6). The quadratic function of the residuals is replaced by an influence function, $\rho(\cdot)$, that reduces the magnitude of the contribution of large residuals to the optimization criterion. An attractive choice is to replace the unbounded quadratic influence function, $\rho(\epsilon) = \epsilon^2$ with a function that has bounded influence; that is, $|\rho(\epsilon)| < c < \infty$. Given a choice for the robust influence function, the original problem is replaced by

$$\hat{r}_\rho(\cdot) = \arg\min_{a(\cdot)} \int \sum_{i=1}^n K_h(x - x_i)\, \rho\big(a(x) - y_i\big)\, dx.$$

This estimator has been discussed by Härdle (1984) and Härdle and Gasser (1984).

A second algorithm that down-weights the influence of large residuals was proposed by Cleveland (1979). The LOWESS procedure described in Section 8.1.2 can be extended for this purpose. Rather than replacing the quadratic influence function of the residuals $\{\epsilon_i\}$, an additional multiplicative factor is introduced term by term based on the relative magnitude of the sample residuals $\{\hat{\epsilon}_i\}$. The factor is small for large residuals. Cleveland suggests computing a robust scale estimate $\hat{\sigma}$ and weights $\{\delta_i\}$ by

$$\hat{\sigma} = \text{median}\{|\hat{\epsilon}_i|\} \quad \Rightarrow \quad \delta_i = \left[1 - \big(\hat{\epsilon}_i/(6\hat{\sigma})\big)^2\right]_+^2.$$

For example, if fitting a local linear polynomial, the fitted polynomial solves

$$(\hat{a}, \hat{b}) = \arg\min_{(a,b)} \sum_{i=1}^n \delta_i \times w_h(x, x_i)\, \{y_i - (a + bx)\}^2.$$

The LOWESS estimate is $\hat{a} + \hat{b}x$. The procedure is applied recursively until the estimates remain unchanged. The computational effort is an order of magnitude greater than the nonrobust version.

8.3.2 Modal Regression

It is ironic that while the original kernel density estimator is quite resistant to outliers, the regression function derived from the kernel estimate is not. Is there an alternative regression estimator derived from the kernel estimator that is resistant? The answer comes from a moment's reflection upon the three measures of center discussed in elementary textbooks: mean, median, and mode. The obvious proposal is to go beyond the conditional mean of the kernel estimator and to consider either its conditional median or its conditional mode. Of course, if the errors are normal or symmetric, all three choices give the same result, asymptotically. A strong argument can be made for summarizing the bivariate data with a trace of the conditional mode. This trace will be called the *modal regression curve* or *trace*. The formal definition is given by

$$\text{Modal regression curve}: \quad \hat{r}(x) = \arg(s) \max_{y} \hat{f}(y|x),$$

where the plural on "arg" indicates *all local maxima*. An equivalent definition is

$$\hat{r}(x) = \arg(s) \max_{y} \hat{f}(x,y).$$

Some examples reveal the practical advantages of the conditional mode compared to the conditional mean estimators, both robust and nonrobust. Consider the Old Faithful eruption prediction dataset displayed in Figure 8.4 together with the LOWESS smoother. Because of the (conditional) multimodal behavior for $x > 3$, the resistant kernel estimate LOWESS is actually rougher than a nonresistant LOWESS estimate. But as a scatterplot smoother and data summarizer, any linear smoother is inadequate to summarize the structure. The right frame of the figure shows the modal regression trace, as well as the bivariate ASH from which it was derived. As a measure of center, the mode summarizes the "most likely" conditional values rather than the conditional average. On the other hand, when the conditional density is symmetric, these two criteria match. In this example, they do not.

Observe that the modal trace is much smoother than the usual regression. This smoothness may seem counterintuitive at first. Compare the two regression estimates

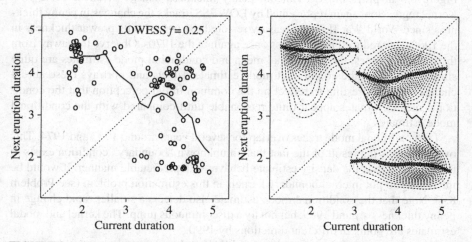

FIGURE 8.4 Conditional mean (LOWESS) and conditional mode smoothers of the lagged Old Faithful duration dataset. The conditional mode is displayed with the symbol "o" when above the 25% contour and with an "x" between 5 and 25%. The 45° line is also shown.

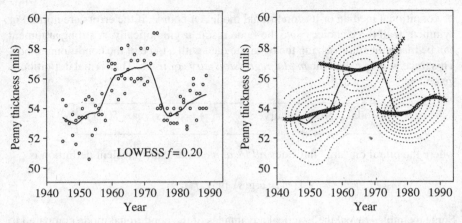

FIGURE 8.5 Conditional mean and conditional mode smoothers of the US penny thickness dataset.

in the interval (1.5, 2.5), in which the conditional densities are unimodal but not symmetric. The LOWESS estimate faithfully follows the conditional mean, which dips down. In the interval (3.5, 5), the conditional densities are bimodal. The bifurcation of the modal regression trace indicates that a single prediction such as the conditional mean misses the structure in the data. The conditional mode is nearly half a minute longer if the previous eruption was short. Observe that the ordinary regression estimate gives predictions in that interval where there is little probability mass, around $y = 3$ min.

A second example provides further evidence for the usefulness of this approach. Figure 8.5 displays the US Lincoln penny thickness dataset given in Table B.4. Kernel regression, again represented by LOWESS, tracks the changes in penny thickness since World War II. The Treasury restored the penny to its prewar thickness in the 1960s, but reduced its thickness once again in the 1970s. Observe that away from the transition times, the conditional mean and conditional mode estimates are quite similar. However, the conditional mode estimate detects and portrays these abrupt changes. It is interesting to reflect on the common initial perception that the conditional mean estimates appear quite reasonable until compared with the conditional mode estimates.

The conditional mode traces overlap for several years around 1957 and 1974. This overlapping is the result of the fact that an unknown "boundary" condition exists in the data, and thus the density estimate behaves in a predictable manner. It would be interesting to put internal boundary kernels in this estimation problem (see Problem 8.4). Note that the conditional mode estimate also detects a smaller rapid change in penny thickness around 1985, but not by a discontinuous jump. The kernel and modal estimates are going in different directions by 1990.

Thus the conditional mode estimate, while introduced as an alternative regression algorithm that is resistant to outliers, has potential beyond that purpose. The problem encountered in the penny dataset is an example of the "change point" problem where there is a jump in the regression curve (see Siegmund (1988)). Displaying the conditional modal trace introduces some interesting design choices. The curves above were computed slice by slice. The estimates are shown with an "x" symbol if the density is less than 25% of the maximum bivariate density value. Other choices could be investigated. In particular, when the data contain many outliers, a consequence is the appearance of an additional modal trace at each of those outliers. Outlier traces will be very short, composed of "x" symbols, and will serve to identify the outliers. The "central" modal traces will be totally unaffected by such outliers; hence, the modal trace is resistant to outliers. Finally, the actual computation of the conditional modes was performed by computing a linear blend of the bivariate ASH estimate using the biweight kernel. Tarter et al. (1990) have also produced software to compute conditional modes. A related idea is discussed in Sager and Thisted (1982).

8.3.3 L_1 Regression

In the general parametric linear model, the L_1 criterion on the residuals leads to a particularly resistant estimate. Consider data lying on a straight line except for one outlier. As the least-squares fit goes through $(\bar{x}, \bar{y})$, it is clear that the fit will be affected by the outlier. On the other hand, the L_1 fit will go through the $n-1$ points on the line, ignoring the outlier. The reason is simple—any movement of the line incurs a cost for each of the $n-1$ points and a gain for only the outlying point. This argument makes it clear that the L_1 fit will be resistant to multiple outliers as well.

Wang (1990) investigated the theoretical and practical aspects of the L_1 criterion in nonparametric regression. The idea is essentially equivalent to the criterion in (8.6), with the squared loss replaced by absolute loss. It is well-known that the solution to this problem is the solution of an ordinary linear programming problem (Wagner, 1959). If the ith residual is $\epsilon_i = y_i - x_{(i)}\beta$ and ϵ_i^+ and ϵ_i^- are defined to be the absolute values of the positive and negative parts of ϵ_i, respectively, then

$$\epsilon_i = \epsilon_i^+ - \epsilon_i^- \quad \text{and} \quad |\epsilon_i| = \epsilon_i^+ + \epsilon_i^- .$$

Thus the problem of minimizing the sum of the absolute residuals becomes

$$\min_{\epsilon^+, \epsilon^-} \sum_{i=1}^{n} (\epsilon_i^+ + \epsilon_i^-)$$

$$\text{s/t} \quad \mathbf{y} = \mathbf{X}\beta + \epsilon^+ - \epsilon^- \quad \text{and} \quad \epsilon^+, \epsilon^- \geq 0.$$

Modern linear programming (LP) packages can handle tens of thousands of points in reasonable time. To make the procedure nonparametric, the L_1 model is applied locally, with kernel weights on the absolute criterion values. The modified problem remains an LP problem. Wang discusses efficient methods for updating the series of linear programs required to define the L_1 nonparametric regression curve.

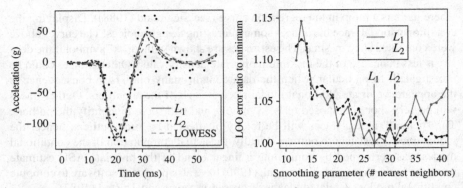

FIGURE 8.6 (Left) Local L_1 and L_2 quadratic fits to the motorcycle data, together with a LOWESS fit with $f = 0.25$. (Right) normalized leave-one-out (LOO) cross-validation criteria, namely, the mean absolute error and the standard deviation for the L_1 and L_2 fits, respectively. The raw values range from $(16.2, 18.6)$ and $(18.5, 20.8)$, respectively. The best L_1 fit occurred with 27 points in each local neighborhood.

As an example, local quadratic fits were computed with the motorcycle accident data using both L_1 and L_2 criteria (see Figure 8.6). A uniform kernel was applied over an interval containing the k nearest neighbors. The leave-one-out cross-validation curves are shown in the right frame of Figure 8.6, with one important modification. Since a robust fit will produce large residuals, using squared loss in the LOO cross-validation function will not be robust itself. Instead, Wang and Scott (1994) suggest using the mean of the absolute residuals for the leave-one-out-cross-validation (LOO-CV) function. This gave a data-based choice $\hat{k} = 27$. The local least-squares estimate was calibrated at $\hat{k} = 31$ using the usual LOO-CV function; however, the risk curve is not rapidly increasing for $k > 31$. A LOWESS curve is included for reference, although a smaller smoothing parameter might be appropriate.

The special feature that makes this proposal interesting is its immediate extension to multiple dimensions. Identifying influential points becomes problematic beyond three dimensions (two predictors) and this approach seems quite promising as no interaction (or iteration) is required to eliminate the effects of outliers. An example depicted in Figure 8.7 with a smooth bivariate surface on a 61×61 mesh contaminated by 2.5% large-variance $N(0, 2.5^2)$ noise or 97.5% $N(0, 0.1^2)$ noise serves to illustrate the power of the approach. The true surface was a portion of a standard bivariate normal density over $[-3, 3]^2$ multiplied by 2π; hence, the surface ranges from 0 to 1. The local polynomial model was quadratic and a uniform kernel was introduced in the criterion. The noise is so large that the vertical scale of the raw data surface is truncated at $(-0.25, 1.25)$, the full vertical range being $(-5.17, 6.58)$. The robust LOO-CV criterion picked a 21×17 rectangular neighborhood. The L_1 fit is unaffected by the clusters of outliers. The local least-squares is noisier even in this data-rich example. If a local linear model is employed (not shown), there will

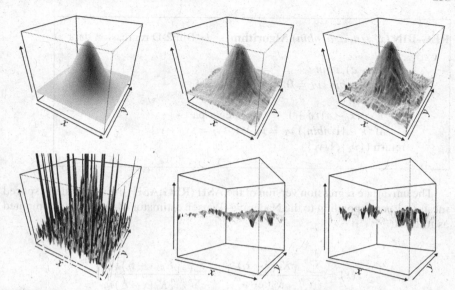

FIGURE 8.7 (Top row) Regression surface that is a $2\pi \times$ bivariate normal on $[-3,3] \times [-3,3]$ on a 61×61 mesh; L_1 local full quadratic nonparametric estimate with $\hat{m}_x = 10$ and $\hat{m}_y = 8$; and L_2 local full quadratic nonparametric estimate with same smoothing parameters. (Bottom row) Surface contaminated with noise from the mixture density $0.975N(0, 0.1^2) + 0.025N(0, 2.5^2)$, and residual surfaces for estimates above. The residual plots are centered at $z = 0$ and the vertical scale is expanded by a factor of 3.

be noticeable bias in the corners (see Wang and Scott (1994) for further details and examples).

8.4 REGRESSION IN SEVERAL DIMENSIONS

8.4.1 Kernel Smoothing and WARPing

The extension of the Nadaraya–Watson estimator to several dimensions is straightforward. The multivariate estimator is again a local average of responses, with the product kernel defining the size of the local neighborhood and the specific weights on the responses.

As with the kernel density estimator, the computational burden can be greatly reduced by considering a binning algorithm similar to the ASH. A moment's consideration suggests that binning need only be done in the predictor space, with only the *sum of all the responses* in each bin recorded. For purposes of notation, the sum of the ν_k bin responses in the kth bin with average response $\bar{y}_k$ may be computed by $\nu_k \bar{y}_k$. In the regression (bivariate) binning algorithm (REG-BIN) the quantity $sy_k = \nu_k \bar{y}_k$ is tabulated, which is the kth bin response sum.

REG-BIN $(x, y, n, a, b, nbin)$ **Algorithm**:   (∗ Bin 2-D regression data ∗)

> $\delta = (b - a)/nbin$
> for $k = 1, nbin \{ \nu_k = 0; sy_k = 0 \}$
> for $i = 1, n \{$
> $\quad k = (x_i - a)/\delta + 1$ (∗ integer part ∗)
> $\quad$ if $(k \in [1, nbin]) \ \nu_k = \nu_k + 1; sy_k = sy_k + y_i \}$
> return $(\{\nu_k\}, \{sy_k\})$

The univariate regression version of the ASH (REG-ASH) over the equally spaced mesh $\{t_k\}$ corresponding to the Nadaraya–Watson estimator (8.3) is easily computed as follows:

$$\hat{r}(x) = \frac{\sum_{i=1}^n K_h(x - x_i) y_i}{\sum_{i=1}^n K_h(x - x_i)} \approx \frac{\sum_{k=1}^{nbin} K_h(x - t_k) \nu_k \bar{y}_k}{\sum_{k=1}^{nbin} K_h(x - t_k) \nu_k}.$$

If the regression estimate is computed on the same grid $\{t_k\}$ as the binned data, then the kernel weights need only be computed once. When a finite support kernel is used to compute the weights $w_m(k)$, then the regression estimate in some bins may equal 0/0. The algorithm detects this condition and returns "not-a-number" (*NaN*). The speedup is comparable to that observed with the ASH.

REG-ASH $(m, \nu, sy, nbin, a, b, n, w_m)$ **Algorithm**:   (∗ 2-D REG-ASH ∗)

> $\delta = (b - a)/nbin; h = m\delta$
> for $k = 1, nbin \{ f_k = 0; r_k = 0; t_k = a + (k - 0.5)\delta \}$
> for $k = 1, nbin \{$
> $\quad$ if $(\nu_k = 0)$ next k
> $\quad$ for $i = \max(1, k - m + 1), \min(nbin, k + m - 1) \{$
> $\qquad f_i = f_i + \nu_k w_m(i - k)$
> $\qquad r_i = r_i + sy_k w_m(i - k) \} \}$
> for $k = 1, nbin \{$ if $(f_k > 0) r_k = r_k/f_k$ else $r_k = NaN \}$
> return $(\mathbf{x} = \{t_k\}, \mathbf{y} = \{r_k\})$ (∗ Bin centers & REG heights ∗)

The ASH and REG-ASH are special cases of a general approach to accelerating kernel-like calculations. The basic idea is to round the data to a fixed mesh and perform computations on those data. The idea of using such rounded data was explored by Scott (1981). The WARPing approach is described by Härdle and Scott (1988). The Fast Fourier Transform (FFT) approach to regression is described by Wand (1994).

8.4.2 Additive Modeling

For more than a few variables, in addition to increased concern about the curse of dimensionality and boundary effects, the kernel method begins to lose its interpretability. There is a need to "constrain" the multivariate kernels in a manner that allows flexible modeling but retains the ease of interpretation of multivariate parametric modeling. *Additive regression models* achieve this goal. The regression surface is modeled by the equation

$$r(\mathbf{x}) = r_0 + \sum_{i=1}^{d} r_i(x_i) + \epsilon_i.$$

The flexibility is achieved by fitting 1-D linear smoothers for r_i in each dimension. The set of multivariate surfaces that can be generated by this model is not very complex, but it is richer than parametric choices such as x_i or x_i^2 for $r_i(x_i)$. Further flexibility can be achieved by adding in terms of the form $r_{ij}(x_i, x_j)$. However, combinatorial explosion quickly sets in. The level of each additive function is arbitrary, although a constant r_0 can be introduced into the model and each additive function constrained to have average 0.

The fitting of such models takes two extremes. On the one hand, Wahba (1990) has advocated a penalty function approach that estimates the d functions simultaneously as splines. A simpler iterative scheme has been proposed by Friedman and Stuetzle (1981) called *backfitting*. The idea is to begin with an initial set of estimates of r_i, perhaps obtained parametrically, and to iterate over each of the d variables, smoothing the residuals not explained by the $d - 1$ predictors remaining. Any bivariate scatterplot smoother may be used. Simple kernel smoothers may be used as well as the smoothing splines. Friedman and Stuetzle (1981) used *super-smoother* in the more general projection pursuit regression (PPR) algorithm. Hastie and Tibshirani (1990) demonstrate how the backfitting algorithm mimics the Gauss–Siedel algorithm for solving linear equations for certain linear smoothers. In any case, the algorithm usually converges quickly to the neighborhood of a reasonable solution.

The backfitting algorithm is easily implemented using the WARPing regression estimator after removing the mean from $\{y_i\}$:

---

1. Initialize $r_j(x_j) = 0$ (∗ Backfitting Algorithm ∗)
2. Loop on j until convergence {

$$\epsilon_i \leftarrow y_i - \sum_{j \neq i} r_j(x_i)$$

$$r_j \leftarrow \text{smooth}(\epsilon_i) \}$$

The REG-ASH additive model (RAM) was applied to a simulation example of Wahba (1990, p. 139) with $\mathbf{x} \in \Re^4, \epsilon_i \sim N(0, 1), n = 100$ design points sampled uniformly over the hypercube $(0, 1)^4$, and true additive surface

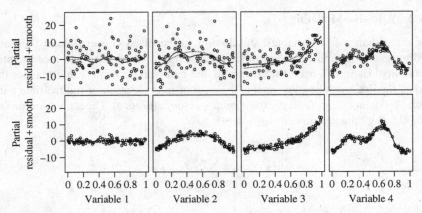

FIGURE 8.8   Additive model iteration for simulated data from (8.23). The true additive function is shown as a dotted line and the estimated additive function as a solid line. The top row gives the initial loop and the bottom row the final iteration (six loops).

$$r(\mathbf{x}) = 10 \sin \pi x_2 + \exp 3x_3 + 10^6 x_4^{11}(1 - x_4)^6 + 10^4 x_4^3 (1 - x_4)^{10}, \qquad (8.23)$$

where the true value of the additive function $r_1(x_1) = 0$. In Figure 8.8, the first row displays plots of (x_i, ϵ_i) for the first iteration together with the computed smooth; in the second row, the final iteration is displayed after five loops through the algorithm. The initial residual variance was 57.2. Even by the end of the first loop, the residual variation had been significantly reduced to 4.7 and was 1.45 at the final solution.

A second example shown in Figure 8.9 comes from the complete set of gas flow data. The meter was tested at seven different pressures with between 24 and 51 flows at each pressure. The first variable is log flow rate rescaled to $(0, 1)$. The second variable is log psia also rescaled to $(0, 1)$. The initial residual variance of 0.27 was reduced to 0.12. The relationship with log psia is almost perfectly linear, whereas the relationship to log flow rate is more complicated. The accuracy of the meter is affected most by the flow rate, but a small effect from the pressure exists as well.

8.4.3 The Curse of Dimensionality

An additive model is still a kernel estimator, although with highly constrained kernels, and subject to the same asymptotic problems. The most common problem remains rank-deficient predictor data. Some theoretical results require that the data become dense over the hyper-rectangle defining the additive model. This assumption is clearly very strong, and one that presumably is required not in practice but for proving theorems. However, Wahba's example in the preceding section is easily modified to illustrate the effect of the curse of dimensionality. Choose $\mathbf{x} \sim U([0, 1]^4)$ but with pairwise correlations of 0.99. In practice, this is accomplished by applying the inverse of the principal components transformation to points sampled uniformly on the hypercube and then rescaled back to the hypercube. The initial and final additive fits are

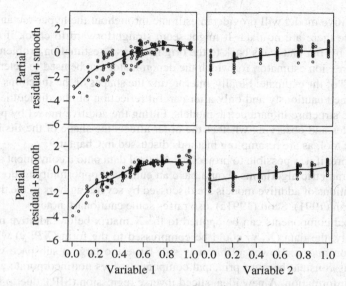

FIGURE 8.9   Additive model iteration for the gas flow dataset. The 214 measurements were taken at seven different pipeline pressures. The additive fits for the initial and final iteration are shown.

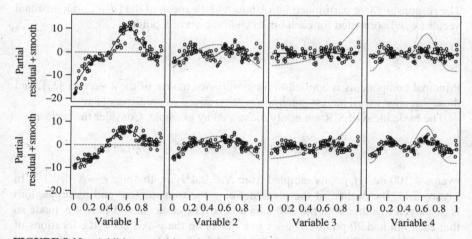

FIGURE 8.10   Additive model iteration for simulated data from (8.23). The pairwise correlations of the design points are all 0.99. The true additive function is shown as a dotted line and the estimated additive function as a solid line.

shown in Figure 8.10. The initial and final residual variances were 6.6 and 2.1. However, as the design points are concentrated along the hyperdiagonal, the individual additive functions are not identifiable since the design space is singular. In fact, only one function is required (set $x_2 = x_3 = x_4$ in (8.23)). In very high dimensions, such singularities are common and the curse of dimensionality applies.

An additive model will provide an estimate throughout the hyper-rectangle base, even where there are no data. It might seem straightforward to check if there are data close by, but that leads back to the original density estimation problem and the kernel regression estimator, for which the denominator can be used to determine the "accuracy" of the estimate. Finally, interpreting the shapes of the functions r_i should be performed cautiously and only after careful reflection on the difficulties of estimating the structure in parametric models. Fitting the additive model by permuting the variables and observing whether the order affects the shape of the fits is a good diagnostic tool, as are resampling methods discussed in Chapter 9.

As before, if it is possible to project the original data onto a convenient subspace, the nonparametric algorithms should have an enhanced opportunity to do their job. The limitations of additive models are discussed by several authors in the discussion of Friedman (1991). Scott (1991a) also raises some cautionary notes.

Principal components can be applied to the $\mathbf{X}$ matrix before additive modeling. Specifically, the data $(\mathbf{X}, \mathbf{y})$ could be compressed to the form $(\mathbf{X}P, \mathbf{y})$ where P is a projection matrix. However, that may still leave too large a subspace where the response is constant because principal components does not incorporate any of the response information. A new idea, sliced inverse regression (SIR), due to Li (1991) incorporates the response information in an approach designed for normal data but like principal components has much greater general applicability. The SIR algorithm is a simple extension of the principal components and informative components ideas. The responses y_i are partitioned into k bins and the mean of the $(d+1)$-dimensional vector $(\mathbf{x}_i, y_i)$ computed for each bin to yield the set of k points

$$(\bar{\mathbf{x}}_j, \bar{y}_j), \quad j = 1, \ldots, k.$$

Principal components is applied to the covariance matrix of the k vectors $\{\bar{\mathbf{x}}_j\}$, and those directions with largest eigenvalues are retained as predictors of y.

The basic ideas of SIR are easily illustrated by example. Consider the surface

$$r(x_1, x_2) = 0.5625 - 0.1375 x_1 - 0.2875 x_2 + 0.0125 x_1 x_2$$

over $n = 100$ design points sampled from $N(0, 0.4^2 I_2)$ with noise $\epsilon \sim N(0, 0.1^2)$. In Figure 8.11, the outline of the plane defining the true surface (and its projection) is displayed together with the 100 data points. The choice $k = 5$ bins was made so that each bin had 20 points. The six tick marks on the y-axis show the locations of the bins. The conditional means for each of the five bins were computed and are displayed as diamonds together with a line down to the projection onto the design space. The covariance matrix of the five points on the design plane was computed and found to have two eigenvalues, given by 0.365 and 0.034. Thus the SIR direction is the eigenvector $(0.667, 0.745)^T$ corresponding to the larger eigenvalue. The ellipse corresponding to the covariance matrix is shown surrounding the five means. Notice that the responses are essentially constant in the direction of the other eigenvector. For higher dimensional data that are not normal, this simple idea has proven a powerful diagnostic and exploratory tool.

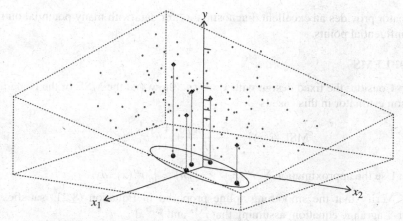

FIGURE 8.11   Example of the SIR dimension reduction technique.

Simple models such as the SIR algorithm have an appeal due solely to the parsimonious solution. A similar simple model is the regression tree model for the regression surface that is piecewise constant

$$\hat{r}(\mathbf{x}) = \sum_{i=1}^{p} c_i I\{\mathbf{x} \in N_i\}, \quad \text{where} \quad \bigcup_{i=1}^{p} N_i = \Re^d \quad \text{and} \quad N_i \cap N_j = \emptyset.$$

The so-called CART regression model has had much success in practice for classification problems (see Breiman et al. (1984)).

8.5 SUMMARY

The choice of kernel in the regression setting is more flexible than in the density estimation setting. In particular, the use of higher order kernels is more attractive because the issue of nonnegativity vanishes. Care must be exercised at the boundaries to avoid anomalies in the estimate. Boundary kernels may induce artifacts as often as they correct artifacts.

The direct use of kernel estimates for regression may or may not be the best choice. Certainly, issues of resistance to outliers are of practical importance. Naive kernel regression algorithms provide no protection against influential points. The use of modal regression is to be encouraged. Additive models provide the most easily understood representation of multidimensional regression surfaces, when those surfaces are not too complex. However, the curse of dimensionality suggests that the additive functions can be quite unreliable if the design data are nearly singular. The use of the L_1 norm as a robust measure for local polynomial fitting can be expected to grow dramatically as the availability of good LP codes increases. At the least, the L_1

estimator provides an excellent diagnostic tool for data with many potential outliers and influential points.

PROBLEMS

8.1   Consider the fixed design with $x_i = i/n$. Show that the MSE of the Nadaraya–Watson estimator in this case is

$$\mathrm{MSE}\{\hat{r}(x)\} = \frac{\sigma_\epsilon^2 R(K)}{nh} + \frac{1}{4}h^4 \sigma_K^2 R(r'').$$

Hint: Use the approximation $\sum r''(x_i)^2 \times (1/n) \approx \int_0^1 r''(x)^2 \, dx$.

8.2   Verify that the smoothing spline proposed in Equation (8.21) satisfies the Euler–Lagrange equation, assuming that $\hat{r}_\lambda^{(1)}$ and $\hat{r}_\lambda^{(2)}$ do.

8.3   Prove that the weight vector $\mathbf{w}_x$ in Equation (8.18) is linear in x.

8.4   Construct an internal boundary modification to the penny dataset displayed in Figure 8.5.

8.5   Consider a bivariate additive model for a function $f(x, y)$, where each additive function is a piecewise constant function. Specifically,

$$f_1(x) = \sum a_i I_{A_i}(x) \quad \text{and} \quad f_2(y) = \sum b_j I_{B_j}(y),$$

where $\{A_i\}$ and $\{B_j\}$ are bins along the x and y axes, respectively.
   (a) Given data $\{(x_i, y_i, z_i), i = 1, n\}$, find the equations that correspond to the least-squares estimates of $\{a_i\}$ and $\{b_j\}$.
   (b) Compute and plot this estimator for the gas-flow-accuracy dataset, for several choices of the mesh.

9

OTHER APPLICATIONS

9.1 CLASSIFICATION, DISCRIMINATION, AND LIKELIHOOD RATIOS

The problems of classification and discrimination are closely related. In each instance, the data, $\mathbf{x}_i \in \Re^d$, are assumed to comprise K clusters. If the number of clusters K is known, and a training sample of data is available from each cluster, then the *discrimination problem* is to formulate rules for assigning new unclassified observations to one of the clusters. If the dataset is made up of an unknown number of clusters, then the *classification problem* is to stratify the unlabeled data into proposed clusters.

The general principles of discrimination may be understood in the simplest case when $K = 2$. Using a notation common in biostatistical applications, the two categories are labeled "positive" or "negative" depending on the presence or absence of a disease or on the finding of a diagnostic test. The *prior probabilities* of the categories are denoted by $P(+)$ and $P(-)$, both known and satisfying $P(+) + P(-) = 1$. The ratio $P(+) : P(-)$ or $[P(+)/P(-)] : 1$ is called the *prior odds* in favor of disease. There is a simple 1:1 relationship between the probability p of an event and the corresponding odds o in favor of the event:

$$o = \frac{p}{1-p} \quad \text{and} \quad p = \frac{o}{1+o}. \tag{9.1}$$

When $p = \frac{1}{2}$ the odds are even or 1:1.

Multivariate Density Estimation, First Edition. David W. Scott.
© 2015 John Wiley & Sons, Inc. Published 2015 by John Wiley & Sons, Inc.

Suppose that a discrete covariate such as race or sex is thought to be associated with the risk of disease. Let the event A denote the presence of a covariate. Then Bayes' theorem states that

$$P(+|A) = P(A|+)P(+)/P(A) \qquad (9.2)$$

and

$$P(-|A) = P(A|-)P(-)/P(A), \qquad (9.3)$$

where $P(+|A)$ and $P(-|A)$ are the *posterior probabilities* given the additional knowledge of the information contained in the event A. The conditional probabilities $P(A|\pm)$ may be estimated through epidemiological studies, or by experiment. Taking the ratio of the two equations in (9.2) and (9.3), we have

$$\frac{P(+|A)}{P(-|A)} = \frac{P(A|+)}{P(A|-)} \times \frac{P(+)}{P(-)}, \qquad (9.4)$$

where $P(A|+)/P(A|-)$ is the *likelihood ratio* (LR). The ratio on the left-hand side is called the *posterior odds* in favor of the disease. In odds form, (9.4) is called *Bayes' rule*:

$$o(+|A) = \frac{P(A|+)}{P(A|-)} o(+).$$

Depending on whether the likelihood ratio is greater than or less than 1, the posterior odds in favor of disease are increased or decreased, respectively.

If the (multivariate) covariate $\mathbf{x}$ is a continuous random variable, then Bayes' rule still holds when the likelihood ratio is replaced with the corresponding ratio of the probability densities at $\mathbf{x}$ for the two groups.

$$\boxed{\text{Bayes' rule}: \qquad o(+|\mathbf{x}) = \frac{f_+(\mathbf{x})}{f_-(\mathbf{x})} o(+).} \qquad (9.5)$$

Again the odds change with the likelihood ratio. An unclassified (or unlabeled) point $\mathbf{z}$ would be assigned to the "+" group if $o(+|\mathbf{z}) > 1$ and "−" otherwise.

In practice, the *shifted-normal model* is the model most commonly assumed for the pair of densities f_+ and f_-. Specifically, in the univariate case

$$f_+(x) = N(\mu_1, \sigma^2) \quad \text{and} \quad f_-(x) = N(\mu_0, \sigma^2).$$

A little algebra reveals that the logarithm of Bayes' rule is

$$\log[o(+|x)] = \frac{\mu_1 - \mu_0}{\sigma} \left[\frac{1}{\sigma} \left(x - \frac{\mu_0 + \mu_1}{2} \right) \right] + \log[o(+)].$$

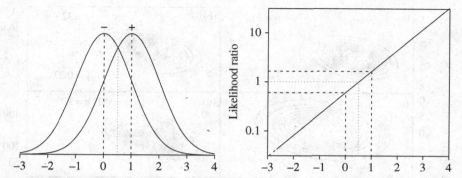

FIGURE 9.1 Shifted-normal model of two populations for a single risk factor or covariate. The log–likelihood ratio is linear in x.

That is, the log–odds change linearly in x. The posterior odds equal the prior odds when the covariate x is midway between the two population means (see Figure 9.1). If x is measured in standard units, then the slope of the log–LR ratio is $(\mu_1 - \mu_0)/\sigma$, which is called the *coefficient of detection*. The greater (in magnitude) the coefficient of detection, the greater the predictive power of Bayes' rule. The multivariate shifted-normal model is considered in Problem 9.2.

Nonparametric discrimination simply involves estimation of the unknown densities in the likelihood ratio using averaged shifted histogram (ASH) estimates. Consider the two populations in the plasma lipid study. By imaging the coronary arteries using angiography, 320 males were found to have significant occlusion (narrowing) while 51 had none. There is no obvious deviation from normality in the group of 51; see frame (a) in Figure 9.2. The normal fit to the 320 males shown in frame (b) is not obviously unsatisfactory, although the superposition of the normal estimate with the ASH estimate in frame (d) indicates the lack-of-fit clearly. Frames (c) and (e) contain the parametric estimates that lead to the contours of the LR (on a $\log_{10}$ scale). If the covariance matrices are assumed to be identical, then the $\log_{10}$-LR surface will be a plane (see Problem 9.2) and the LR contours a series of parallel lines. Looking only at areas with significant probability mass, the logarithmic LR estimates range from about -0.5 to 1.0, which suggests moderate predictive power.

The nonparametric LR shown in frame (f) used the ASH estimate in the numerator and the normal fit in the denominator. As ASH estimates do not extrapolate well, the LR is plotted only over the region where the ASH estimate was greater than 5% of the modal value. The LR surface is not monotonic increasing, reflecting the multi-modal nature of the data. The parametric LR estimates are never grossly incorrect; however, the parametric fit seems to overestimate the importance of high cholesterol and underestimate the importance of high triglyceride concentrations. (Fitting a single covariance matrix to the data results in heavier tails along the cholesterol axis because of the bimodality in the data along that axis.) In the nonparametric LR plot, the relative contributions to risk of the two lipids seem nearly equal. Similar complex interactions were observed in a much larger follow-up study by Scott et al. (1980).

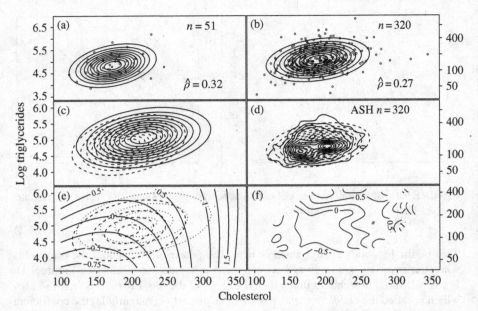

FIGURE 9.2 Risk analysis of plasma lipid dataset. (Upper left) Normal fit "−" group. (Upper right) Normal fit "+" group. (Middle left) Overlay of two normal fits. (Middle right) Overlap of ASH of f_+ [biweight kernel $h = (21.7, 0.33)$] and normal fit to "+" group. (Lower left) Contours of parametric $\log_{10}(\text{LR})$. (Lower right) $\log_{10}(\text{LR})$ nonparametric estimate.

That study included age, smoking behavior, and history of hypertension as covariates. For various combinations of levels of the three additional covariates, the lipid dataset exhibited bimodality, and the risk contours suggested that the predictive power of triglycerides was again underestimated in a shifted-normal model.

In the author's opinion, bivariate contour plots of surfaces that are not monotone or uncomplicated are much less satisfactory than perspective plots. Perspective plots of the two $\log_{10}$-LR surfaces in Figure 9.2 are shown in Figure 9.3. The parametric LR estimates around the border of the plot are clearly driven by the normal model, not by the presence of significant data. Caution should be exercised when extrapolating outside the range of the data.

Prediction and classification in $\Re^d$ using nonparametric techniques have a long history of success. Habbema et al. (1974) reported excellent results using a multivariate normal kernel estimator for classification. Titterington et al. (1981) used mixed variables. The classification depends only on the contour where the LR equals $1 [\log_{10}(\text{LR}) = 0]$. As in the univariate case, a new observation is classified as positive if $\text{LR}(\mathbf{x}) > 1$ and as negative otherwise. Thus classification is a relatively easy task for kernel estimation, as only a relatively small fraction of the data is near the boundary where the LR is 1. This ease seems to hold over a wide range of smoothing parameters. For example, Hand (1982, p. 88) makes a connection with the nearest-neighbor

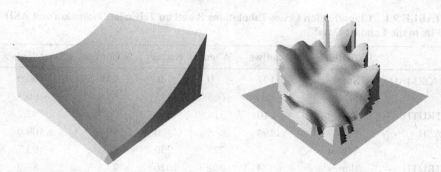

FIGURE 9.3 Perspective plots of the $\log_{10}$ likelihood ratio surfaces in Figure 9.2. The range of the vertical axes is $(-0.91, 0.91)$ in both frames, corresponding to a range of odds in favor of disease from 0.15:1 to 8.1:1.

classification rule, which assigns a new point **x** to a category based on the category of the training sample point closest to it in Euclidean norm. This behavior is precisely that of a Gaussian kernel classifier by noting that if the distance between **x** and the nearest data point is δ and to the second nearest point $\delta + \epsilon$, then the relative contribution to the kernel estimate $\hat{f}(\mathbf{x})$ is

$$\frac{\phi[-(\delta+\epsilon)^2/(2h^2)]/h}{\phi[-\delta^2/(2h^2)]/h} \to 0 \quad \text{as} \quad h \to 0.$$

Therefore, for a very small choice of bandwidth, the Gaussian kernel estimates for the two training samples are dominated by the nearest sample point in each, with the classification going to the nearest one.

This connection may be extended to very large bandwidths as $h \to \infty$. As h dwarfs the span of the data, each kernel estimate converges pointwise to a single kernel centered on the sample mean. Therefore, the two kernel estimates at the new observation reflect only the distance to the sample mean of each cluster, the classification being made to the closer sample mean. This classification rule is known as average linkage. Thus for extreme values of the bandwidth, the kernel classifier mimics two well-known classification algorithms known to work reasonably well in practice. Of course, for bandwidths in between, the classifier benefits from a good estimate. Therefore, the kernel classifier should perform well in almost all situations.

For the Landsat data, the actual agricultural use for each pixel was determined on the ground by trained observers. The primary crops observed were sunflower (1200 pixels), wheat (1010), and barley (1390). Scott (1985c) compared the observed misclassification rates of the trivariate normal fits and the ASH fits to the training sample data from these three clusters. The results are summarized in Table 9.1. Visually, the ASH estimates were egg-shaped, slightly longer on one end of the egg. The result of the small skewness is a bias in the location of the parametric means, and asymmetric error rates for wheat and barley compared to the ASH estimates. Improved error rates were obtained by considering not only the classification prediction of each pixel but

TABLE 9.1   Classification Cross-Tabulations Based on Trivariate Gaussian and ASH Fits to the Landsat Data[a]

	PREDICTED	Sunflwr	Wheat	Barley	% Correct	Smoothed (%)
NORMAL	Sunflwr	1191	9	0	99.3	100.0
	Wheat	10	665	335	65.8	80.3
TRUTH:	Barley	10	314	1066	76.7	93.7
ASH	Sunflwr	1194	5	0	99.5	100.0
	Wheat	7	773	230	76.5	93.7
TRUTH:	Barley	3	361	1026	73.8	89.9

[a]The first three columns summarize the predictions of the classifier using the training data. The last column summarizes the rates using a classification rule based on a majority rule of a pixel and its eight neighbors.

also the predictions for its eight neighbors. This modification takes advantage of the obvious spatial correlation among neighboring pixels. The final smoothed classification was based on the majority prediction in each block of nine pixels. The resulting spatial classification appears much smoother visually. The improvement in prediction is significant, as shown in the final column of Table 9.1.

For smaller samples, a jackknife or cross-validation procedure should be used (Efron, 1982). Each sample in the training set is removed from the density estimator before classification. The reuse procedure removes the somewhat overoptimistic error rates obtained when all the data in the training sample are used to form the density estimates and then those same data are used to test the classification performance.

An alternative approach when there are two clusters is to treat the problem as a nonparametric regression problem, where the outcome Y is 0 or 1, corresponding to the "−" and "+" clusters. Thus the data are $\{\mathbf{x}_i, y_i\}$. Now,

$$\mathrm{E}Y_i = 0 \cdot \Pr\{Y_i = 0\} + 1 \cdot \Pr\{Y_i = 1\} = \Pr(+|\mathbf{X} = \mathbf{x}_i),$$

which can be estimated from the odds estimate in Equation (9.5) via Equation (9.1). A kernel regression estimate of Y will fall between 0 and 1 if a positive kernel is used. If the sizes of the training samples reflect the relative frequency of the two classes and the same smoothing parameter is used for both classes, then

$$\hat{y} = \hat{r}(\mathbf{x}) = \frac{\sum_{i=1}^{n} y_i K_h(\mathbf{x} - \mathbf{x}_i)}{\sum_{i=1}^{n} K_h(\mathbf{x} - \mathbf{x}_i)} = \frac{n_1 \hat{f}_+(\mathbf{x})}{n_0 \hat{f}_-(\mathbf{x}) + n_1 \hat{f}_+(\mathbf{x})},$$

which gives identical predictions to the nonparametric likelihood ratio estimate using the same smoothing parameter [as $p = o/(1 + o)$ and $o = \hat{f}_+/\hat{f}_-$]. The obvious disadvantage of the regression approach is the use of the same smoothing parameter for both populations. If $P(+) \neq P(-)$, then the sample sizes will not be equal and h_+ and h_- should not be equal either. Likewise, the same conclusion holds if the shapes of the two densities are not identical. For example, with the lipid dataset, neither condition supports the use of the regression approach. Hall and Wand (1988b) observed that the classification rule $o(+|x) > 1$, together with (9.5), is equivalent to the inequality $P(+)f_+(x) - P(-)f_-(x) > 0$ and investigated a single optimal bandwidth for this weighted density difference.

9.2 MODES AND BUMP HUNTING

9.2.1 Confidence Intervals

The asymptotic distribution of the sample mode provides the basis for finding a confidence region for the true mode. The same approach may be extended to finding confidence regions for the modal regression curves.

Kernel estimates are actually sums of rows in a triangular array of random variables, $\{Y_{n1}, Y_{n2}, \ldots, Y_{nn}; n \geq 1\}$, as is demonstrated later. Consider conditions under which sums of triangular arrays converge to normality. Suppose $Y_{nj} \sim F_{nj}$ (a cdf) with $\mu_{nj} = \mathrm{E}Y_{nj}$ and

$$A_n = \sum_{j=1}^n \mu_{nj} \quad \text{and} \quad B_n^2 = \mathrm{Var}\left\{ \sum_{j=1}^n Y_{nj} \right\}.$$

Then Serfling (1980) proves the following proposition: If

$$\sum_{j=1}^n \mathrm{E}(Y_{nj} - \mu_{nj})^4 = o(B_n^4), \quad \text{then} \quad \sum_{j=1}^n Y_{nj} \sim \mathrm{AN}(A_n, B_n^2). \tag{9.6}$$

From Equation (6.20), the kernel estimate of the derivative is a sum of random variables from a triangular array:

$$\hat{f}'(x) = \sum_{j=1}^n Y_{nj} \quad \text{where} \quad Y_{nj} = \frac{1}{nh^2} K'\left(\frac{x - x_j}{h} \right). \tag{9.7}$$

The asymptotic normality of the mode follows from the asymptotic normality of $\hat{f}'(x)$. The first four noncentral moments of Y_{nj} are $f'(x)/n + h^2 \sigma_K^2 f'''(x)/(2n)$, $f(x)R(K')/(n^2 h^3)$, $-f'(x)R((K')^{3/2})/(n^3 h^4)$, and $f(x)R((K')^2)/(n^4 h^7)$ (see Problem 9.4). Therefore,

$$A_n = \sum_{j=1}^n \mu_{nj} \approx f'(x) + h^2 \sigma_K^2 f'''(x)/2$$

$$B_n^2 = n \mathrm{Var}\{Y_{nj}\} \approx f(x)R(K')/(nh^3).$$

Now the fourth noncentral and central moments are the same to first order, so

$$\sum_{j=1}^{n} \mathrm{E}(Y_{nj} - \mu_{nj})^4 \approx \frac{f(x)R((K')^2)}{n^3 h^7} \quad \text{and} \quad B_n^4 \approx \frac{f(x)^2 R(K')^2}{n^2 h^6} ;$$

their ratio is $O((nh)^{-1})$, which vanishes if $nh \to \infty$ as $n \to \infty$. Thus the conditions of Serfling's proposition hold and

$$\pm \{\hat{f}'(x) - [f'(x) + h^2 \sigma_K^2 f'''(x)/2]\} \sim \mathrm{AN}(0, f(x)R(K')/(nh^3)).$$

Let $\hat{\theta}_n$ be a sequence of sample modes converging to a true mode at $x = \theta$. Then $f'(\theta) = 0, \hat{f}'(\hat{\theta}_n) = 0, f'(\hat{\theta}_n) \approx f'(\theta) + (\hat{\theta}_n - \theta)f''(\theta) = (\hat{\theta}_n - \theta)f''(\theta)$, and $f'''(\hat{\theta}_n) \approx f'''(\theta)$. Hence,

$$\pm \{0 - [(\hat{\theta}_n - \theta)f''(\theta) + h^2 \sigma_K^2 f'''(\theta)/2]\} \sim \mathrm{AN}(0, f(\theta)R(K')/(nh^3))$$

or

$$(\hat{\theta}_n - \theta) \sim \mathrm{AN}\left(-\frac{h^2 \sigma_K^2 f'''(\theta)}{2f''(\theta)}, \frac{f(\theta)R(K')}{f''(\theta)^2 nh^3}\right),$$

ignoring a $O(1/nh^3)$ term in the bias that comes from the next term in the Taylor's series, $(\hat{\theta}_n - \theta)^2 f'''(\theta)/2$. The variance is $O(1/nh^3)$ and the squared bias is $O(h^4)$; therefore, optimally, $h = O(n^{-1/7})$ with corresponding MSE $= O(n^{-4/7})$, which are the same rates as for the derivative estimator itself. However, if $h = O(n^{-1/5})$, then the same result holds, but the bias is asymptotically negligible and the MSE is dominated by the variance, which is of order $n^{-2/5}$. Hence a $100(1 - \alpha)\%$ confidence interval in this case is

$$\theta \in \left(\hat{\theta}_n \pm z_{\alpha/2}[f(\theta)R(K')/(f''(\theta)^2 nh^3)]^{1/2}\right).$$

Replacing $f''(\theta)$ with $\hat{f}''(\hat{\theta}_n)$ is often adequate; otherwise, an auxiliary bandwidth must be introduced. Compare this result to the FP approximation obtained in Section 4.1.4.

For the modal regression problem, let $\hat{\theta}_{x,n}$ denote a conditional mode of $\hat{f}(x, y)$ that converges to θ_x. Then defining

$$Y_{nj} = K((x - x_j)/h_x)K'((y - y_j)/h_y)/(nh_x h_y^2),$$

a similar approach shows that Y_{nj} is asymptotically normal and that

$$\mathrm{E}Y_{nj} \approx f_y(x, y) + \frac{1}{2}h_y^2 \sigma_K^2 f_{yyy}(x, y); \quad \mathrm{Var}\, Y_{nj} \approx \frac{f(x, y)R(K)R(K')}{nh_x h_y^3}$$

$$\theta_x \in \hat{\theta}_{x,n} \pm z_{\alpha/2}\left[f(x, \theta_x)R(K)R(K')/(f_{yy}(x, \theta_x)^2 nh_x h_y^3)\right]^{1/2},$$

where the subscripts on f indicate partial derivatives.

9.2.2 Oversmoothing for Derivatives

Sample modes that are not "real" tend to have confidence intervals that overlap with neighboring sample modes. If a larger bandwidth of optimal order $n^{-1/7}$ is chosen, then the bias becomes significant. Choosing a bandwidth for the derivative could be done by cross-validation (Härdle and Marron, 1988), but this approach is even less reliable than for the density estimate itself. Fortunately, the oversmoothing approach can be extended to f' and f'' as well. The oversmoothed bandwidths can be used as a point of reference for cross-validation or to provide a conservative estimator for modes and bumps. From Theorem 6.2, the functionals to minimize are $R(f''')$ and $R(f^{\mathrm{iv}})$ for the first and second derivatives, respectively, subject to the constraint $\int x^2 f = 1$. The solutions are

$$f_3^{\mathrm{OS}}(x) = \frac{315}{256}(1-x^2)_+^4 \quad \text{and} \quad f_4^{\mathrm{OS}}(x) = \frac{693}{512}(1-x^2)_+^5,$$

leading to the inequalities

$$R(f''') \geq 14{,}175\,\sigma^{-7}11^{-9/2} \quad \text{and} \quad R(f^{\mathrm{iv}}) \geq 1{,}091{,}475\,\sigma^{-9}13^{-11/2}.$$

Substituting into the expressions for the optimal bandwidths in Theorem 6.2 for the choice $K = N(0,1)$ leads to

$$
\begin{aligned}
&\text{Oversmoothed } f': \quad h_{f'}^* \leq 1.054\,\sigma\,n^{-1/7} \equiv h_{\mathrm{OS}}(f') \\
&\text{Oversmoothed } f'': \quad h_{f''}^* \leq 1.029\,\sigma\,n^{-1/9} \equiv h_{\mathrm{OS}}(f'').
\end{aligned}
\tag{9.8}
$$

These formulas are nearly optimal for estimating the derivatives when the data are normal. For a sample size of $n = 100, h_{\mathrm{OS}}(f), h_{\mathrm{OS}}(f')$, and $h_{\mathrm{OS}}(f'')$ equal 0.455, 0.546, and 0.617. When $n = 1000$ the values are 0.287, 0.393, and 0.478 [which is only 1.66 times $h_{\mathrm{OS}}(f)$]. These values may be rescaled for use with other more computationally tractable kernels using an equivalent kernel rule similar to that in Equation (6.29), but with the rth derivative of the kernel (compare h^* in Theorems 6.1 and 6.2) (see Problem 9.5).

9.2.3 Critical Bandwidth Testing

Bump hunting falls precisely into the classification category. Given a density estimate, the modes and bumps are easily located. Some of those modes and bumps may be spurious and should be smoothed away locally. Good and Gaskins (1980) proposed an iterative procedure for computing the odds in favor of the existence of a bump's "reality" over an interval (a_i, a_{i+1}) with their maximum penalized likelihood estimator. Assume that the data are bin counts. Briefly, the counts in the bins covering (a_i, a_{i+1}) were reduced to match the probability mass in those bins computed from a maximum penalized likelihood (MPL) density estimate; then all the bin counts were rescaled so that the sum of the bin counts was again n. (The MPL criterion can easily accommodate the notion of noninteger bin counts.) As all density estimates are biased downward at bumps, the effect was to effectively eliminate the bump by "surgery" in the authors' terminology. Usually, a dozen iterations were sufficient to completely eliminate any hint of the bump. To evaluate the odds in favor of the bump being real, the penalized log-likelihood of the original data for the MPL estimate was computed with and without the bump. Each bump is surgically removed and tested individually. For the Lawrence Radiation Laboratory (LRL) dataset discussed in Section 3.5, Good and Gaskins found that the loss in likelihood when removing bumps 3, 8, 11, and 12 seemed small (numbering the bumps from the left). The authors found that the exact choice of smoothing parameter within a narrow range did not affect the estimated odds very much. As bumps may or may not contain a mode, the evidence in favor of a mode is indirect.

For univariate data, Silverman (1981) suggests a conservative test of the null hypothesis that the unknown density has at most k modes. The role played by the bandwidth is important, because the number of sample modes in the density estimate can be anything between 1 and n, as h ranges over $(0, \infty)$. Surprisingly, the number of modes need not be monotonic increasing as $h \to 0$. For example, Hart (1985) shows that the property of monotonicity need not hold even if the kernel K is unimodal. Silverman (1981) showed that the property of monotonicity does hold for the normal kernel.

Starting with a large value of h and a normal kernel, clearly there exists a bandwidth h_1 such that $\hat{f}$ is unimodal for all $h > h_1$ but not for $h < h_1$. Similarly, $\hat{f}$ is bimodal for $h_1 > h > h_2$ but not if $h < h_2$, and so on. Silverman calls the bandwidths where $h = h_k$ the critical bandwidths, corresponding to a density estimate which has exactly k modes and 1 saddle point. The rationale is simple. If the true density has fewer than k modes, then the density estimate The test is performed by computing an estimate of the p-value by resampling from $\hat{f}_{h_k}$, which is the same smoothed bootstrap sample used in cross-validation by Taylor (1989) in Section 6.5.1.4. For each bootstrap sample, the normal kernel estimate is computed using the bandwidth $h = h_k$ and the number of modes counted. The p-value is the observed fraction of estimates where the number of modes exceeds k.

The test is supported by theory that suggests that the critical bandwidth converges to 0 if k is less than or equal to the true number of modes, while the critical bandwidth does not converge to 0 otherwise. Izenman and Sommer (1988) have used the test to examine differences in thicknesses of new issues of stamps.

Matthews (1983) has experimented with this test and concluded "that when the underlying density is k-modal, the p-values for less than k modes may be large and stable, but that the p-values for k modes or more are often highly variable." With a distinctly bimodal mixture density $0.5[\phi(x) + \phi(x - 3.2)]$, he computed the number of times the test concluded that the density was exactly bimodal with sample sizes of 40, 200, and 1000. In 10 simulations each, the test reached the correct conclusion 30, 40, and 60% of the time, respectively. Further research directed toward the use of adaptive bandwidths would be interesting.

9.2.4 Clustering via Mixture Models and Modes

Clustering is the dual of discrimination and a much more difficult problem. Typically, the class labels are unknown, not even the number of classes K (Kaufman and Rousseeuw, 2005). The most powerful clustering technique may simply be the parametric general normal mixture model (Titterington et al., 1985). Karl Pearson (1894) first considered the estimation of two general normal mixtures based on estimates of the first five sample moments and the solution of a ninth-degree polynomial. More recent references include Day (1969), Aitkin and Wilson (1980), Hathaway (1985), and McLachlan and Krishnan (2007).

One parametric approach is to treat the class labels as missing data, which leads to the expectation-maximization (EM) algorithm (Dempster et al., 1977). For computational and conceptual simplicity, especially in higher dimensions, the mixing density is commonly chosen to the be Gaussian (Titterington et al., 1985). Overlapping mixture components can be difficult to interpret as clusters. For example, in one dimension, densities of the form $0.5\phi(-\mu, 1) + 0.5\phi(\mu, 1)$ are not bimodal unless $|\mu| > 1$. Roeder (1990) used the EM algorithm with some data pertaining to the distribution of galaxies and reported many practical problems using a mixture model with general parameters, such as delta spikes at isolated data points. It is also well-known that a mixture of K Gaussians may or may not display K modes. We explore this further in the Section 9.2.4.1.

The parallel nonparametric approach to clustering is to construct a kernel estimate and locate the modes. A cluster is then associated with each mode. Now this is conservative in the sense that pairs of adjacent clusters might manifest as a single mode and a single bump in the kernel estimate. In an extreme case, there may be only a single mode and no bump. But clustering is a highly exploratory activity and such limitations are to be expected. One may hope that adding more informative variables would lead to further separation of the clusters in the higher dimensional feature space, with some clusters becoming bumps, some bumps becoming modes, and new bumps appearing.

Finally, there are many clustering algorithms that do not rely on a parametric or nonparametric probability of the data. The most commonly used is the hierarchical clustering algorithm, which is an iterative distance-based approach (see Izenman (2009), Chapter 12). The results of the algorithm are displayed as a tree. An example is given below in Section 9.2.4.2. The most popular nonhierarchical clustering algorithm is K-means (MacQueen, 1967) that iteratively updates the means of points currently assigned to the K groups, reallocates points to the closest mean, and stops when no further changes occur.

9.2.4.1 Gaussian Mixture Modeling
A Gaussian mixture model with K components may be written most generally as

$$\hat{f}_K(x) = \sum_{k=1}^{K} w_k \phi(\mathbf{x}|\boldsymbol{\mu}_k, \Sigma_k),$$ (9.9)

where the nonnegative weights sum to 1. The full model is relatively parameter intense: $K - 1$ parameters in the weight vector, $K \cdot d$ parameters in the mean vectors, and $K \cdot d(d+1)/2$ parameters in the covariance matrices. Thus one fundamental challenge is choosing initial values for the parameters as input to a maximum likelihood algorithm (such as EM). A simplifying assumption is that there is a common covariance matrix, $\Sigma_k = \Sigma$, which greatly reduces the number of parameters at the cost of generality and perhaps accuracy. There are less severe assumptions about $\{\Sigma_k\}$, and the R package *mclust* by Fraley et al. (2014) explores 10 different assumptions. Furthermore, they evaluate every model using a Bayesian Information Criterion (BIC) (see Banfield and Raftery (1993)). When applied to the $n = 320$ log-lipid dataset, BIC selects $K = 1$, a density that is depicted in the third frame of Figure 9.2. In Figure 9.4, there is an intriguing trimodal fit when $K = 10$ that mirrors the kernel estimate; however, the weight of evidence according to BIC is very small. (The fit shown in the left frame of Figure 9.4 was obtained by re-running MLE using as initial parameter estimates a solution provided by an earlier version of Mclust, which was available in 2005.) The four components in the middle have nearly 92% of the weight. The remaining six components generally have weights less than 1 or 2% and "cover" points in the surrounding low-density region.

A more problematic difficulty of the general covariance assumption is that if the determinant of any one of the Σ_k approaches 0, the maximum likelihood will approach $+\infty$. Happily, there are sound theoretical results that show there are many local maxima in the likelihood function that provide useful and desirable estimates (Hathaway, 1985). Thus trying a number of different initializations is recommended.

Usefulness of Mixture Weights as Surrogates for Cluster Size.   The most common use of mixture models is in one dimension, where there may be considerable interest in the relative size of the components. The estimated weights $\{\hat{w}_k\}$ provide a natural

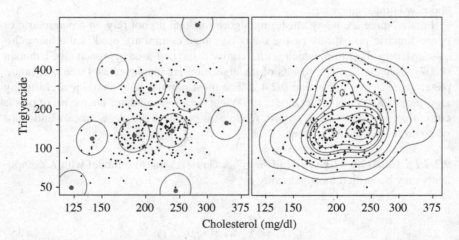

FIGURE 9.4   Mclust (2005 version) applied to log-lipid dataset ($n = 320$).

choice for this purpose. However, since the Gaussian density does not provide an orthonormal basis for density functions, caution should be exercised. The good news is that the mixture model is dense in the L_2 function space; however, the true K may be quite large (or even infinite).

If the K clusters are well-separated and are each approximately normal, then the weights should reasonably capture the size of each cluster. However, if a cluster density is skewed, then it is not so clear what the cluster weights are capturing.

To illustrate this point, consider the skewed density, $f \sim \chi^2(7)$, which is shown in Figure 9.5. How close can (9.9) approximate this density for finite K? The integrated squared error was minimized numerically over the $3K - 1$ mixture parameters. The K components and the resulting mixture are displayed in Figure 9.5. When K is 3 or 4, the two densities are almost indistinguishable to the naked eye. However, no finite K will result in zero ISE.

These best-fitting mixtures correspond to what might be expected if fitted by the EM algorithm with a very large sample ($n = \infty$). If we focus on the $K = 3$ mixture, a social scientist might wish to interpret the three clusters as low, middle, and upper class groups. The clusters weights are $c(0.210, 0.446, 0.344)$. How robust are these weights in two senses: are the best parameters sensitive (or insensitive), and how noisy are the parameter estimates with finite samples?

The Hessian matrix at the best solution defines a hyperellipse whose contours are nearly equal values of the integrated squared error (ISE) criterion. In Figure 9.6, six examples of parameters inside the one-sigma ellipse are displayed. The root ISE ranges from 15 to 66% larger, but the fits are nearly as good as the original in Figure 9.5 to the naked eye. However, a close look at the size of the cluster components reveals that the middle cluster has weights ranging from 27.6 to 60.0%, while

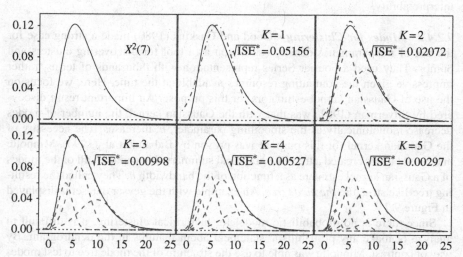

FIGURE 9.5 Numerically best Gaussian mixture approximations to a $\chi^2(7)$ density for $1 \le K \le 5$. The root optimal ISE is shown in each figure.

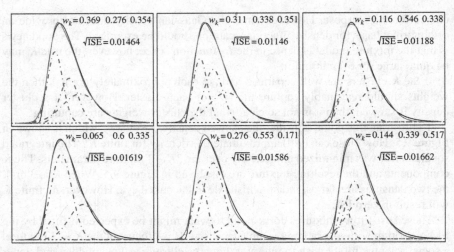

FIGURE 9.6 Six examples of nearby mixture solutions when $K = 3$ that are within a one-sigma confidence hyperellipse. The weights and criterion value are shown in each frame.

the lower cluster ranges from 6.5 to 36.9%. This lack of stability or robustness is perhaps not as well known as it should be. The variance of the high cluster is so large that its support overlaps the other two clusters in every example. If one tacks on statistical noise due to a finite sample size, then it is clear that taking the values of $\{\hat{w}_k\}$ as an appropriate size of a cluster is very hit-or-miss. Again, the lack of orthonormality of Gaussian mixtures is at the heart of its lack of robustness and reliable interpretability.

9.2.4.2 Modes for Clustering

Good and Gaskins (1980) made a strong case for the use of nonparametric density estimation as a tool for discovering clusters and bumps. They used a Fourier Series representation with thousands of terms, rather impressive given the computing resources available at the time. Here, we focus on the use of Gaussian kernel estimators for this purpose. An important result discovered by Silverman (1981) was that with the Gaussian kernel, the number of modes decrease monotonically as the smoothing parameter, h, increases. The necessity of the Gaussian kernel for this purpose was proven by Babaud et al. (1994). Minnotte and Scott (1993) created a tree-like graphical summary that captured all of the modes of a Gaussian kernel estimate as a function of the bandwidth h. They called the resulting tree-like structure *the mode tree*. An example with the geyser dataset is displayed in Figure 9.7.

Since there is no probabilistic model for hierarhical clustering, it is difficult to analyze or make any probabilistic statement about features in the dendrogram. By way of contrast, Minnotte was able to use the structure of the mode tree to test modes individually (see Minnotte and Scott 1993; Minnotte 1997). This is important since no single choice of h is likely to show all the potential modes at the same time.

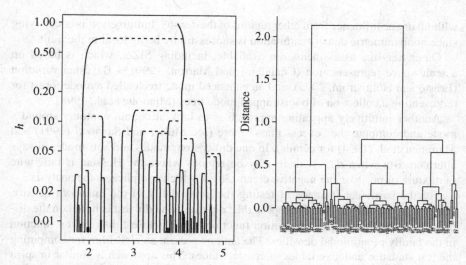

FIGURE 9.7 A mode tree and dendrogram of the geyser eruption times. The dendrogram is the hierarchical clustering tree based on average linkage.

Testing the Number of Modes or Individual Modes. Assessing the modes or the number of clusters is a challenging problem. Thus the bootstrap approach has natural appeal for its generality and lack of specific assumptions. For example, Gaussian mixtures can provide a useful representation of the density but are less useful nailing down the exact number of clusters due to the nonuniqueness of the representation.

Likewise, the lack of a probability model makes the dendrogram tree representation difficult to assess across bootstrap samples, since it is not obvious how to compare pairs of trees due to the change in labels. However, the mode tree is well-suited for this purpose.

As mentioned in Section 9.2.3, Silverman (1981) identified what he called "critical bandwidths" where an additional mode is about to appear when the bandwidth is decreased further. These are exactly where the horizontal dashed lines appear in the mode tree in Figure 9.7. Silverman suggested counting the number of modes at those critical bandwidths across many bootstrap samples, assessing the veracity of that mode count by the distribution of the mode count in the resamples. For complex densities, this approach requires some care, as noted by Matthews (1983). Specifically, if there is a true mode at a relatively low height (that requires a small bandwidth to properly resolve), other taller modes may split into many noisy modes at that critical bandwidth, making it much more likely to mask the smaller mode across bootstrap samples. Likewise, outliers have an influence on a strict counting of modes, since outliers appear in the bootstrap sample with probability $1 - (1 - 1/n)^n \approx 1 - e^{-1}$ or 63.2% of the time.

Minnotte (1998) was able to show that testing individual modes only at critical bandwidths along that branch of the tree is sufficient for appropriate evaluation locally without undue influence from other regions of the density. Intuitively, this is satisfying since nonparametric density estimation is successful by being local in the limit.

Other tree-like assessments are available, including SiZer, which is based on a scale-space representation (Chaudhuri and Marron, 1999), a Bayesian variation (Erästö and Holmström, 2005), and an enhanced mode tree called a mode forest for representing a collection of bootstrapped mode trees (Minnotte et al., 1998).

Another intuitively appealing approach is to look at certain contours around a mode and compute the "excess mass" there (see Müller and Sawitzki (1991) and Mammen et al. (1994) for details). In one dimension, modes and anti-modes always alternate. But when $d > 1$, there may be regions where the Hessian is indefinite (a mixture of positive and negative eigenvalues), which complicates the analysis.

For the particular instance of testing the unimodality of the unknown density, Hartigan and Hartigan (1985) propose the "dip test." The dip test is based on the distance between the sample distribution function and the closest distribution function in the family of unimodal densities. The authors present an algorithm for computing the test statistics and give tables of critical values. This approach is similar in spirit to the "bump surgery" of Good and Gaskins (1980), who essentially attempted to project the density locally on another density without the bump.

Is the Number of Modes Monotone in $\Re^d$?   Given the sufficiency and necessity of a Gaussian kernel for monotonicity in one dimension, a similar result in the multivariate setting would require a spherical Gaussian kernel by consideration of cross sections. Now Lindsay (1983) and Ray and Lindsay (2005) have studied the geometry and presentation of multivariate Gaussian mixtures by focusing on the ridge line, which connects all modes. They show a very simple equal mixture of two bivariate normals with different covariance matrices that form an L-shape with a third mode at the juncture.

Finally, a simple counterexample was provided by Scott and Szewczyk (1997) (see Figure 9.8). Three points at the vertices of an equilateral triangle give rise to three modes if $0 < h < 2.8284$ and one mode if $h > 2.9565$. In the narrow range of bandwidths in between, $2.8284 < h < 2.9565$, a fourth mode appears at the origin. The last frame in Figure 9.8 shows a blowup of the center region with the four modes

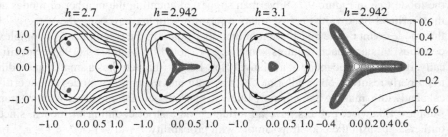

FIGURE 9.8 Contours of a bivariate Gaussian kernel density estimator with $n = 3$ points (black dots) on the unit circle forming an equilateral triangle. A highly nonlinear set of contour levels are displayed, so that the contours near the modes are emphasized.

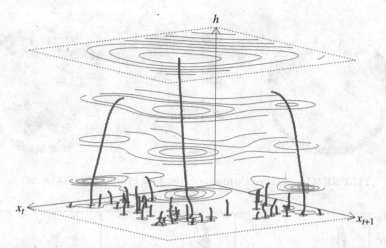

FIGURE 9.9 Bivariate mode tree of the lagged geyser dataset. Contours of the Gaussian kernel estimates are shown for $h \in (0.05, 0.20, 0.45, 0.70, 1.00)$.

clearly visible. This example uses a circular covariance matrix, which corresponds to a mixture of three equally weighted Gaussian densities. The practical consequences of this result do not seem too severe, but the bivariate mode tree might have some (short) orphan limbs. It should be mentioned that in $\Re^3$, four points at the vertices of a regular tetrahedron give rise to five modes over a wider interval of bandwidths.

Multivariate versions of the mode tree are discussed by Minnotte and Scott (1993) and Klemelä (2008). An example for the lagged geyser dataset is shown in Figure 9.9. The basic trimodal feature is readily apparent.

Using Modes to Perform Image Segmentation.   In k-means clustering, features are assigned to clusters by finding the nearest cluster center. With density estimation, each point is assigned to a mode following a hill-climbing algorithm (see Fukunaga and Hostetler (1975), Boswell (1983), and Carreira-Perpiñán and Williams (2003)).

This idea is illustrated on an MRI example, where the clusters provide a rough guide to anatomy and function, the so-called image segmentation problem (Duda et al., 2001). The three features of a 256×256 MRI image are shown in Figure 9.10, slightly enlarged. Of the 65,536 pixels, 63.7% are outside the skull. Marking and annotating such images can be a time-consuming task. Clustering algorithms can be of great help.

Focusing on the two-spin relaxation time variables, an ASH is displayed in Figure 9.11. The hill-climbing algorithm was applied to the data. A line segment is drawn between each data point and its mode. Note how the segmentation is apparent from the white space within the support of the density. The rightmost frame shows the 70 modes, most of which represent only a few pixels.

Six of the modes in Figure 9.11 are highlighted (circled) and represent four regions of interest. Starting at the bottom left corner and proceeding clockwise, the pixels in

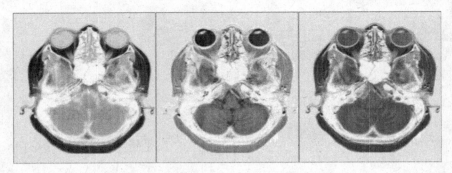

FIGURE 9.10   Gray-scale images of the three MRI variables (t_1, t_2, sd).

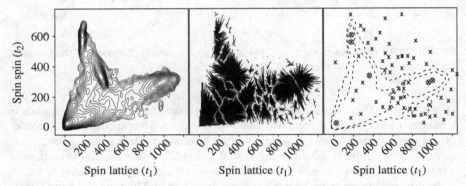

FIGURE 9.11   (Left) ASH contours of (t_1, t_2) of an MRI image with 24,476 pixels. (Center) Hill-climbing of individual pixel values to the nearest mode. (Right) The 70 modes found are superimposed on two contours.

the six clusters are shown in the first four frames of Figure 9.12. No spatial information was used to define the clusters, but not surprisingly, the clusters demonstrate a lot of spatial homogeneity. The second frame displays two adjacent modes (left and right eyes). The fourth cluster displays two adjacent modes (bone). The final two images come from a different MRI image in a patient with a tumor. The tumor is clearly distinguished in the fifth frame, and seems to have surrounded a portion of the white matter (see Scott and Hutson, 1992). Adding more features and hill-climbing in higher dimensions can improve the segmentation.

From Kernels to Mixtures.   Since a kernel estimate is an equal-weight mixture model with $K = n$ and $w_k = 1/n$, a natural question is how to obtain a more parsimonious mixture approximation other than adopting the binning approach. Scott and Szewczyk (2001) proposed a greedy algorithm that starts with an n-component mixture, then iteratively replaces a pair of adjacent components with a single component that makes the least change. For this purpose, they investigated the several criterion, including L_2E and Hellinger. The so-called iterative pairwise replacement

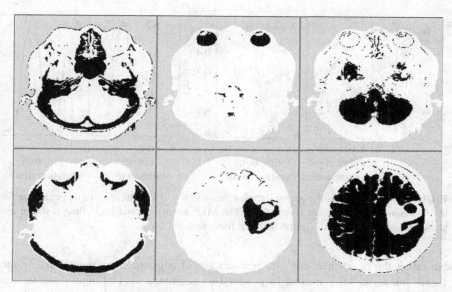

FIGURE 9.12   MRI images and subsets.

algorithm (IPRA) can be extended to be less local in its consideration of components to combine.

At each step, an adjacent pair of components is to be replaced by a single component, which may be illustrated by

$$w_k \phi(x|\mu_k, \sigma_k^2) + w_{k+1} \phi(x|\mu_{k+1}, \sigma_{k+1}^2) \quad \longleftarrow \quad w_{k'} \phi(x|\mu_{k'}, \sigma_{k'}^2),$$

where $w_{k'}$ is chosen to be $w_k + w_{k+1}$ so that the total area remains 1, and $\mu_{k'}$ and $\sigma_{k'}$ are chosen to minimize one of several criteria. The simplest choice is just the method-of-moments matching approach. Let $f_k = w_k/(w_k + w_{k+1})$; then take

$$\mu_{k'} = f_k \mu_k + (1 - f_k) \mu_{k+1} \qquad \text{and}$$
$$\sigma_{k'}^2 = f_k \sigma_k^2 + (1 - f_k) \sigma_{k+1}^2 + f_k (1 - f_k)(\mu_k - \mu_{k+1})^2.$$

At each iteration, the choice of a new mixture density was determined by the minimizer of the Hellinger distance between the original Gaussian kernel estimator and the alternative available mixture densities constructed by the greedy algorithm. In Figure 9.13, the IPRA approach is illustrated on the geyser dataset. For $K = 14$ (filled triangle) and beyond, the mixture densities are indistinguishable from the kernel estimate to the naked eye, and the Hellinger distance is less than 2.2×10^{-5}. Since there are 36 duplicate values in the geyser dataset, the Hellinger distance is exactly 0 for $K \geq 71$. $K = 70$ is depicted in the first frame by an open circle; the 0 values are not plotted on the log scale for $71 \leq K \leq 107$.

When $K = 8$ (second frame and filled circle in first frame), there are two intervals near the antimode and right mode where a small gap is apparent. Note that the two-component mixture in the third frame was not the result of an optimization algorithm,

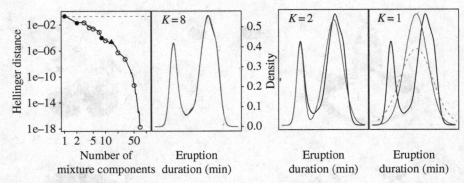

FIGURE 9.13 IPRA of the geyser duration dataset ($n = 107$). The smoothing parameter for the Gaussian kernel estimate is $h = 0.20$. The MLE normal fit and its Helling distance are shown as dashed lines in the fourth and first frames, respectively.

but rather the backwards stepwise replacement of adjacent component pairs. In the fourth frame, an MLE normal fit is added for reference. Its Hellinger distance is a little better than the MOM estimate. Other approaches, some optimization based, are discussed in Scott and Szewczyk (2001).

A related Gaussian-mixture-modeling problem considered by Marchette et al. (1994, 1996), and Priebe and Marchette (2000) is well-suited to a streaming environment. At each step, a new data point is received. The data point is tentatively assigned to one of the current components by a Bayesian procedure. Before the point is assigned to that component (and the parameters updated), an auxiliary decision is made whether to create a *new* component with the data point as its center instead. The result is a mixture that adapts well to the unknown density. The authors show that if a good filter is chosen, the resulting density is usually superior to a good fixed bandwidth estimate (in several dimensions).

9.3 SPECIALIZED TOPICS

9.3.1 Bootstrapping

Bootstrapping provides a means of estimating statistical error by resampling. The ordinary bootstrap resamples from the empirical density function, while the smoothed bootstrap resamples from the kernel density estimate $\hat{f}_h(\mathbf{x})$ given in Equation (6.39) or (6.45). In practice, bootstrap samples are generated by treating the kernel estimator as a mixture density; that is, the kernel estimate is viewed as the mixture of n equally probable kernels. To generate a smoothed bootstrap sample of the same size as the original multivariate sample $\{\mathbf{x}_i, i = 1, \ldots, n\}$:

1. Generate $\{j_1, j_2, \ldots, j_n\} \sim U(1, 2, \ldots, n)$ with replacement.
2. Generate $\{\mathbf{t}_1, \mathbf{t}_2, \ldots, \mathbf{t}_n\}$ from the scaled multivariate kernel.
3. Return $\mathbf{x}_i^* = \mathbf{x}_{j_i} + \mathbf{t}_i, \quad i = 1, \ldots, n.$

where $U(1, 2, \ldots, n)$ is the discrete uniform density and $\mathbf{x}^*$ is a bootstrap sample. The precise details for step 2 depend on the form of the multivariate estimator being used. The most common case is the product kernel in (6.39) with univariate kernel K and smoothing parameters $\{h_1, \ldots, h_d\}$. Then step 2 becomes

2a. Generate $\{u_{i1}, \ldots, u_{id}, i = 1, \ldots, n\}$ from $K(u)$.
2b. $\{\{t_{ij} = h_j u_{ij}, j = 1, \ldots, d\}, i = 1, \ldots, n\}$.

For the ordinary bootstrap, which corresponds to $h_j = 0$, only the first and third steps are required as $\mathbf{t}_i = \mathbf{0}$. The above algorithm is easily modified for adaptive density estimates.

Generating pseudo-random samples from specific kernels may be accomplished by the probability sampling approach $F_K^{-1}(U)$ where $U \sim U(0, 1)$. Packages are widely available for generating normal samples. The biweight kernel may be sampled by the following transformation of six uniform samples:

$$\log(u_1 u_2 u_3 / u_4 u_5 u_6) \div \log(u_1 u_2 u_3 u_4 u_5 u_6), \tag{9.10}$$

although the computations should not be organized in this fashion due to underflow (see Problem 9.7).

Generating Monte Carlo samples from $\hat{f}$ is identical to the bootstrap. Certain density estimates are not amenable to resampling. For example, what does it mean to resample from a kernel that is negative? How should a sample be drawn from the nearest-neighbor estimator when its integral is not finite? Ad hoc procedures may be envisioned, but some thought should be given to what the corresponding density estimate is after any modification.

9.3.2 Confidence Intervals

The difficulty with constructing confidence intervals for nonparametric estimates is the presence of bias in the estimates. Eliminating that bias is not possible, but on average the variance dominates the MSE. A practical compromise is to estimate the variance pointwise and construct two standard error bars around the estimate (see Problem 9.3). Such an interval is not a confidence interval for the unknown density function but rather a confidence interval for the nonparametric estimate.

The bootstrap may be used to obtain pointwise error bands for the positive kernel estimate $\hat{f}(x; h, \{x_i\})$. If samples from the original (unsmoothed) bootstrap are used to compute $\hat{f}(x; h, \{x_i^*\})$, then sample percentiles of these bootstrapped kernel density estimates may be superimposed on the original density estimate. In Figure 9.14, error bars (based on the 10th and 90th percentiles) of 200 bootstrap resamples from the silica dataset ($n = 22$) are shown, with the biweight kernel and $h = 3$. The error bars are shown on the square-root-density scale, which is the variance stabilizing transformation. Except near the boundaries, this technique gives reasonable answers.

The use of the original bootstrap does not address the problem of bias in the kernel estimate and does not help with the question of whether the bumps are spurious. The smoothed bootstrap will result in error bars that reflect this bias. In other

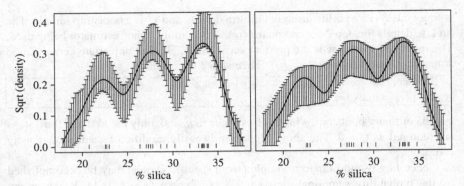

FIGURE 9.14 Bootstrap error bars for the unsmoothed and smoothed bootstrap resamples for the silica dataset. The 90th and 10th sample percentiles from 500 bootstrap estimates are shown.

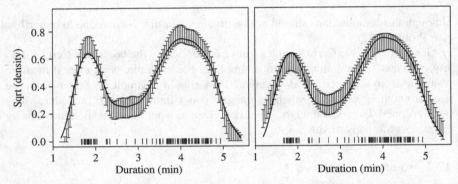

FIGURE 9.15 Bootstrap error bars for the unsmoothed and smoothed bootstrap resamples for the Old Faithful geyser dataset. The 10th and 90th sample percentiles from 200 bootstrap estimates are shown.

words, the bias between $\hat{f}(x; h, \{x_i\})$ and $f(x)$ should be similar to the bias between $\hat{f}(x; h, \{x_i^*\})$ and $\hat{f}(x; h, \{x_i\})$. The right frame in Figure 9.14 displays the error bars of 200 smoothed bootstrap kernel estimates. The downward bias at peaks and upward bias at dips is evident. The trimodal structure seems much less plausible based on the smoothed bootstrap error bars, as there are many unimodal density curves that exist in within the bands. Based on the analysis by Taylor (1989) discussed in Section 6.5.1.4, the smoothed bootstrap error bars will be somewhat wider than necessary because the smoothed bootstrap inflates the variance about 10%. The temptation is to pivot the error bars around the density estimate to reflect the estimated bias correctly.

The same bootstrap procedures were applied to the larger geyser dataset ($n = 107$) in Figure 9.15. The biweight kernel with $h = 0.5$ was chosen. The error bars are much narrower and the bimodal structure apparent in all the bootstrap resamples.

For regression problems, either smoothed or original bootstrap samples may be drawn and the regression estimate computed from those data. This idea was demonstrated by McDonald (1982). Alternatively, bootstrap samples can be based on the residuals rather than the data (Härdle and Bowman, 1988). The extension from pointwise to uniform confidence intervals for both density estimation and regression has been discussed by Hall and Titterington (1988) and first by Bickel and Rosenblatt (1973). Hall et al. (1989) recommend the use of the smoothed bootstrap asymptotically. Finally, the so-called wild bootstrap works in a more general setting (see Wu (1986) and Mammen (1993) for details).

9.3.3 Survival Analysis

Watson and Leadbetter (1964) considered nonparametric estimation of the hazard function

$$h(t) = \frac{f(t)}{1 - F(t)} = -\frac{d}{dt} \log[1 - F(t)],$$

which measures the instantaneous force of mortality given survival to time t. Given a sample of failure times $\{0 \le t_1 \le t_2 \le \ldots \le t_n\}$, the empirical cumulative distribution function may be substituted to provide an estimate of the hazard

$$\tilde{h}(t) = -\frac{d}{dt} \log[1 - F_n(t)]$$

$$= -\sum_{i=1}^{n} \left[\log\left(1 - \frac{i}{n}\right) - \log\left(1 - \frac{i-1}{n}\right) \right] \delta(t - t_i),$$

which is undefined for $t \ge t_n$ as that term involves $\log(0)$. Usually, the ecdf is multiplied by a factor such as $n/(n+1)$ to avoid that difficulty. With this modification and convolving the rough estimate with a kernel, a smooth estimate of the hazard function is obtained:

$$\hat{h}(t) = \sum_{i=1}^{n} \log\left[\frac{n-i+2}{n-i+1} \right] K_\lambda(t - t_i),$$

denoting the smoothing parameter by λ rather than h to avoid confusion.

Often survival data are subject to censoring, which occurs when the actual failure time of a subject is not observed either due to loss of follow-up or due to the end of the observation period. In this case, t_i is either the time of death (or failure) or the length of time the subject was observed before being lost (or the survival time at the end of the study). Suppose n_i is the number of individuals still under observation at time t_i, including the individual who died at time t_i. Then the Kaplan and Meier (1958) product-limit estimator replaces the ecdf:

$$\tilde{F}_n(t_i) = 1 - \prod_{j=1}^{i}\left(\frac{n_j - 1}{n_j}\right).$$

The authors prove that this is a consistent estimate of $F(t)$, assuming that the censoring mechanism is random. The kernel estimate for censored data follows in a similar fashion (see Problem 9.8).

9.3.4 High-Dimensional Holes

A challenge in higher dimensions is recognizing high-dimensional structure. A simple example is a hole in the data, or more generally, regions of lower density surrounded by regions of higher density, rather than the more common vice versa. A bivariate example was given in Figure 7.1. A trivariate sample of 5000 data points from a symmetric density with a hole was constructed. Its pairwise scatter is shown in Figure 9.16. The hole is not clearly visible as was the case with the bivariate example. In fact, applications of interactive tools such as rotation and brushing to the point cloud do not reveal the hole.

The first two frames in Figure 9.17 show a single α-shell with two nested contour surfaces (the inner shell, which is visible through the semitransparent outer shell, has radius about a fourth as long). Again, the appearance of two *nested* contours at the same α level is the *signature* of a hole in the density. (Two *nonnested* contours at the same α level is the signature of a bimodal density.) This figure might easily have resulted from displaying α-shells at two different α levels with ordinary normal data. At a slightly higher value of α than used before, the inner and outer contour surfaces join together, as shown in the third frame of Figure 9.17. The shading of the two contours at the same α-level indicates that the orientation of the two nested contours is reversed, compared to contours that are nested because they represent different α-levels. (The color version of this figure available online more clearly indicates the

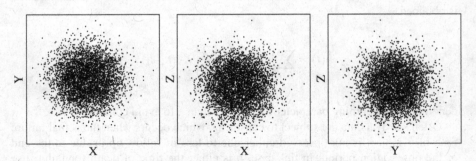

FIGURE 9.16   Pairwise scatterplots of 5000 trivariate simulated points with a hole. The hole is actually a region of lower density rather than a region around the origin with no data.

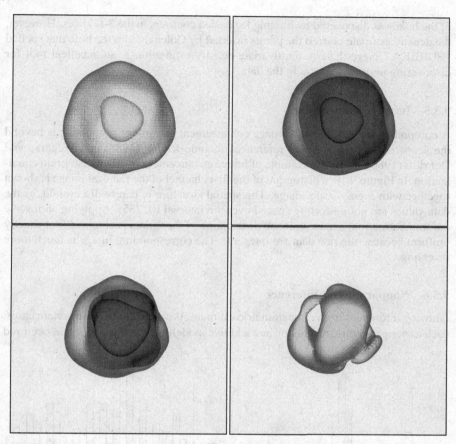

FIGURE 9.17 Contour shells derived from averaged shifted histogram estimate of a pseudo-random sample of 5000 points from a trivariate density with a "hole" in the middle. (First frame) The single α-level contour displayed is a pair of nested and nearly spherical shells. At values of α lower than shown, the inner shell shrinks and then vanishes. (Second frame) Same as the first frame, with the outer shell pealed away. The contour surfaces in a darker shade point to the higher density regions. In this case, the higher region is between the nested shells. Theoretically, the mode is a sphere, although finite samples will not achieve this exactly. (Third frame) At a slightly higher α-level, the outer shell has shrunk and the inner shell has merged, and in fact, they have merged in the back. (Fourth frame) At an even higher α-level, the shells have broken apart, although theoretically they should be converging to a sphere (the mode).

orientation.) Finally, in the fourth frame, with an α-level near 1, the contour surface approaching the sample mode becomes quite complex. These data happen to contain three sample modes located near the theoretical mode, which is the surface of a sphere.

In a poster session sponsored by the ASA Statistical Graphics section in Chicago in 1986, David Coleman created a 5-D dataset with a true hole carved in it. The presence of the hole was discovered by looking for nested contours in the 3-D slices. However, the density estimate blurred the points inserted by Coleman into the hole that spelled "EUREKA." Nevertheless, multivariate density estimation is an excellent tool for discovering unusual features in the data.

9.3.5 Image Enhancement

A comprehensive treatment of image enhancement and image processing is beyond the scope of this book. Some references include Ripley (1988) and Wegman and DePriest (1986). A simple example of image enhancement is called histogram equalization. In Figure 9.18, a histogram of the first channel of the Landsat image is shown together with a gray-scale image. The spatial structure is barely discernible, as the data values are not uniformly spaced over the interval (0, 255). Applying an inverse cdf transformation, a more uniform distribution of these data is obtained (not exactly uniform because the raw data are integers). The corresponding image is much more revealing.

9.3.6 Nonparametric Inference

Drawing inferences from nonparametric estimates is a growing field. In certain cases, such as nonparametric regression and additive modeling, much progress has occurred

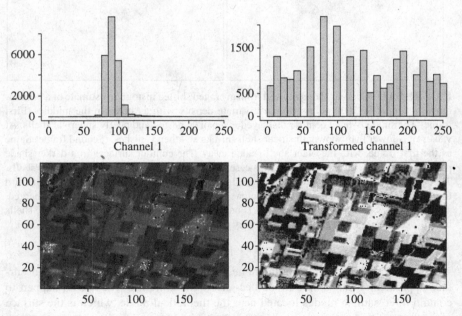

FIGURE 9.18 Histograms of raw data from Landsat scene and transformed data that are more nearly uniform. The increased dynamic range in the gray scale images may be observed.

because many of these procedures are linear smoothers, which represent a hybrid of parametric and nonparametric ideas. The difficult inference is how many terms should be in the model. Conditional on that information, the fitting often follows other linear parametric inference procedures. A few other examples could be given, but much work remains, particularly in the multivariate case.

Consider Pearson's goodness-of-fit hypothesis test:

$$H_0 : f = f_0 \quad \text{vs.} \quad H_1 : f \neq f_0.$$

Given a sample $\{x_i\}$, a bootstrap test may be constructed by introducing a measure of discrepancy such as

$$d(f_0, \hat{f}_h) = \int [\hat{f}_h(x) - f_0(x)]^2 \, dx.$$

A p-value can be determined by sampling from f_0, computing a kernel estimate, and determining the fraction of such samples where the discrepancy exceeds that for the original sample.

While the basis for the test is sound, the discrepancy measures the bias more than anything else because the kernel estimate scale is inflated due to smoothing. Thus the power can be improved by redefining the discrepancy to be

$$d_K(f_0, \hat{f}_h) = \int \left[\hat{f}_h(x) - [f_0 * K_h](x) \right]^2 \, dx.$$

Bowman (1992) reports that the power does not seem to be influenced by the choice of smoothing parameter. If both the null hypothesis and kernel are normal, then $f_0 * K_h$ is $N(0, 1 + h^2)$, so that the computations are relatively easy.

Other authors have considered tests of the adequacy of parametric fits based on nonparametric alternatives. In the regression context, see Cox et al. (1988) and Eubank and Speckman (1990).

9.3.7 Final Vignettes

9.3.7.1 Principal Curves and Density Ridges When bivariate data are not of the regression type, a lower dimensional summary may still be desired. Ordinary scatterplot smoothing is not appropriate as there is no response variable. Hastie and Stuetzle (1989) discuss an iterative procedure that moves points toward the "principal curve." The principal curve can be thought of as a local conditional mean. Alternatively, a similar concept may be developed based on the bivariate density surface. The *density ridge* should be more than just those points for which $\nabla f(x, y) = 0$ and the Hessian $Hf(x, y)$ is negative semidefinite. By the spectral representation,

$Hf(x,y) = \lambda_1 \mathbf{v}_1 \mathbf{v}_1^T + \lambda_2 \mathbf{v}_2 \mathbf{v}_2^T$. Now the curvature of the surface at (x, y) in the direction $\eta = (\eta_1, \eta_2)$ is given by

$$f''(x,y)[\eta] = \eta^T \nabla^2 f(x,y)\eta = \eta^T (\lambda_1 \mathbf{v}_1 \mathbf{v}_1^T + \lambda_2 \mathbf{v}_2 \mathbf{v}_2^T)$$
$$= \lambda_1 \cos^2 \theta_1 + \lambda_2 \cos^2 \theta_2 = \lambda_1 \cos^2 \theta_1 + \lambda_2 \sin^2 \theta_1,$$

where θ_1 and θ_2 are the angles between η and the two eigenvectors, respectively, and $\theta_2 = \theta_1 \pm \pi/2$, so that $\cos^2 \theta_2 = \sin^2 \theta_1$. A point is on the density ridge when the maximum negative curvature is perpendicular to the gradient. Thus a point is on the density ridge if $\nabla f(x,y)$ is an eigenvector and the *other* eigenvalue is negative (and is more negative if both eigenvalues are negative). Scott (1991b) investigated the possibility of identifying such structure in the density function.

9.3.7.2 Time Series Data   Special attention may be paid to time series data when there is substantial autocorrelation. Negative autocorrelation generally is helpful while positive autocorrelation clearly can be difficult in practice. The theory of optimal smoothing and cross-validation changes. For some recent results, see Hart (1984, 1991), Altman (1990), Hart and Vieu (1990), and Chiu (1989) as well as the monograph by Györfi et al. (1989).

9.3.7.3 Inverse Problems and Deconvolution   This class of problems is the most difficult in smoothing. Spline methods have been the favorite technique applied in this area. Some recent applications include estimation of yearly infection rates of AIDS (Brookmeyer, 1991). Excellent summaries are available in O'Sullivan (1986) and Wahba (1990).

9.3.7.4 Densities on the Sphere   See Watson (1985) and Fisher et al. (1987).

PROBLEMS

9.1   Suppose that $\mu_0 = 0$ and $\mu_1 > 0$ in a univariate shifted-normal model and that the prior odds are 1:1. By varying σ^2 (and hence the coefficient of detection), examine how the posterior odds vary at $x = \mu_0$ and $x = \mu_1$.

9.2   The multivariate shifted-normal model is $f_+(\mathbf{x}) = N(\mu_1, \Sigma)$ and $f_-(\mathbf{x}) = N(\mu_0, \Sigma)$. Compute the log–likelihood ratio and describe the level sets of constant likelihood ratio. In particular, describe what happens at $\mathbf{x} = (\mu_0 + \mu_1)/2$.

9.3   Using Serfling's proposition in (9.6), show that the kernel estimate $\hat{f}(x) = AN(f(x) + h^2\sigma_K^2 f''(x)/2, f(x)R(K)/(nh))$.

9.4   Compute the asymptotic approximations to the first four noncentral moments of the random variables Y_{nj} in Equation (9.7).

9.5   Find an equivalent bandwidth rule for the first- and second-derivative kernel estimators.

9.6   Consider a bootstrap sample size n from a dataset with n points. Show that the probability that a particular point is *not* in the resample is approximately e^{-1}. Comment on the effect that a single outlier has on the estimated p-values in Silverman's multimodality test.

9.7   In order to generate samples X from the biweight kernel, show that $X = 2Y - 1$ where $Y \sim \text{Beta}(3,3)$; show that $Y = V_1/(V_1 + V_2)$ where $V_i \sim \text{Gamma}(3)$; show that $V_i = W_1 + W_2 + W_3$ where $W_i \sim \text{Gamma}(1)$; and show that $W_i = -\log(U_i)$ where $U_i \sim U(0,1)$. Prove formula (9.10).

9.8   Investigate the Kaplan–Meier product-limit estimator for censored data. Construct a kernel estimator based on it.

APPENDIX A

COMPUTER GRAPHICS IN $\Re^3$

A.1   BIVARIATE AND TRIVARIATE CONTOURING DISPLAY

A.1.1   Bivariate Contouring

Bivariate contouring algorithms on rectangular grids present some subtle and interesting challenges, beyond considerations of speed. The discussion will be limited to a square mesh with only piecewise linear interpolation and piecewise linear contours. The contour is an approximation to the set $\{\mathbf{x} \in \Re^2 : f(\mathbf{x}) = c\}$. The algorithm should be local; that is, the algorithm is applied bin by bin and is based solely on the four function values at the corners. No other information, such as gradient values or values of the function outside the square, is permitted. (An alternative approach is to trace each individual contour (Dobkin et al., 1990); however, as this assumes that a starting point is given, some form of exhaustive search over the entire mesh cannot be avoided.)

A vertex $\mathbf{v}$ will be labeled "+" if $f(\mathbf{v}) > c$ and "$-$" if $f(\mathbf{v}) \leq c$ (this convention for handling the case $f(\mathbf{v}) = c$ is discussed later). Using linear interpolation along the sides of the square bins and connecting the points of intersection gives approximate contours as shown in Figure A.1. In the left frame, the contour is drawn bin by bin, one line per bin. However, in the next two frames, two lines are drawn in the center bin, as each side of that square has a point where $f(\mathbf{x}) = c$. Depending on the manner in which the four points are connected (assuming that the contour lines cannot cross), one of the shown contour patterns will be chosen.

Multivariate Density Estimation, First Edition. David W. Scott.
© 2015 John Wiley & Sons, Inc. Published 2015 by John Wiley & Sons, Inc.

The ambiguity could be resolved if a finer mesh were available, but by assumption it is not. Alternatively, a triangular mesh can be superimposed on the square mesh, as a triangle can have at most two points of intersection on its boundary when using linear interpolation along the three sides of the triangle. Save rotations, there are four distinct cases, as shown in Figure A.2. Reexamining the center bin in Figure A.1, the preference in a triangular mesh is determined by the orientation of the triangles in that bin. Therefore, the decision is by fiat rather than appeal to any other preference.

The question of how to handle equalities at the vertices is a practical problem. In statistics, very often a contour is attempted at the zero level, with disastrous results if the function has large regions where $f(\mathbf{x}) = 0$. Here, a simple idea is advanced that handles the equality problem neatly. The approach is summarized in Figure A.3. Contrast the second and fourth triangles. A contour line is drawn along the 0–0 edge when

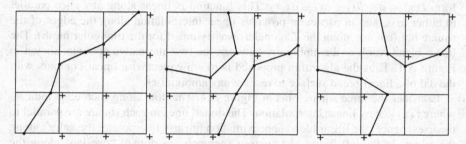

FIGURE A.1   Three examples of portions of a contour mesh. The unlabeled vertices are "−", that is, $f(\mathbf{v}) \leq c$. The right two meshes have identical values on the vertices, but different contours.

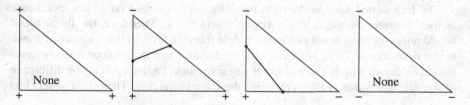

FIGURE A.2   Examples of function values on triangular mesh elements.

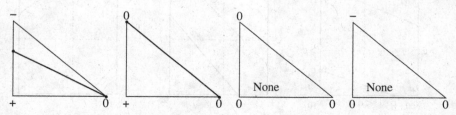

FIGURE A.3   Examples of function values with equality on triangular mesh elements.

the other corner is at a higher level, but a contour line is not drawn if the third corner satisfies $f(\mathbf{x}) \leq c$. The rationale is that such an edge may be drawn in the adjoining triangle, if appropriate. When a large region satisfies $f(\mathbf{x}) = c$, the only contour lines drawn are at the boundary where the function finally begins to increase. This choice seems more correct than drawing all 0–0 edges. Mathematically, the contour is being drawn at the level $f(\mathbf{x}) = c + \epsilon$ as $\epsilon \searrow 0$ and not at the exact level $f(\mathbf{x}) = c$.

A triangular mesh is attractive for its simplicity and its speed. Yet the multiplicity of choices suggests further consideration of the original square meshes. Of the 2^4 cases (depending as $f(\mathbf{v}) > c$ or $f(\mathbf{v}) \leq c$ at each of the 4 vertices), only the 2 cases illustrated in Figure A.1 result in any ambiguity. The cases occur when the sequence of signs along the four corners is $+ - + -$. One approach is to draw the exact contours of an approximation function defined on each square. The simplest function that interpolates the four values at the corners is the linear blend, defined by the bilinear form $f(\mathbf{x}) = a + bx_1 + cx_2 + dx_1x_2$. This function is linear along any slice parallel to either axis, and in particular provides linear interpolation along the edges of the square bin (but not along the diagonal as was assumed for the triangular mesh). The linear blend resolves the ambiguity between the two alternative contours shown in Figure A.1. Thus the algorithm proposed is to draw piecewise linear contours with the aid of a linear blend surface to resolve any ambiguities.

Consider the three square bins in Figure A.4. The dots along each edge indicate where $f(\mathbf{x}) = c$ by linear interpolation. The dotted lines in each square are parallel to the axes; hence, the linear blend approximation linearly interpolates the values along the edges. In the left frame, the contour segments (as drawn) "overlap" along the x-axis but not along the y-axis. It is left to the reader to prove that this configuration is impossible. For example, the vertical dotted line takes on values at its endpoints that are less than c, but takes on values greater than c in the middle. Either there is overlap in both directions or no overlap at all.

The four dotted lines in the right pair of square bins take on values less than c as they intersect the edges in the region where $f < c$. Therefore, the linear blend would reject the contours drawn in the middle frame in favor of the contours as drawn in the right frame. There is an easy test to decide when the contour line segments should be drawn with positive slope or negative slope. Define v_{ij} to be the difference of the function value at the vertex less the contour value c. Then the contour line

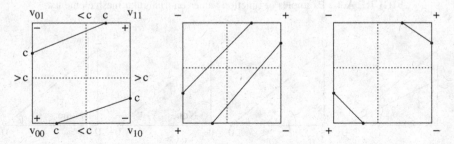

FIGURE A.4 Using the linear blend to resolve contour ambiguities on a square mesh.

segments with positive slope are drawn if $v_{00}v_{11} < v_{01}v_{10}$ and with negative slope if $v_{00}v_{11} > v_{01}v_{10}$. Equality may occur in two distinct cases. The first occurs when $v_{ij} = 0$ and all four function values at the corners equal c. By convention, no contour lines are drawn in this square bin. When equality occurs but $v_{ij} \neq 0$, the linear blend offers no guidance. (Verify that the contours of the linear blend in the square bin cross in this rare situation.)

A.1.2 Trivariate Contouring

The detailed description of bivariate contouring carries over to trivariate contouring over a cubical mesh. Assume that the array of function values f_{ijk} is given. In each cube, a collection of triangular patches is computed that defines the contour surface. As in the bivariate case, there are ambiguities when contouring the cubical bins directly. The analog of the triangular mesh is to divide the cube into six regions, for example, $x_1 < x_2 < x_3$. The three cases of interest with the contour patches are illustrated in Figure A.5. Notice that in the middle frame, a pair of triangular patches are drawn connecting the four points on the edges, and that the triangles are not unique. Since the orientation of the six tetrahedra in the cube is not unique, different contours can result just as in the bivariate case.

In general, fewer triangular patches are required if the cubical bins are contoured directly. A few of the cases have been illustrated in Chapter 1. As in the bivariate case, saddle points provide the greatest challenge. One special case that requires an unusual solution is illustrated in Figure A.6. The pair of patches leaves a situation on the closest face of the cube with overlap in both directions that would not be permitted with the (bivariate) linear blend fix. In fact, a "hole" in the 3-D contour surface remains and should be filled in during this step.

More visually pleasing surfaces can be drawn through enhancement of the *marching cubes* algorithm (Lorensen and Cline, 1987) by local smoothing over the triangular patches. The difference is illustrated in the Farin (1987), for example. Completely eliminating any noise from the surface is to be discouraged generally, in order to remind the user that these are estimated surfaces and subject to noise. A rough measure of the accuracy of the estimate can be obtained by examining the noise in the contours. These are subjective matters and viewing programs such as MinneView

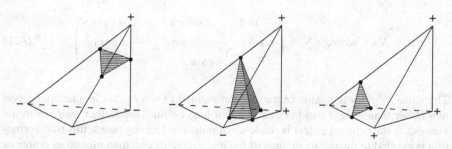

FIGURE A.5 Trivariate contour patches with tetrahedron bins.

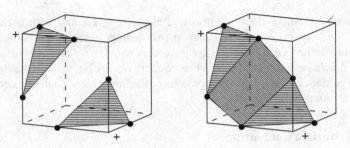

FIGURE A.6   Special case of cubical contouring.

allow toggling between different viewing options. For further details, see Foley and Van Dam (1982), Diamond (1982), Ripley (1981), Farin (1987), and Dobkin et al. (1990).

A.2   DRAWING 3-D OBJECTS ON THE COMPUTER

In this section, a simplified description of the algorithms for presenting 3-D objects on a computer screen is presented, both with and without perspective. Suppose that the object of interest has been positioned and scaled to fit in the cube $[-1, 1]^3$. Let $\mathbf{e} \in \Re^3$ be the Euclidean coordinates of the viewer and imagine that the computer screen is positioned halfway between the viewer and the origin.

The viewer's position is first transformed to spherical coordinates (r, ϕ, θ):

$$r = \left[\sum_{i=1}^{3} e_i^2 \right]^{1/2}; \quad \phi = \cos^{-1}(e_3/r); \quad \theta = \tan^{-1}(e_2, e_1),$$

where $\tan^{-1}(e_1, e_2)$ is a version of $\tan^{-1}(e_2/e_1)$ that returns a value between $(-\pi, \pi)$ rather than $(-\pi/2, \pi/2)$. Then given $\mathbf{x} = (x_1, x_2, x_3)^T$ in the cube, the screen coordinates are (x_1', x_2'), where $\mathbf{x}' = (x_1', x_2', x_3')^T$ is given by

$$\mathbf{x}' = \mathbf{V}\mathbf{x} \quad \text{where} \quad \mathbf{V} = \begin{pmatrix} -\sin\theta & -\cos\theta\cos\phi & -\cos\theta\sin\phi \\ \cos\theta & -\sin\theta\cos\phi & -\sin\theta\sin\phi \\ 0 & \sin\phi & -\cos\phi \end{pmatrix}. \qquad \text{(A.1)}$$

The value x_3' gives the value of the axis orthogonal to screen, but in a left-hand system rather than a right-hand system (so that larger values are farther away from the viewer). If the original object is made up of points or line segments, this transformation is applied to the points or ends of the line segments and then plotted as points or line segments on the plane.

Adding perspective to the equations is not difficult. A "1" is appended to each point $\mathbf{x}$ and the matrix $\mathbf{V}$ becomes the 4×4 matrix

$$\mathbf{V} = \begin{pmatrix} -\sin\theta & -\cos\theta\cos\phi & -\cos\theta\sin\phi & 0 \\ \cos\theta & -\sin\theta\cos\phi & -\sin\theta\sin\phi & 0 \\ 0 & \sin\phi & -\cos\phi & 0 \\ 0 & 0 & r & 1 \end{pmatrix}. \quad (A.2)$$

After computing $\mathbf{x}' = \mathbf{V}[\mathbf{x}^T\ 1]^T$, the screen coordinates (x_1'', x_2'') are given by

$$x_1'' = \frac{x_1'}{2} \times \frac{r}{x_3'} \quad \text{and} \quad x_2'' = \frac{x_2'}{2} \times \frac{r}{x_3'}. \quad (A.3)$$

APPENDIX B

DATASETS

B.1 US ECONOMIC VARIABLES DATASET

Annual economic variables for the United States between 1925 and 1940. The selection of variables for Chernoff faces in Figure 1.7 is also indicated.

Year	GNP	WPI	CPI	Income	Banks	Unemployment	Fuel	House	Suicide	Homicide
1925	1.794	5.33	5.25	1.274	2.8442	3.2	4.0014	9.37	12.0	8.3
1926	1.900	5.16	5.30	1.274	2.7742	1.8	4.1342	8.49	12.6	8.4
1927	1.898	4.93	5.20	1.274	2.6650	3.3	4.2492	8.10	13.2	8.4
1928	1.909	5.00	5.13	1.274	2.5798	4.2	4.3020	7.53	13.5	8.6
1929	2.036	4.91	5.13	1.274	2.4970	3.2	4.9039	5.09	13.9	8.4
1930	1.835	4.46	5.00	1.167	2.3679	8.7	4.7544	3.30	15.6	8.8
1931	1.693	3.76	4.56	1.108	2.1654	15.9	4.3954	2.54	16.8	9.2
1932	1.442	3.36	4.09	0.949	1.8734	23.6	3.4489	1.34	17.4	9.0
1933	1.415	3.40	3.88	0.921	1.4207	24.9	3.5274	0.93	15.9	9.7
1934	1.543	3.86	4.01	0.981	1.5348	21.7	3.9367	1.26	14.9	9.5
1935	1.695	4.13	4.11	1.068	1.5488	20.1	4.0797	2.21	14.3	8.3
1936	1.930	4.17	4.15	1.198	1.5329	16.9	5.0144	3.19	14.3	8.0
1937	2.032	4.45	4.30	1.236	1.5094	14.3	5.3560	3.36	15.0	7.6
1938	1.929	4.05	4.22	1.153	1.4867	19.0	4.8560	4.06	15.3	6.8
1939	2.094	3.98	4.16	1.232	1.4667	17.2	5.7958	5.15	14.1	6.4
1940	2.272	4.05	4.20	1.303	1.4534	14.6	6.2942	6.03	14.4	6.3

GNP	–	1958 prices; $$ 10**8	Area of face
WPI	–	Wholesale Price Index	Shape of face
CPI	–	Consumer Price Index	Length of nose
Income	–	Personal 1958 prices	Location of mouth
Banks	–	Number of commercial banks	Curve of smile
Unemployment	–	Percent civilian	Width of mouth
Fuel	–	Electric utilities cost	Location of eyes
House	–	New starts (000s)	Separation of eyes
Suicides	–	Rate/100,000	Angle of eyes
Homicide	–	Rate/100,000	Shape of eyes

Source: US Department of Commerce, Bureau of the Census, "Historical Statistics of the United States: Colonial Times to 1970," Washington, D.C., 1975.

B.2   UNIVERSITY DATASET

Characteristics of 28 selected universities around 1984.

College	$/F	S/F	G/U	Tuit	Bks	$$/F	NMP	$R&D	Bk/F
Amherst	0.59	9.7	0.00	81.5	2.4	1.2	0.8	3.	3.8
Brown	0.27	13.6	0.22	82.0	4.2	0.9	0.7	34.	3.8
Cal Tech	0.70	5.8	1.05	75.0	2.0	0.4	3.8	110.	1.3
Carnegie-M	0.29	12.9	0.38	63.0	2.2	0.6	0.4	51.	1.2
Chicago	0.38	7.5	1.73	70.7	6.8	0.5	3.3	51.	4.4
Columbia	0.52	12.3	2.49	78.9	7.2	0.4	0.3	56.	3.5
Dartmouth	0.89	17.1	0.16	81.9	3.9	3.1	0.8	36.	5.1
Duke	0.11	7.5	0.43	62.1	5.6	0.2	0.7	31.	2.1
Emory	0.90	21.2	1.34	62.0	4.1	0.9	1.2	53.	5.2
Harvard	0.70	6.6	1.46	81.9	10.2	0.5	4.1	36.	4.3
J. Hopkins	0.91	11.0	0.47	67.0	4.9	1.6	1.2	1260.	8.3
MIT	0.63	9.4	1.04	87.0	4.4	0.8	1.9	146.	1.9
Northwest.	0.27	8.5	0.66	80.8	5.4	0.4	1.4	24.	2.1
Notre Dame	0.29	12.4	0.26	59.5	3.7	3.2	0.4	19.	2.0
Oberlin	0.44	16.7	0.38	75.1	3.2	0.6	0.3	1.	4.4
U. Penn.	0.06	4.7	0.97	80.0	4.8	0.2	0.6	22.	0.7
Princeton	1.28	8.5	0.32	83.8	5.8	1.5	3.3	34.	4.9
Rice	0.95	9.3	0.31	35.0	3.3	1.7	5.5	20.	3.0
USC	0.11	16.6	1.13	68.4	4.7	0.5	0.3	30.	1.4
SMU	0.03	20.3	0.74	50.0	4.4	1.2	0.3	4.	4.3
Southwest.	0.66	17.5	0.00	35.0	1.0	3.3	0.2	0.	1.8
Stanford	0.66	10.1	0.79	82.2	7.0	0.8	1.2	90.	4.2
Swarthmore	0.56	7.6	0.02	71.3	2.4	0.7	1.5	2.	3.4
Texas A&M	0.02	10.6	0.26	12.0	3.7	0.1	0.6	12.	0.5
U. Texas	0.96	24.5	0.35	12.0	6.9	0.2	0.3	23.	2.5
Tulane	0.10	9.7	0.35	59.5	3.7	0.5	0.7	13.	2.1
Vanderbilt	0.14	7.0	0.63	61.0	3.9	0.2	0.3	21.	1.2
Yale	0.47	6.4	0.97	81.9	8.8	0.7	2.7	46.	4.8

1. Endowment per faculty ($millions)
2. Total students per faculty
3. Ratio graduate/undergraduate students
4. Tuition ($100s)
5. Square root (# Library Books/100,000)
6. Fund drive per faculty ($10,000s)

7. Percentage National Merit undergraduates
8. Federal R&D funds per faculty ($1000s)
9. Library books (1000s)/faculty

Source: Rice University Self-Study, 1985.

B.3 BLOOD FAT CONCENTRATION DATASET

Concentration of plasma cholesterol and plasma triglycerides (mg/dl) in 371 patients evaluated for chest pain. The data are listed sequentially for each patient $\{x_1, y_1, x_2, y_2, \ldots, x_{51}, y_{51}\}$ for the first group and similarly for the second group.

Data for 51 males with no evidence of heart disease:

```
195 348 237 174 205 158 201 171 190  85 180  82 193 210 170  90 150 167 200 154
228 119 169  86 178 166 251 211 234 143 222 284 116  87 157 134 194 121 130  64
206  99 158  87 167 177 217 114 234 116 190 132 178 157 265  73 219  98 266 486
190 108 156 126 187 109 149 146 147  95 155  48 207 195 238 172 168  71 210  91
208 139 160 116 243 101 209  97 221 156 178 116 289 120 201  72 168 100 162 227
207 160
```

Data for 320 males with narrowing of the arteries:

```
184 145 263 142 185 115 271 128 173  56 230 304 222 151 215 168 233 340 212 171
221 140 239  97 168 131 231 145 221 432 131 137 211 124 232 258 313 256 240 221
176 166 210  92 251 189 175 148 185 256 184 222 198 149 198 333 208 112 284 245
231 181 171 165 258 210 164  76 230 492 197  87 216 112 230  90 265 156 197 158
230 146 233 142 250 118 243  50 175 489 200  68 240 196 185 116 213 130 180  80
208 220 386 162 236 152 230 162 188 220 200 101 212 130 193 188 230 158 169 112
181 104 189  84 180 202 297 232 232 328 150 426 239 154 178 100 242 144 323 196
168 208 197 291 417 198 172 140 240 441 191 115 217 327 208 262 220  75 191 115
119  84 171 170 179 126 208 149 180 102 254 153 191 136 176 217 283 424 253 222
220 172 268 154 248 312 245 120 171 108 239  92 196 141 247 137 219 454 159 125
200 152 233 127 232 131 189 135 237 400 319 418 171  78 194 183 244 108 236 148
260 144 254 170 250 161 196 130 298 143 306 408 175 153 251 117 256 271 285 930
184 255 228 142 171 120 229 242 195 137 214 223 221 268 204 150 276 199 165 121
211  91 264 259 245 446 227 146 197 265 196 103 193 170 211 122 185 120 157  59
224 124 209  82 223  80 278 152 251 152 140 164 197 101 172 106 174 117 192 101
221 179 283 199 178 109 185 168 181 119 191 233 185 130 206 133 210 217 226  72
219 267 215 325 228 130 245 257 186 273 242  85 201 297 239 137 179 126 218 123
```

279 317 234 135 264 269 237 88 162 91 245 166 191 90 207 316 248 142 139 173
246 87 247 91 193 290 332 250 194 116 195 363 243 112 271 89 197 347 242 179
175 246 138 91 244 177 206 201 191 149 223 154 172 207 190 120 144 125 194 125
105 36 201 92 193 259 262 88 211 304 178 84 331 134 235 144 267 199 227 202
243 126 261 174 185 100 171 90 222 229 231 161 258 328 211 306 249 256 209 89
177 133 165 151 299 93 274 323 219 163 233 101 220 153 348 154 194 400 230 137
250 160 173 300 260 127 258 151 131 61 168 91 208 77 287 209 308 260 227 172
168 126 178 101 164 80 151 73 165 155 249 146 258 145 194 196 140 99 187 390
171 135 221 156 294 135 167 80 208 201 208 148 185 231 159 82 222 108 266 164
217 227 249 200 218 207 245 322 242 180 262 169 169 158 204 84 184 182 206 148
198 124 242 248 189 176 260 98 199 153 207 150 206 107 210 95 229 296 232 583
267 192 228 149 187 115 304 149 140 102 209 376 198 105 270 110 188 148 160 125
218 96 257 402 259 240 139 54 213 261 178 125 172 146 198 103 222 348 238 156
273 146 131 96 233 141 269 84 170 284 149 237 194 272 142 111 218 567 194 278
252 233 184 184 203 170 239 38 232 161 225 240 280 218 185 110 163 156 216 101

Source: Scott et al. (1978).

B.4 PENNY THICKNESS DATASET

Thickness in *mils* of a sample of 90 US Lincoln pennies dated from 1945 to 1989.
Two pennies were measured for each year. 1 *mil* = 0.001 inch.

1945	1946	1947	1948	1949	1950	1951	1952	1953	1954
51.8	53.2	53.0	53.6	53.2	53.4	53.4	53.4	50.6	52.2
54.6	54.2	53.0	52.2	51.6	51.0	54.0	54.6	54.8	54.6
1955	1956	1957	1958	1959	1960	1961	1962	1963	1964
52.6	54.0	53.6	53.2	56.8	57.0	56.6	54.6	58.0	55.0
54.2	53.6	55.0	53.2	57.2	55.2	57.0	56.0	56.2	56.0
1965	1966	1967	1968	1969	1970	1971	1972	1973	1974
56.6	57.2	57.6	57.2	58.0	57.2	58.0	57.0	59.0	59.0
58.2	56.0	55.8	56.0	56.0	55.6	55.8	57.0	57.6	54.0
1975	1976	1977	1978	1979	1980	1981	1982	1983	1984
54.0	53.2	54.0	53.0	54.0	53.6	54.8	54.4	52.8	55.2
54.2	53.2	53.2	53.6	53.0	53.0	54.0	54.6	52.6	54.0
1985	1986	1987	1988	1989					
55.6	55.0	55.0	54.0	56.2					
56.0	53.2	54.0	55.6	54.0					

Source: Bradford S. Brown, Registered Professional Engineer, Houston, Texas.

B.5　GAS METER ACCURACY DATASET

Accuracy of a gas meter as a function of flow rate and pressure. Alternating rows give the flow rate and the accuracy (100% is perfect).

44.217 psia:

35.00	54.00	69.0	71.00	88.0	107.00	179.00	252.00	314.00	341.00
96.36	97.33	97.7	97.86	97.7	98.17	98.25	98.54	98.83	99.05

360.00	360.0	376.00	395.00	721.00	722.00	1076.00	1077.00
99.16	99.1	99.21	99.25	99.87	99.99	100.41	100.32

2243.00	2354.00	3265.00	3614.00	3625.00	3772.00
100.35	100.38	100.45	100.73	100.85	100.87

74.609 psia:

58.00	74.00	74.00	92.00	112.0	187.00	205.0	258.00	327.00	346.00
97.25	97.67	97.72	97.95	98.3	99.61	98.9	98.92	99.47	99.68

369.00	372.00	392.00	402.00	402	438.00	438.00	486.00	486.00
99.93	99.92	99.72	100.08	100	99.88	99.85	99.93	99.91

538.00	650.00	650.00	728.00	729.00	813.00	814.00	1328.00
99.97	100.05	99.99	100.03	99.97	100.25	100.24	100.27

1331.00	2326.00	2348.00	3682.00	3686.00	3899.00
100.19	100.25	100.28	100.59	100.63	100.52

134.59 psia:

54.00	72.00	73.00	92.00	111.00	181.00	186.00	257.00	271.00
97.93	98.54	98.71	98.82	98.75	99.31	99.12	99.51	99.72

303.00	335.00	344.00	348.00	360.00	361.00	405.00	741.00	748.00
100.41	99.69	99.92	99.91	100.05	100.08	100.32	100.51	100.46

1079.0	1079.00	2343.00	2368.00	3614.00	3614.0	3815.0
100.8	100.78	100.47	100.46	100.58	100.5	100.3

194.6 psia:

37.00	55.00	73.00	74.00	92.00	111.00	183.00	183.00	188.00
97.68	98.75	98.88	98.69	99.03	98.98	99.37	99.27	99.19
189.00	265.00	328.00	342.00	358.00	360.00	385.00	408.00	658.00
99.44	100.14	99.97	100.17	100.34	100.36	100.64	100.63	100.53
681.00	1094.0	1101.0	2386.00	2391.00	3596.00	3694.00	3755.00	
100.49	100.8	100.7	100.49	100.51	100.64	100.69	100.57	

314.58 psia:

25.00	38.00	56.00	73.00	74.00	92.00	109.00	183.00	184.00	251.00
96.65	98.33	99.22	99.53	99.48	99.76	99.79	99.89	99.76	100.05
251.00	324.00	330.00	330.00	345.00	351.0	355.00	356.00	372.00	
99.88	100.04	100.14	100.04	100.19	100.3	100.29	100.17	100.71	
380.00	380.0	383.00	401.00	402.00	482.0	539.00	637.00	638.00	
100.72	100.8	100.83	100.85	100.75	100.8	100.92	100.69	100.75	
640.0	712.00	725.00	739.00	746.00	887.00	901.00	902.00	1066.00	
100.7	100.71	100.59	100.74	100.74	100.68	100.63	100.91	100.95	
1069.00	1082.00	1093.00	1117.00	1126.00	1194.00	2149.00	2323.00		
100.87	100.94	100.77	100.88	100.89	100.85	100.66	100.74		
2348.00	2403.00	2669.00	3608.00	3632.00	3781.00				
100.86	100.68	100.64	100.46	100.37	100.51				

434.6 psia:

19.00	36.00	56.00	74.00	74.00	90.00	108.00	178.00	183.00
96.27	98.78	99.31	99.58	99.58	99.85	99.94	99.85	99.74
187.00	257.00	257.00	328.00	345.00	362.00	363.00	364.00	407.00
99.89	100.02	99.88	100.15	100.22	100.99	100.35	100.34	100.81
726.00	726.00	1085.0	1116.00	1131.00	2291.00	2293.00	3627.00	
100.62	100.64	100.7	100.68	100.72	100.46	100.44	100.43	
3653.00	3767.00							
100.38	100.59							

614.7 psia:

19.00	36.00	56.00	71.00	71	73.0	74.0	95.00	110.00	180.00
97.17	98.75	99.32	100.05	100	99.8	99.7	100.01	100.17	100.06
181.00	251.0	327.00	342.00	362.00	362.00	377.00	400.00	750.00	
100.03	100.5	100.37	100.38	100.33	100.32	101.04	100.95	100.75	
757.00	1073.00	1082.00	2118.00	2177.00	3557.00	3577.00	3742.00		
100.72	100.74	100.79	100.92	100.88	100.39	100.34	100.57		

B.6   OLD FAITHFUL DATASET

Duration in minutes of 107 nearly consecutive eruptions of the Old Faithful geyser. A dash indicates missing observations in the sequence.

4.37	3.87	4.00	4.03	3.50	4.08	2.25	4.70	1.73	4.93
1.73	4.62	3.43	—	4.25	1.68	3.92	3.68	3.10	4.03
1.77	4.08	1.75	3.20	1.85	4.62	1.97	—	4.50	3.92
4.35	2.33	3.83	1.88	4.60	1.80	4.73	1.77	4.57	1.85
3.52	—	4.00	3.70	3.72	4.25	3.58	3.80	3.77	3.75
2.50	4.50	4.10	3.70	3.80	3.43	—	4.00	2.27	4.40
4.05	4.25	3.33	2.00	4.33	2.93	4.58	1.90	3.58	3.73
3.73	—	1.82	4.63	3.50	4.00	3.67	1.67	4.60	1.67
4.00	1.80	4.42	1.90	4.63	2.93	—	3.50	1.97	4.28
1.83	4.13	1.83	4.65	4.20	3.93	4.33	1.83	4.53	2.03
—	4.18	4.43	4.07	4.13	3.95	4.10	2.72	4.58	1.90
4.50	1.95	4.83	4.12						

Source: Weisberg (1985, table 9.1, p. 213).

B.7   SILICA DATASET

Percentage of silica in 22 chondrites meteors.

20.77	22.56	22.71	22.99	26.39	27.08	27.32	27.33	27.57	27.81
28.69	29.36	30.25	31.89	32.88	33.23	33.28	33.40	33.52	33.83
33.95	34.82								

Source: Ahrens (1965) and Good and Gaskins (1980).

B.8   LRL DATASET

Bin counts from a particle physics experiment. There are 172 bins with centers ranging from 285 to 1995 with bin width of 10 MeV. The sample size is 25,752.

5	11	17	21	15	17	23	25	30	22	36	29	33
43	54	55	59	44	58	66	59	55	67	75	82	98
94	85	92	102	113	122	153	155	193	197	207	258	305
332	318	378	457	540	592	646	773	787	783	695	774	759
692	559	557	499	431	421	353	315	343	306	262	265	254
225	246	225	196	150	118	114	99	121	106	112	122	120
126	126	141	122	122	115	119	166	135	154	120	162	156
175	193	162	178	201	214	230	216	229	214	197	170	181
183	144	114	120	132	109	108	97	102	89	71	92	58
65	55	53	40	42	46	47	37	49	38	29	34	42
45	42	40	59	42	35	41	35	48	41	47	49	37
40	33	33	37	29	26	38	22	27	27	13	18	25
24	21	16	24	14	23	21	17	17	21	10	14	18
16	21	6										

Source: Good and Gaskins (1980).

B.9   BUFFALO SNOWFALL DATASET

Annual snowfall in Buffalo, NY, 1910–1972, in inches.

126.4	82.4	78.1	51.1	90.9	76.2	104.5	87.4	110.5	25.0
69.3	53.5	39.8	63.6	46.7	72.9	79.6	83.6	80.7	60.3
79.0	74.4	49.6	54.7	71.8	49.1	103.9	51.6	82.4	83.6
77.8	79.3	89.6	85.5	58.0	120.7	110.5	65.4	39.9	40.1
88.7	71.4	83.0	55.9	89.9	84.8	105.2	113.7	124.7	114.5
115.6	102.4	101.4	89.8	71.5	70.9	98.3	55.5	66.1	78.4
120.5	97.0	110.0							

Source: Carmichael (1976) and Parzen (1979).

APPENDIX C

NOTATION AND ABBREVIATIONS

C.1 GENERAL MATHEMATICAL AND PROBABILITY NOTATION

I_A	indicator function for the set A	37
$E(X)$	expectation of X	37
$Var(X)$	variance of X	37
$Cov(X, Y)$	covariance of X and Y	102
$X \sim f$	X has density f	31
$C.V.$	coefficient of variation	179
Φ	cdf of the $N(0, 1)$ density	91
ϕ	pdf of the $N(0, 1)$ density	43
$\chi^2(n)$	pdf of the chi-squared density	31
$Pr(\cdot)$	probability of an event	31
$B(n, p)$	binomial pdf	37
$U(a, b)$	uniform pdf	7
$P(\lambda)$	Poisson pdf	91
$B(\mu, \nu)$	Beta function $[= \Gamma(\mu)\Gamma(\nu)/\Gamma(\mu + \nu)]$	29
$\Gamma(x)$	Gamma function	30
$Beta(\mu, \nu)$	Beta pdf	60
$N(\boldsymbol{\mu}, \Sigma)$	multivariate normal pdf	21

Multivariate Density Estimation, First Edition. David W. Scott.
© 2015 John Wiley & Sons, Inc. Published 2015 by John Wiley & Sons, Inc.

IQR	interquartile range	59
$R(g)$	$\int g(x)^2\, dx$ ("roughness" of g)	57
arg min	argument that minimizes a function	47
arg max	argument that maximizes a function	91
s/t	subject to (some constraints)	76
$f * g$	convolution of f and g	140
$\delta(t)$	Dirac delta function	39
$\delta_{\mu\nu}$	Kronecker delta function	140
$\mathrm{tr}(A)$	trace of the matrix A	165
I_d	$d \times d$ identity matrix	21
$|H|$	determinant of the matrix H	45
∇f	gradient of f $(\partial f/\partial x_i)$	165
$\nabla^2 f$	Hessian of $[\partial^2 f/(\partial x_i \partial x_j)]$	165
$(1 - x^2)_+$	$= \max(0, 1 - x^2)$	72
$a_n = O(b_n)$	$\Leftrightarrow a_n/b_n \to c$ as $n \to \infty$ ("big O")	42
$a_n = o(b_n)$	$\Leftrightarrow a_n/b_n \to 0$ as $n \to \infty$ ("little o")	56
AN	asymptotically normal	178
$V_d(a)$	volume of sphere in $\Re^d$ of radius a	30
$\lfloor x \rfloor$	floor(x), greatest integer $\leq x$	210
vech$A$	half-vectorization operation	167

C.2 DENSITY ABBREVIATIONS

cdf	cumulative distribution function F	37
ecdf	cumulative distribution function F_n	39
pdf	probability density function f	37
ASFP	averaged shifted frequency polygon	125
ASH	averaged shifted histogram	125
FP	frequency polygon	100
FP-ASH	frequency polygon interpolant of the ASH	125
LBFP	linear blend of a frequency polygon	111
WARP	weighting average of rounded points	135
$\hat{f}$	probability density estimator	21
$\hat{f}_{-i}$	probability density estimator omitting x_i	80
k–NN	kth nearest neighbor (density estimator)	167
B_k	kth bin	53
$K_h(t)$	$= K(t/h)/h$	137
S_α	α-level contour surface	21

C.3 ERROR MEASURE ABBREVIATIONS

AAISB	asymptotic adaptive integrated squared bias	197
AAMISE	asymptotic adaptive MISE	71
AAMSE	asymptotic adaptive mean squared error	71
AB	asymptotic bias	202
AISB	asymptotic integrated squared bias	58
AIV	asymptotic integrated variance	58
AMIAE	asymptotic mean absolute integrated error	223
AMISE	asymptotic mean integrated squared error	43
AMSE	asymptotic mean squared error	107
ASB	asymptotic squared bias	196
AV	asymptotic variance	196
Bias	bias of an estimator	40
BMISE	bootstrap mean integrated squared error	181
BMSE	bootstrap mean squared error	181
IMSE	integrated mean squared error	41
ISB	integrated squared bias	57
ISE	integrated squared error	41
IV	integrated variance	56
MAE	mean absolute error	222
MISE	mean integrated squared error	41
MSE	mean squared error	40
RCV	root coefficient of variation	221
RRMSE	relative root mean squared error	221
$(\cdot)^*$	the optimal value of $(\cdot)$, for example, MISE*	55

C.4 SMOOTHING PARAMETER ABBREVIATIONS

BCV	biased cross-validation	79
CV	cross-validation	79
OS	oversmoothed	77
PI	plug-in	184
UCV	unbiased cross-validation	80
h	smoothing parameter (bin or kernel width)	53
h^*	optimal smoothing parameter	55
h_{MISE}	optimal smoothing parameter w.r.t. MISE	173
$\hat{h}$	data-based choice for smoothing parameter	59
$\hat{h}_{BCV}$	data-based optimal choice w.r.t. BCV	80

REFERENCES

Abramson, I.S. (1982a). "On Bandwidth Variation in Kernel Estimates—A Square Root Law" *Ann. Statist.* **10** 1217–1223.

Abramson, I.S. (1982b). "Arbitrariness of the Pilot Estimator in Adaptive Kernel Methods" *J. Multivariate Analysis* **12** 562–567.

Ahrens, L.H. (1965). "Observations on the Fe–Si–Mg Relationship in Chondrites" *Geochimica et Cosmochimica Acta* **29** 801–806.

Aitkin, M. and Wilson, G.T. (1980). "Mixture Models, Outliers, and the EM Algorithm" *Technometrics* **22** 325–331.

Allen, D.M. (1974). "The Relationship Between Variable Selection and Data Augmentation and a Method for Prediction" *Technometrics* **16** 125–127.

Altman, N.S. (1990). "Kernel Smoothing of Data with Correlated Errors" *J. Amer. Statist. Assoc.* **85** 749–759.

Anderson, T.W. (2003). *An Introduction to Multivariate Statistical Analysis*, 3rd Edition. John Wiley & Sons, Inc., Hoboken, NJ.

Andrews, D.F. (1972). "Plots of High Dimensional Data" *Biometrics* **28** 125–136.

Asimov, D. (1985). "The Grand Tour: A Tool for Viewing Multidimensional Data" *SIAM J. Sci. Statist. Comp.* **6** 128–143.

Babaud, J., Witkin, A.P., Baudin, M., and Duda, R.O. (1994). "Uniqueness of the Gaussian Kernel for Scale-Space Filtering" *Pattern Analysis and Machine Intelligence, IEEE Trans. on* **1** 26–33.

Badhwar, G.D., Carnes, J.G., and Austin, W.W. (1982). "Use of Landsat-Derived Temporal Profiles for Corn-Soybean Feature Extraction and Classification" *Remote Sensing of Environment* **12** 57–79.

Banchoff, R.F. (1986). "Visualizing Two-Dimensional Phenomena in Four-Dimensional Space: A Computer Graphics Approach." In *Statistical Image Processing and Graphics*, E.J. Wegman and D.J. DePriest (eds.), pp. 187–202. Marcel Dekker, New York.

Banfield, D.F. and Raftery, A.E. (1993). "Model-Based Gaussian and Non-Gaussian Clustering" *Biometrics* **49** 803–821.

Bartlett, M.S. (1963). "Statistical Estimation of Density Functions" *Sankhyā Ser. A* **25** 245–254.

Basu, A., Harris, I.R., Hjort, N.L., and Jones, M.C. (1998). "Robust and efficient estimation by minimising a density power divergence" *Biometrika* **85** 549–559.

Becker, R.A., Chambers, J.M., and Wilks, A.R. (1988). *The New S Language.* Wadsworth & Brooks/Cole, Pacific Grove, CA.

Becker, R.A. and Cleveland, W.S. (1987). "Brushing Scatterplots" *Technometrics* **29** 127–142.

Bellman, R.E. (1961). *Adaptive Control Processes.* Princeton University Press, Princeton, NJ.

Bickel, P.J. and Rosenblatt, M. (1973). "On Some Global Measures of the Deviations of Density Function Estimates" *Ann. Statist.* **1** 1071–1095.

Blackman, R.B. and Tukey, J.W. (1958). *Measurement of Power Spectra*, Dover Publications, New York.

Boswell, S.B. (1983). "Nonparametric Mode Estimation for Higher Dimensional Densities." Ph.D. thesis, Department of Mathematical Sciences, Rice University.

Bowman, A.W. (1984). "An Alternative Method of Cross-Validation for the Smoothing of Density Estimates" *Biometrika* **71** 353–360.

Bowman, A.W. (1985). "A Comparative Study of Some Kernel-Based Nonparametric Density Estimators" *J. Statist. Comp. Simul.* **21** 313–327.

Bowman, A.W. (1992). "Density Based Tests for Goodness-of-Fit" *J. Statist. Comp. Simul.* **40** 1–13.

Bowyer, A. (1980). "Experiments and Computer Modelling in Stick-Slip." Ph.D. thesis, University of London, England.

Box, G.E.P. and Cox, D.R. (1964). "An Analysis of Transformations" *J. Roy. Statist. Soc. B* **26** 211–243.

Breiman, L., Friedman, J.H., Olshen, A., and Stone, C.J. (1984). *CART: Classification and Regression Trees.* Wadsworth, Belmont, CA.

Breiman, L., Meisel, W., and Purcell, E. (1977). "Variable Kernel Estimates of Multivariate Densities and Their Calibration" *Technometrics* **19** 135–144.

Brookmeyer, R. (1991). "Reconstruction and Future Trends of the AIDS Epidemic in the United States" *Science* **253** 37–42.

Cacoullos, T. (1966). "Estimation of a Multivariate Density" *Ann. Inst. Statist. Math.* **18** 178–189.

Carmichael, J.-P. (1976). "The Autoregressive Method: A Method for Approximating and Estimating Positive Functions." Ph.D. thesis, Statistical Science Division, SUNY, Buffalo, NY.

Carreira-Perpiñán, M.Á. and Williams, C.K.I. (2003). "On the Number of Modes of a Gaussian Mixture" *Scale Space Methods in Computer Vision* Springer, pp. 625–640.

Carr, D.B., Littlefield, R.J., Nicholson, W.L., and Littlefield, J.S. (1987). "Scatterplot Matrix Techniques for Large N" *J. Amer. Statist. Assoc.* **83** 596–610.

Carr, D.B. and Nicholson, W.L. (1988). "EXPLOR4: A Program for Exploring Four-Dimensional Data Using Stereo-Ray Glyphs, Dimensional Constraints, Rotation, and Masking." In *Dynamic Graphics for Statistics*, W.S. Cleveland and M.E. McGill (eds.), pp. 309–329. Wadsworth & Brooks/Cole, Pacific Grove, CA.

Carr, D.B., Nicholson, W.L., Littlefield, R.J., and Hall, D.L. (1986). "Interactive Color Display Methods for Multivariate Data." In *Statistical Image Processing and Graphics*, E.J. Wegman and D.J. DePriest (eds.), pp. 215–250. Marcel Dekker, New York.

Cencov, N.N. (1962). "Evaluation of an Unknown Density from Observations" *Soviet Mathematics* **3** 1559–1562.

Chamayou, J.M.F. (1980). "Averaging Shifted Histograms" *Computer Physics Communications* **21** 145–161.

Chambers, J.M., Cleveland, W.S., Kleiner, B., and Tukey, P.A. (1983). *Graphical Methods for Data Analysis*. Wadsworth, Belmont, CA.

Chaudhuri, P. and Marron, J.S. (1999). "SiZer for Exploration of Structures in Curves" *J. Amer. Statist. Assoc.* **94** 807–823.

Chernoff, H. (1973). "The Use of Faces to Represent Points in k-Dimensional Space Graphically" *J. Amer. Statist. Assoc.* **68** 361–368.

Chiu, S.T. (1989). "Bandwidth Selection for Kernel Estimates with Correlated Noise" *Statist. Prob. Lett.* **8** 347–354.

Chiu, S.T. (1991). "Bandwidth Selection for Kernel Density Estimation" *Ann. Statist.* **19** 1883–1905.

Choi, E. and Hall, P. (1999). "Data sharpening as a prelude to density estimation" *Biometrika* **86** 941–947.

Chow, Y.S., Geman, S., and Wu, L.D. (1983). "Consistent Cross-Validated Density Estimation" *Ann. Statist.* **11** 25–38.

Clark, R.M. (1975). "A Calibration Curve for Radiocarbon Dates" *Antiquity* **49** 251–266.

Cleveland, W.S. (1979). "Robust Locally Weighted Regression and Smoothing Scatterplots" *J. Amer. Statist. Assoc.* **74** 829–836.

Cleveland, W.S. and McGill, M.E. (eds.) (1988). *Dynamic Graphics for Statistics*. Wadsworth & Brooks/Cole, Pacific Grove, CA.

Cox, D.D., Koh, E., Wahba, G., and Yandell, B. (1988). "Testing the (Parametric) Null Model Hypothesis in (Semiparametric) Partial and Generalized Spline Models" *Ann. Statist.* **16** 113–119.

Davis, K.B. (1975). "Mean Square Error Properties of Density Estimates" *Ann. Statist.* **3** 1025–1030.

Day, N.E. (1969). "Estimating the Components of a Mixture of Normal Distributions" *Biometrika* **56** 463–474.

Deheuvels, P. (1977a). "Estimation non paramétrique de la densité par histogrammes généralises" *Revue de Statistique Appliquée, v.15* **25/3** 5–42.

Deheuvels, P. (1977b). "Estimation non paramétrique de la densité par histogrammes généralises II" *Publications de l'Institute Statistique de l'Université Paris* **22** 1–23.

Dempster, A.P., Laird, N.M., and Rubin, D.B. (1977). "Maximum Likelihood Estimation from Incomplete Data Via the EM Algorithm" *J. Roy. Statist. Soc. B* **39** 1–38.

Devroye, L. (1987). *A Course in Density Estimation*. Birkhäuser, Boston.

Devroye, L. and Györfi, L. (1985). *Nonparametric Density Estimation: The L_1 View*. John Wiley, New York.

Diaconis, P. and Freedman, D. (1984). "Asymptotics of Graphical Projection Pursuit" *Ann. Statist.* **12** 793–815.

Diaconis, P. and Friedman, J.H. (1983). "M and N Plots." In *Recent Advances in Statistics: Papers in Honor of Herman Chernoff on His Sixtieth Birthday*, M.H. Rizvi, J.S. Rustagi, D. Siegmund (eds.), pp. 425–447. Academic Press, New York.

Diamond, R. (1982). "Two Contouring Algorithms." In *Computational Crystallograph*, D. Sayre (ed.), pp. 266–272. Clarendon Press, Oxford, England.

DIW (1983). "Das Sozio-Ökonomische Panel." Deutsches Institut für Wirtschaftsforschung, Berlin.

Doane, D.P. (1976). "Aesthetic Frequency Classifications" *Amer. Statist.* **30** 181–183.

Dobkin, D.P., Levy, S.V.F., Thurston, W.P., and Wilks, A.R. (1990). "Contour Tracing by Piecewise Linear Approximations" *ACM Trans. Graphics* **9** 389–423.

Donoho, A.W., Donoho, D.L., and Gasko, M. (1988). "MacSpin: Dynamic Graphics on a Desktop Computer." In *Dynamic Graphics for Statistics*, W.S. Cleveland and M.E. McGill (eds.), pp. 331–351. Wadsworth & Brooks/Cole, Pacific Grove, CA.

Donoho, D.L. (1988). "One-Sided Inference About Functionals of a Density" *Ann. Statist.* **16** 1390–1420.

Donoho, D.L. and Liu, R.C. (1988). "The 'Automatic' Robustness of Minimum Distance Functional" *Ann. Statist.* **16** 552–586.

Duan, N. (1991). "Comment on "Transformations in Density Estimation (with Discussion)" by Wand, M.P., Marron, J.S., and Ruppert, D." *J. Amer. Statist. Assoc.* **86** 355–356.

Duda, R.O. and Hart, P.E. (1973). *Pattern Classification and Scene Analysis*. John Wiley, New York.

Duda, R.O., Hart, P.E., and Stork, D.G. (2001). *Pattern Classification*. Second edition. John Wiley, New York.

Duin, R.P.W. (1976). "On the Choice of Smoothing Parameter for Parzen Estimators of Probability Density Functions" *IEEE Trans. Comp.* **C-25** 1175–1179.

Duong, T. (2004). "Bandwidth Selectors for Multivariate Kernel Density Estimation" Ph.D. thesis, School of Mathematics and Statistics, University of Western Australia.

Duong, T. (2014). "R Package 'ks' Version 1.9.2" *http://CRAN.R-project.org/package=ks*.

Duong, T. and Hazelton, M.L. (2005). "Cross-validation Bandwidth Matrices for Multivariate Kernel Density Estimation" *Scand. J. Statist.* **32** 485–506.

Duong, T. and Hazelton, M.L. (2005). "Convergence Rates for Unconstrained Bandwidth Matrix Selectors in Multivariate Kernel Density Estimation" *J. Multivariate Analysis* **93** 417–433.

Efron, B. (1982). *The Jackknife, Bootstrap, and Other Resampling Plans*. SIAM, Philadelphia.

Emerson, J.D. and Hoaglin, D.C. (1983). "Stem-and-Leaf Displays." In *Understanding Robust and Exploratory Data Analysis*, D.C. Hoaglin, F. Mosteller, and J.W. Tukey (eds.), pp. 7–32. John Wiley, New York.

Epanechnikov, V.K. (1969). "Non-Parametric Estimation of a Multivariate Probability Density" *Theory Prob. Appl.* **14** 153–158.

Erästö, P. and Holmström, L. (2005). "Bayesian Multiscale Smoothing for Making Inferences About Features in Scatterplots" *J. Comp. Graph. Statist.* **14** 569–589.

Eubank, R.L. (1988). *Spline Smoothing and Nonparametric Regression*. Marcel Dekker, New York.

Eubank, R.L. and Speckman, P. (1990). "Curve Fitting by Polynomial-Trigonometric Regression" *Biometrika* **77** 1–10.

Everitt, B.S. and Hand, D.J. (1981). *Finite Mixture Distributions*. Chapman and Hall, London.

Fan, J.Q. (1992). "Design-Adaptive Nonparametric Regression" *J. Amer. Statist. Assoc.* **87** 998–1004.

Farin, G.E., ed. (1987). *Geometric Modeling: Algorithms and New Trends*. SIAM, Philadelphia.

Farrell, R.H. (1972). "On the Best Obtainable Asymptotic Rates of Convergence in Estimation of the Density Function at a Point" *Ann. Math. Statist.* **43** 170–180.

Fienberg, S.E. (1979). "Graphical Methods in Statistics" *Amer. Statist.* **33** 165–178.

Fisher, N.I., Lewis, T., and Embleton, J.J. (1987). *Statistical Analysis of Spherical Data*. Cambridge University Press, Cambridge, England.

Fisher, R.A. (1922). "On the Mathematical Foundations of Theoretical Statistics" *Philosophical Trans. Royal Society London (A)* **222** 309–368.

Fisher, R.A. (1932). *Statistical Methods for Research Workers*, Fourth Edition. Oliver and Boyd, Edinburgh.

Fisherkeller, M.A., Friedman, J.H., and Tukey, J.W. (1974). "PRIM-9: An Interactive Multidimensional Data Display and Analysis System." SLAC-PUB-1408, Stanford Linear Accelerator Center, Stanford, CA.

Fix, E. and Hodges, J.L., Jr. (1951). "Nonparametric Discrimination: Consistency Properties." Report Number 4, USAF School of Aviation Medicine, Randolph Field, Texas.

Flury, B. and Riedwyl, H. (1981). "Graphical Representation of Multivariate Data by Means of Asymmetrical Faces" *J. Amer. Statist. Assoc.* **76** 757–765.

Foley, J.D. and Van Dam, A. (1982). *Fundamentals of Interactive Computer Graphics*. Addison-Wesley, Reading, MA.

Fraley, C. and Raftery, A.E. (2002). "Model-based clustering, discriminant analysis, and density estimation" *J. Amer. Statist. Assoc.* **97** 611–631.

Fraley, C., Raftery, A.E., and Scrucca, L. (2014). "R Package 'mclust' Version 4.3" *http://CRAN.R-project.org/package=mclust*.

Freedman, D. and Diaconis, P. (1981). "On the Histogram as a Density Estimator: L_2 Theory" *Zeitschrift für Wahrscheinlichkeitstheorie und verwandte Gebiete* **57** 453–476.

Friedman, J.H. (1984). "A Variable Span Smoother." Technical Report 5, Department of Statistics, Stanford University.

Friedman, J.H. (1991). "Multivariate Adaptive Regression Splines (with Discussion)" *Ann. Statist.* **19** 1–141.

Friedman, J.H. and Stuetzle, W. (1981). "Projection Pursuit Regression" *J. Amer. Statist. Assoc.* **76** 817–23.

Friedman, J.H., Stuetzle, W., and Schroeder, A. (1984). "Projection Pursuit Density Estimation" *J. Amer. Statist. Assoc.* **79** 599–608.

Friedman, J.H. and Tukey, J.W. (1974). "A Projection Pursuit Algorithm for Exploratory Data Analysis" *IEEE Trans. in Computers* **C-23** 881–890.

Fryer, M.J. (1976). "Some Errors Associated with the Non-Parametric Estimation of Density Functions" *J. Inst. Maths. Applics.* **18** 371–380.

Fukunaga, K. and Hostetler, L.D. (1975). "The Estimation of the Gradient of a Density Function, With Applications in Pattern Recognition" *Information Theory, IEEE Trans. on* **21** 32–40.

Galton, F. (1886). "Regression Towards Mediocrity in Hereditary Stature" *J. Anthropological Institute* **15** 246–263.

Gasser, T. and Engel, J. (1990). "The Choice of Weights in Kernel Regression Estimation" *Biometrika* **77** 377–381.

Gasser, T. and Müller, H.G. (1979). "Kernel Estimation of Regression Functions." In *Smoothing Techniques for Curve Estimation*, Lecture Notes in Mathematics 757, pp. 23–68. Springer-Verlag, Berlin.

Gasser, T., Müller, H.G., and Mammitzsch, V. (1985). "Kernels for Nonparametric Curve Estimation" *J. Roy. Statist. Soc. B* **47** 238–252.

Glenn, N.D. (2005). *With this ring: A national survey on marriage in America*, National Fatherhood Organization, Gaithersburg, MD.

Good, I.J. and Gaskins, R.A. (1972). "Global Nonparametric Estimation of Probability Densities" *Virginia Journal of Science* **23** 171–193.

Good, I.J. and Gaskins, R.A. (1980). "Density Estimation and Bump-Hunting by the Penalized Likelihood Method Exemplified by the Scattering and Meteorite Data (with Discussion)" *J. Amer. Statist. Assoc.* **75** 42–73.

Graunt, J. (1662). *Natural and Political Observations Made upon the Bills of Mortality*. Martyn, London.

Gross, A.J. and Clark, V.A. (1975). *Survival Distributions: Reliability Applications in the Biomedical Sciences*. John Wiley, New York.

Györfi, L., Härdle, W., Sarda, P., and Vieu, P. (1989). *Nonparametric Curve Estimation from Time Series*. Springer-Verlag, Berlin.

Habbema, J.D.F., Hermans, J., and Van Der Broek, K. (1974). "A Stepwise Discriminant Analysis Program Using Density Estimation," *COMPSTAT 1974, Proceedings in Computational Statistics*, G. Bruckman (ed.), pp. 101–110, Physica-Verlag, Vienna.

Hald, A. (1990). *A History of Probability and Statistics and Their Application Before 1750*. John Wiley, New York.

Hall, P. (1984). "Central Limit Theorem for Integrated Square Error of Multivariate Density Estimators" *J. Multivariate Analysis* **14** 1–16.

Hall, P., DiCiccio, T.J., and Romano, J.P. (1989). "On Smoothing and the Bootstrap" *Ann. Statist.* **17** 692–704.

Hall, P. and Marron, J.S. (1987a). "On the Amount of Noise Inherent in Bandwidth Selection for a Kernel Density Estimator" *Ann. Statist.* **15** 163–181.

Hall, P. and Marron, J.S. (1987b). "Extent to Which Least-Squares Cross-Validation Minimises Integrated Square Error in Nonparametric Density Estimation" *Prob. Theory Related Fields* **74** 567–581.

Hall, P. and Marron, J.S. (1987c). "Estimation of Integrated Squared Density Derivatives" *Statist. Prob. Lett.* **6** 109–115.

Hall, P. and Marron, J.S. (1988). "Variable Window Width Kernel Estimates of Probability Densities" *Prob. Theory Related Fields* **80** 37–49.

Hall, P. and Marron, J.S. (1991). "Lower Bounds for Bandwidth Selection in Density Estimation" *Prob. Theory Related Fields* **90** 149–173.

Hall, P., Marron, J.S., and Park, B.U. (1992). "Smoothed cross-validation" *Prob. Theory Related Fields* **92** 1–20.

Hall, P. and Minnotte, M.C. (2002). "High order data sharpening for density estimation" *J. Roy. Statist. Soc. B* **64** 141–157.

Hall, P., Sheather, S.J., Jones, M.C., and Marron, J.S. (1991). "On Optimal Data-Based Bandwidth Selection in Kernel Density Estimation" *Biometrika* **78** 263–270.

Hall, P. and Titterington, D.M. (1988). "On Confidence Bands in Nonparametric Density Estimation and Regression" *J. Multivariate Analysis* **27** 228–254.

Hall, P. and Wand, M.P. (1988a). "Minimizing L_1 Distance in Nonparametric Density Estimation" *J. Multivariate Analysis* **26** 59–88.

Hall, P. and Wand, M.P. (1988b). "On Nonparametric Discrimination Using Density Differences" *Biometrika* **75** 541–547.

Hall, P. and Wand, M.P. (1996). "On the accuracy of binned kernel density estimators" *J. Multivariate Analysis* **56** 165–184.

Hampel, F.R. (1974). "The Influence Curve and Its Role in Robust Estimation" *J. Amer. Statist. Assoc.* **69** 383–393.

Hand, D.J. (1982). *Kernel Discriminant Analysis*. Research Studies Press, Chichester, England.

Härdle, W. (1984). "Robust Regression Function Estimation" *J. Multivariate Analysis* **14** 169–180.

Härdle, W. (1990). *Applied Nonparametric Regression*. Cambridge University Press, Cambridge, England.

Härdle, W. and Bowman, A.W. (1988). "Bootstrapping in Nonparametric Regression: Local Adaptive Smoothing and Confidence Bands" *J. Amer. Statist. Assoc.* **83** 102–110.

Härdle, W. and Gasser, T. (1984). "Robust Nonparametric Function Fitting" *J. Roy. Statist. Soc. B* **46** 42–51.

Härdle, W., Hall, P., and Marron, J.S. (1988). "How Far are Automatically Chosen Regression Smoothing Parameters from their Optimum? (with Discussion)" *J. Amer. Statist. Assoc.* **83** 86–101.

Härdle, W. and Scott, D.W. (1988). "Smoothing in Low and High Dimensions by Weighted Averaging Using Rounded Points" *Comp. Statist.* **7** 97–128.

Hart, J.D. (1984). "Efficiency of a Kernel Density Estimator Under an Autoregressive Dependence Model" *J. Amer. Statist. Assoc.* **79** 110–117.

Hart, J.D. (1985). "A Counterexample to a Claim Concerning the Convolution of Multimodal Distributions." *Comm. Statist.-Theor. Meth.* **14** 2943–2945.

Hart, J.D. (1991). "Kernel Regression Estimation with Time Series Errors" *J. Roy. Statist. Soc. B* **53** 173–187.

Hart, J.D. and Vieu, P. (1990). "Data-Driven Bandwidth Choice for Density Estimation Based on Dependent Data" *Ann. Statist.* **18** 873–890.

Hartigan, J.A. and Hartigan, P.M. (1985). "The Dip Test of Unimodality" *Ann. Statist.* **13** 70–84.

Hastie, T.J. and Stuetzle, W. (1989). "Principal Curves" *J. Amer. Statist. Assoc.* **84** 502–516.

Hastie, T.J. and Tibshirani, R.J. (1990). *Generalized Additive Models.* Chapman and Hall, London.

Hathaway, R.J. (1985). "A Constrained Formulation of Maximum-Likelihood Estimation for Normal Mixture Distributions", *Ann. Statist.* **13** 795–800.

Hazelton, M.L. (1998). "Bias Annihilating Bandwidths for Kernel Density Estimation at a Point" *Statist. Prob. Lett.* **38** 305–309.

Hiebert-Dodd, K.L. (1982). "An Evaluation of Mathematical Software That Solves Systems of Nonlinear Equations" *ACM Trans. Math. Soft.* **8** 5–20.

Hjort, N.L. (1986). "On Frequency Polygons and Averaged Shifted Histograms in Higher Dimensions." Tech Report 22, Stanford University.

Hodges, J.L. and Lehmann, E.L. (1956). "The Efficiency of Some Nonparametric Competitors of the *t*-test" *Ann. Math. Statist.* **27** 324–335.

Hoerl, A.E. and Kennard, R.W. (1970). "Ridge Regression: Biased Estimation for Non-Orthogonal Problems" *Technometrics* **12** 55–67.

Huber, P.J. (1964). "Robust Estimation of a Location Parameter" *Ann. Math. Statist.* **33** 73–101.

Huber, P.J. (1985). "Projection Pursuit (with Discussion)" *Ann. Statist.* **13** 435–525.

Hüsemann, J.A. and Terrell, G.R. (1991). "Optimal Parameter Choice for Error Minimization in Bivariate Histograms" *J. Multivariate Analysis* **37** 85–103.

IMSL (1991). "Fortran Subroutine Library and Exponent Graphics Package Manuals." Houston, TX. Currently RogueWave.com offers IMSL Numerical Libraries.

Inselberg, A. (1985). "The Plane with Parallel Coordinates" *The Visual Computer* **1** 69–91.

Izenman, G. (2009). *Modern Multivariate Statistical Techniques: Regression, Classification, and Manifold Learning* Springer, New York.

Izenman, A.J. and Sommer, C.J. (1988). "Philatelic Mixtures and Multimodal Densities" *J. Amer. Statist. Assoc.* **83** 941–953.

Jee, J.R. (1985). "A Study of Projection Pursuit Methods." Ph.D. thesis, Department of Mathematical Sciences, Rice University.

Jee, J.R. (1987). "Exploratory Projection Pursuit Using Nonparametric Density Estimation," *Proceedings of the Statistical Computing Section*, pp. 335–339, American Statistical Association, Alexandria, VA.

Jing, J., Koch, I., and Naito, K. (2012). "Polynomial histograms for multivariate density and mode estimation" *Scand. J. Statist.* **39** 75–96.

Johnson, N.L. (1949). "Systems of Frequency Curves Generated by Methods of Translation" *Biometrika* **36** 149–176.

Johnson, R.A. and Wichern, D.W. (1982). *Applied Multivariate Statistical Analysis.* Prentice Hall, Englewood Cliffs, NJ.

Jones, M.C. (1989). "Discretized and Interpolated Kernel Density Estimates" *J. Amer. Statist. Assoc.* **84** 733–741.

Jones, M.C. (1990). "Variable Kernel Density Estimates" *Austral. J. Statist.* **32** 361–371.

Jones, M.C. and Kappenman, R.F. (1992). "On a Class of Kernel Density Estimate Bandwidth Selectors" *Scand. J. Statist.* **19** 337–349.

Jones, M.C. and Sibson, R. (1987). "What Is Projection Pursuit? (with Discussion)" *J. Roy. Statist. Soc. A* **150** 1–36.

Jones, M.C. and Signorini, D.F. (1997). "A Comparison of Higher-Order Bias Kernel Density Estimators" *J. Amer. Statist. Assoc.* **92** 1064–1073.

Kaplan, E.L. and Meier, P. (1958). "Nonparametric Estimation from Incomplete Observations" *J. Amer. Statist. Assoc.* **53** 457–481.

Kaufman, L. and Rousseeuw, P.J. (2005). *Finding groups in data: an introduction to cluster analysis*, John Wiley & Sons, Hoboken, NJ.

Kendall, M.G. (1961). *A Course in the Geometry of n Dimensions.* Griffin's Statistical Monographs and Courses.

Kent, S.T. and Millett, L.I. (eds.) (2002). *IDs — Not That Easy: Questions About Nationwide Identity Systems* National Academy Press, Washington, DC.

Klemelä, J. (2008). "Mode trees for multivariate data" *J. Comp. Graph. Statist.* **17** 860–869.

Klonias, V.K. (1982). "Consistency of Two Nonparametric Maximum Penalized Likelihood Estimators of the Probability Density Function" *Ann. Statist.* **10** 811–824.

Kogure, A. (1987). "Asymptotically Optimal Cells for a Histogram" *Ann. Statist.* **15** 1023–1030.

Kronmal, R.A. and Tarter, M.E. (1968). "The Estimation of Probability Densities and Cumulatives by Fourier Series Methods" *J. Amer. Statist. Assoc.* **63** 925–952.

Kruskal, J.B. (1969). "Toward a Practical Method Which Helps Uncover the Structure of a Set of Multivariate Observations by Finding the Linear Transformation That Optimizes a New Index of Condensation." In *Statistical Computation*, R.C. Milton and J.A. Nelder (eds.), pp. 427–440. Academic Press, New York.

Kruskal, J.B. (1972). "Linear Transformation of Multivariate Data to Reveal Clustering." In *Multidimensional Scaling: Theory and Applications in the Behavioral Sciences, Vol. 1, Theory*, R.N. Shepard, A.K. Romney, and S.B. Nerlove (eds.), pp. 179–191. Seminar Press, London.

Kullback, S. and Leibler, R.A. (1951). "On information and sufficiency" *Ann. Math. Statist.* **22** 79–86.

Lehmann, E.L. and Casella, G. (1998). *Theory of point estimation, Second Edition*, Springer, New York.

Li, K.-C. (1991). "Sliced Inverse Regression for Dimension Reduction (with Discussion)" *J. Amer. Statist. Assoc.* **86** 316–342.

Lindsay, B.G. (1983). "The geometry of mixture likelihoods: a general theory" *Ann. Statist.* **11** 86–94.

Loftsgaarden, D.O. and Quesenberry, C.P. (1965). "A Nonparametric Estimate of a Multivariate Density Function" *Ann. Math. Statist.* **36** 1049–1051.

Lorensen, W.E. and Cline, H.E. (1987). "Marching Cubes: A High Resolution 3D Surface Construction Algorithm" *Computer Graphics* **21** 163–169.

Mack, Y.P. and Müller, H.G. (1989). "Convolution Type Estimators for Nonparametric Regression" *Statist. Prob. Lett.* **7** 229–239.

MacQueen, J.B. (1967). "Some Methods for classification and Analysis of Multivariate Observations" *Proceedings of the 5th Berkeley Symposium on Mathematical Statistics and Probability 1* University of California Press. pp. 281–297.

Mammen, E. (1993). "Bootstrap and Wild Bootstrap for High Dimensional Linear Models". *Ann. Statist.* **21** 255–285.

Mammen, E. Marron, J.S., and Fisher, N.I. (1994) "Asymptotics for Multimodality Tests Based on Kernel Density Estimates" *Prob. Theory Related Fields* **91** 115–132.

Marchette, D.J., Priebe, C.E., Rogers, G.W., and Solka, J.L. (1994). "Filtered Kernel Density Estimation" DTIC Technical Report, George Mason, Dept of Computational Statistics.

Marchette, D.J., Priebe, C.E., Rogers, G.W., and Wegman, E.J. (1996). "Filtered Kernel Density Estimation" *Comp. Statist.* **11** 112.

Marron, J.S. and Nolan, D. (1988). "Canonical Kernels for Density Estimation" *Statist. Prob. Lett.* **7** 195–199.

Marron, J.S. and Wand, M.P. (1992). 'Exact Mean Integrated Squared Error" *Ann. Statist.* **20** 712–736.

Marshall, A.W. and Olkin, I. (1985). "A Family of Bivariate Distributions Generated by the Bivariate Bernoulli Distribution" *J. Amer. Statist. Assoc.* **80** 332–338.

Mathematica (2012). Version 9.0, Wolfram Research, Inc. , Champaign, IL.

Matthews, M.V. (1983). "On Silverman's Test for the Number of Modes in a Univariate Density Function." Honors Bachelor's thesis, Harvard University.

McDonald, J.A. (1982). "Interactive Graphics for Data Analysis." ORION Technical Report 011, Stanford University.

McLachlan, G. and Krishnan, T. (2007). *The EM Algorithm and Extensions* John Wiley & Sons, Hoboken, NJ.

Minnotte, M.C. (1996). "The bias-optimized frequency polygon" *Comp. Statist.* **11** 35–48.

Minnotte, M.C. (1997). "Nonparametric Testing of the Existence of modes" *Ann. Statist.* **25** 1646–1660.

Minnotte, M.C. (1998). "Achieving higher-order convergence rates for density estimation with binned data" *J. Amer. Statist. Assoc.* **93** 663–672.

Minnotte, M.C., Marchette, D.J. and Wegman, E.J. (1998). "The Bumpy Road to the Mode Rorest" *J. Comp. Graph. Statist.* **7** 239–251.

Minnotte, M.C. and Scott, D.W (1993). "The Mode Tree: A Tool for Visualization of Nonparametric Density Features" *J. Comp. Graph. Statist.* **2** 51–68.

de Montricher, G.F., Tapia, R.A., and Thompson, J.R. (1975). "Nonparametric Maximum Likelihood Estimation of Probability Densities by Penalty Function Methods" *Ann. Statist.* **3** 1329–1348.

Müller, H.G. (1988). *Nonparametric Regression Analysis of Longitudinal Data.* Springer-Verlag, Berlin.

Müller, D.W. and Sawitzki, G. (1991). "Excess Mass Estimates and Tests for Multimodality" *J. Amer. Statist. Assoc.* **86** 738–746.

Nadaraya, E.A. (1964). "On Estimating Regression" *Theory Prob. Appl.* **15** 134–137.

Neudecker, H. and Magnus, J.R. (1988). *Matrix Differential Calculus with Applications in Statistics and Econometrics*, John Wiley, New York.

Nezames, D. (1980). "Some Results for Estimating Bivariate Densities Using Kernel, Orthogonal Series, and Penalized-Likelihood Procedures." Ph.D. thesis, Department of Mathematical Sciences, Rice University.

O'Sullivan, F. (1986). "A Statistical Perspective on Ill-Posed Inverse Problems" *Statist. Sci.* **1** 502–527.

Papkov, G.I. and Scott, D.W. (2010). "Local-Moment Nonparametric Density Estimation of Pre-Binned Data" *Comp. Stat. & Data Analysis* **54** 3421–3429.

Park, B.U. and Marron, J.S. (1990). "Comparison of Data-Driven Bandwidth Selectors" *J. Amer. Statist. Assoc.* **85** 66–72.

Parzen, E. (1962). "On Estimation of Probability Density Function and Mode" *Ann. Math. Statist.* **33** 1065–1076.

Parzen, E. (1979). "Nonparametric Statistical Data Modeling" *J. Amer. Statist. Assoc.* **74** 105–131.

Pearson, E.S. (1938). *Karl Pearson: An Appreciation of Some Aspects of His Life and Work.* Cambridge University Press, Cambridge, England.

Pearson, K. (1894). "Contributions to the Mathematical Theory of Evolution" *Philosophical Trans. Royal Society London (A)* **185** 71–110.

Pearson, K. (1902a). "On the Systematic Fitting of Curves to Observations and Measurements, I" *Biometrika* **1** 265–303.

Pearson, K. (1902b). "On the Systematic Fitting of Curves to Observations and Measurements, II" *Biometrika* **2** 1–23.

Prakasa Rao, B.L.S. (1983). *Nonparametric Functional Estimation.* Academic Press, Orlando, FL.

Priebe, C.E. and Marchette, D.J. (2000). "Alternating Kernel and Mixture Density Estimates" *Comp. Stat. & Data Analysis* **35** 43–65.

R Core Team (2014). *R: A Language and Environment for Statistical Computing*. R Foundation for Statistical Computing, Vienna, Austria. URL http://www.R-project.org/.

Ray, S. and Lindsay, B.G. (2005). "The Topography of Multivariate Normal Mixtures" *Ann. Statist.* **33** 2042–2065.

Redner, R.A. and Walker, H.F. (1984). "Mixture Densities, Maximum Likelihood and the EM Algorithm" *SIAM Review* **26** 195–202.

Reinsch, C.H. (1967). "Smoothing by Spline Functions" *Numerische Mathematik* **10** 177–183.

Rice, J.A. (1984a). "Boundary Modification for Kernel Regression" *Comm. Statist.* **13** 893–900.

Rice, J.A. (1984b). "Bandwidth Choice for Nonparametric Kernel Regression" *Ann. Statist.* **12** 1215–1230.

Ripley, B.D. (1981). *Spatial Statistics*. John Wiley, New York.

Ripley, B.D. (1988). *Statistical Inference for Spatial Processes*. Cambridge University Press, Cambridge, England.

Roeder, K. (1990). "Density Estimation with Confidence Sets Exemplified by Superclusters and Voids in the Galaxies" *J. Amer. Statist. Assoc.* **85** 617–624.

Romano, J.P. and Siegel, A.F. (1986). *Counterexamples in Probability and Statistics*. Wadsworth & Brooks/Cole, Pacific Grove, CA.

Rosenblatt, M. (1956). "Remarks on Some Nonparametric Estimates of a Density Function" *Ann. Math. Statist.* **27** 832–837.

Rosenblatt, M. (1969). "Conditional Probability Density and Regression Estimates." In *Multivariate Analysis II*, P.R. Krishnaiah (ed.), pp. 25–31. Academic Press, New York.

Rousseeuw, P.J. and Leroy, A.M. (1987). *Robust Regression and Outlier Detection*. John Wiley, New York.

Rudemo, M. (1982). "Empirical Choice of Histograms and Kernel Density Estimators" *Scand. J. Statist.* **9** 65–78.

S-PLUS (1990). "User's Manual." StatSci, Inc., Seattle, WA.. Currently TIBCO.com offers S-PLUS 2014 through Spotfire.

Sagae, M., Noro, T., and Scott, D.W. (2009). "The multi-dimensional non-parametric probability density estimation by Multivariate Polynomial Histogram Density Estimation," *J. Japanese Statist. Soc.* **39** 265–298.

Sagae, M., Scott, D.W. and Kusano, N. (2006). "A multivariate polynomial histogram by the method of local moments" *Proceedings of the 8th Workshop on Nonparametric Statistical Analysis and Related Area*, pp. 14–33, Tokyo.

Sager, T.W. and Thisted, R.A. (1982). "Maximum Likelihood Estimation of Isotonic Modal Regression" *Ann. Statist.* **10** 690–707.

Sain, S.R. (1994). "Adaptive Kernel Density Estimation." Ph.D. thesis, Department of Statistics, Rice University.

Sain, S.R. (2002). "Multivariate locally adaptive density estimation" *Comp. Stat. & Data Analysis* **39** 165–186.

Sain, S.R. (2003). "A New Characterization and Estimation of the Zero–bias Bandwidth" *Austral. & New Zealand J. Statist.* **45** 29–42.

Sain, S.R., Baggerly, K.A., and Scott, D.W. (1994). "Cross-Validation of Multivariate Densities" *J. Amer. Statist. Assoc.* **89** 807–817.

Sain, S.R. and Scott, D.W. (2002). "Zero-Bias Locally Adaptive Density Estimators" *Scand. J. Statist.* **29** 441–460.

Samiuddin, M. and El-Sayyad, G.M. (1990). "On nonparametric kernel density estimates" *Biometrika* **77** 865–874.

Schoenberg, I. (1964). "On Interpolation by Spline Functions and Its Minimum Properties" *Internat. Ser. Numer. Analysis* **5** 109–129.

Schucany, W.R. (1989). "Locally Optimal Window Widths for Kernel Density Estimation with Large Samples" *Statist. Prob. Lett.* **7** 401–405.

Schucany, W.R. (1995). "Adaptive Bandwidth Choice for Kernel Regression." *J. Amer. Statist. Assoc.* **90** 535–540.

Schucany, W.R. and Sommers, J.P. (1977). "Improvement of Kernel Density Estimators" *J. Amer. Statist. Assoc.* **72** 420–423.

Schuster, E.F. and Gregory, G.G. (1981). "On the Nonconsistency of Maximum Likelihood Nonparametric Density Estimators," *Proceedings of the Thirteenth Interface of Computer Science and Statistics*, W.F. Eddy (ed.) pp. 295–298. Springer-Verlag, New York.

Schwartz, S.C. (1967). "Estimation of a Probability Density by an Orthogonal Series" *Ann. Math. Statist.* **38** 1262–1265.

Scott, D.W. (1976). "Nonparametric Probability Density Estimation by Optimization Theoretic Techniques." Ph.D. thesis, Department of Mathematical Sciences, Rice University.

Scott, D.W. (1979). "On Optimal and Data-Based Histograms" *Biometrika* **66** 605–610.

Scott, D.W. (1980). "Comment on a Paper by Good and Gaskins" *J. Amer. Statist. Assoc.* **75** 61–62.

Scott, D.W. (1981). "Using Computer-Binned Data for Density Estimation," *Proceedings of the Thirteenth Interface of Computer Science and Statistics*, W.F. Eddy (ed.), pp. 292–294. Springer-Verlag, New York.

Scott, D.W. (1983). "Nonparametric Probability Density Estimation for Data Analysis in Several Dimensions," *Proceedings of the Twenty-Eighth Conference on the Design of Experiments in Army Research Development and Testing*, U.S. Army Research Office, pp. 387–397.

Scott, D.W. (1984). "Multivariate Density Function Representation," *Proceedings of the Sixth Annual National Computer Graphics Association Conference*, Volume II, pp. 794–801.

Scott, D.W. (1985a). "Frequency Polygons" *J. Amer. Statist. Assoc.* **80** 348–354.

Scott, D.W. (1985b). "Averaged Shifted Histograms: Effective Nonparametric Density Estimators in Several Dimensions" *Ann. Statist.* **13** 1024–1040.

Scott, D.W. (1985c). "Classification Using Multivariate Nonparametric Density Estimation," *Proceedings of the Sixth Annual National Computer Graphics Association Conference*, Volume III, pp. 715–718.

Scott, D.W. (1988a). "A Note on Choice of Bivariate Histogram Bin Shape" *J. Official Statistics* **4** 47–51.

Scott, D.W. (1988b). "Comment on "How Far Are Automatically Chosen Regression Smoothing Parameters from Their Optimum" by W. Härdle, P. Hall, and J.S. Marron" *J. Amer. Statist. Assoc.* **83** 96–98.

Scott, D.W. (1990). "Statistics in Motion: Where Is It Going?," *Proceedings of the Statistical Graphics Section*, pp. 17–22, American Statistical Association, Alexandria, VA.

Scott, D.W. (1991a). "On Estimation and Visualization of Higher Dimensional Surfaces." In *IMA Computing and Graphics in Statistics*, Volume 36 in IMA Volumes in Mathematics and its Applications, P. Tukey and A. Buja (eds.), pp. 187–205. Springer-Verlag, New York.

Scott, D.W. (1991b). "On Density Ridges." Technical Report, Rice University.

Scott, D.W. (1991c). "Comment on "Transformations in Density Estimation (with Discussion)" by Wand, M.P., Marron, J.S., and Ruppert, D." *J. Amer. Statist. Assoc.* **86** 359.

Scott, D.W. (2001). "Parametric statistical modeling by minimum integrated square error" *Technometrics* **43** 274–285.

Scott, D.W. and Factor, L.E. (1981). "Monte Carlo Study of the Three Data-Based Nonparametric Density Estimators" *J. Amer. Statist. Assoc.* **76** 9–15.

Scott, D.W., Gorry, G.A., Hoffmann, R.G., Barboriak, J.J., and Gotto, A.M. (1980). "A New Approach for Evaluating Risk Factors in Coronary Artery Disease: A Study of Lipid Concentrations and Severity of Disease in 1847 Males" *Circulation* **62** 477–484.

Scott, D.W., Gotto, A.M., Cole, J.S., and Gorry, G.A. (1978). "Plasma Lipids as Collateral Risk Factors in Coronary Artery Disease: A Study of 371 Males with Chest Pain" *J. Chronic Diseases* **31** 337–345.

Scott, D.W. and Hall, M.R. (1989). "Interactive Multivariate Density Estimation in the S Language," *Proceedings of the Twentieth Interface of Computer Science and Statistics*, E.J. Wegman (ed.), pp. 241–245, American Statistical Association, Alexandria, VA.

Scott, D.W. and Hutson, R.K. (1992). "Nonparametric Density Estimation as an Aid to Tissue Characterization in Magnetic Resonance Imaging" Poster, 3rd Keck Symposium on Computational Biology, Houston, November 1–3, 1992.

Scott, D.W. and Jee, J.R. (1984). "Nonparametric Analysis of Minnesota Spruce and Aspen Tree Data and Landsat Data," *Proceedings of the Second Symposium on Mathematical Pattern Recognition and Image Analysis*, pp. 27–49, NASA & Texas A&M University.

Scott, D.W. and Sagae, M. (1997). "Adaptive Density Estimation With Massive Data Sets," *Proceedings of the Statistical Computing Section*, pp. 104–108, ASA.

Scott, D.W. and Schmidt, H.-P. (1988). "Calibrating Histograms with Applications to Economic Data" *Empirical Economics* **13** 155–168.

Scott, D.W. and Schmidt, H.-P. (1989). "Calibrating Histograms with Applications to Economic Data." Reprinted in *Semiparametric and Nonparametric Economics*, A. Ullah (ed.), pp. 33–46. Physica-Verlag, Heidelberg.

Scott, D.W. and Scott, W.R. (2008). "Smoothed Histograms for Frequency Data on Irregular Intervals" *Amer. Statist.* **62** 256–261.

Scott, D.W. and Sheather, S.J. (1985). "Kernel Density Estimation with Binned Data." *Comm. Statist.* **14** 1353–1359.

Scott, D.W. and Szewczyk, W.F (1997). "Bumps Along the Road Towards Multivariate Mode Trees" NSF Workshop on Bumps, Jumps, Clustering and Discrimination, May 11–14, 1997, Houston.

Scott, D.W. and Szewczyk, W.F. (2001). "From Kernels to Mixtures," *Technometrics* **43** 323–335.

Scott, D.W., Tapia, R.A., and Thompson, J.R. (1977). "Kernel Density Estimation Revisited" *J. Nonlinear Analysis Theory Meth. Appl.* **1** 339–372.

Scott, D.W. and Terrell, G.R. (1987). "Biased and Unbiased Cross-Validation in Density Estimation" *J. Amer. Statist. Assoc.* **82** 1131–1146.

Scott, D.W. and Thompson, J.R. (1983). "Probability Density Estimation in Higher Dimensions," *Proceedings of the Fifteenth Interface of Computer Science and Statistics*, J.E. Gentle (ed.), pp. 173–179, North-Holland, Amsterdam.

Scott, D.W. and Wand, M.P. (1991). "Feasibility of Multivariate Density Estimates" *Biometrika* **78** 197–206.

Scott, D.W. and Wilks, A.R. (1990). "Animation of 3 and 4 Dimensional ASH Surfaces." Videotape, U. Minnesota Geometry Project.

Serfling, R.J. (1980). *Approximation Theorems of Mathematical Statistics.* John Wiley, New York.

Sheather, S.J. (1983). "A Data-Based Algorithm for Choosing the Window Width When Estimating the Density at a Point" *Comp. Stat. & Data Analysis* **1** 229–238.

Sheather, S.J. (2004). "Density Estimation" *Statist. Sci.* **19** 588–597.

Sheather, S.J. and Jones, M.C. (1991). "A Reliable Data-Based Bandwidth Selection Method for Kernel Density Estimation" *J. Roy. Statist. Soc. B* **53** 683–690.

Shibata, R. (1981). "An Optimal Selection of Regression Variables" *Biometrika* **68** 45–54.

Siegmund, D. (1988). "Confidence Sets in Change-Point Problems" *Internat. Statist. Review* **56** 31–48.

Silverman, B.W. (1978a). "Choosing the Window Width When Estimating a Density" *Biometrika* **65** 1–11.

Silverman, B.W. (1978b). "Density Ratios, Empirical Likelihood and Cot Death" *Appl. Statist.* **27** 26–33.

Silverman, B.W. (1981). "Using Kernel Density Estimates to Investigate Multimodality" *J. Roy. Statist. Soc. B* **43** 97–99.

Silverman, B.W. (1982). "Algorithm AS176. Kernel Density Estimation Using the Fast Fourier Transform" *Appl. Statist.* **31** 93–99.

Silverman, B.W. (1984). "Spline Smoothing: The Equivalent Variable Kernel Method" *Ann. Statist.* **12** 898–916.

Silverman, B.W. (1986). *Density Estimation for Statistics and Data Analysis.* Chapman and Hall, London.

Silverman, B.W. and Jones, M.C. (1989). "E. Fix and JL Hodges (1951): An important contribution to nonparametric discriminant analysis and density estimation: Commentary on Fix and Hodges (1951)" *Internat. Statist. Review* **57** 233–238.

Simonoff, J.S. (1996) *Smoothing Methods in Statistics* Springer, New York.

Speckman, P. (1987). "Kernel Smoothing in Partial Linear Models" *J. Roy. Statist. Soc. B* **50** 413–436.

Staniswalis, J.G. (1989). "Local Bandwidth Selection for Kernel Estimates" *J. Amer. Statist. Assoc.* **84** 276–283.

Staniswalis, J.G., Messer, K., and Finston, D.R. (1993). "Kernel Estimators for Multivariate Regression" *J. Nonp. Statist.* **3** 103–121.

Stein, C.M. (1956). "Inadmissibility of the Usual Estimator for the Mean of a Multivariate Normal Distribution," *Proceedings of the Third Berkeley Symposium Math. Statist. Prob.*, Vol. 1, pp. 197–206, U. California Press, Berkeley, CA.

Stigler, S.M. (1986). *The History of Statistics*. Harvard University Press, Cambridge, MA.

Stone, C.J. (1977). "Consistent Nonparametric Regression (with Discussion)" *Ann. Statist.* **5** 595–645.

Stone, C.J. (1985). "Additive Regression and Other Nonparametric Models" *Ann. Statist.* **13** 689–705.

Stone, M. (1974). "Cross-Validatory Choice and Assessment of Statistical Predictions (with Discussion)" *J. Roy. Statist. Soc. B* **36** 111–147.

Stuart, A. and Ord, J.K. (1987). *Kendall's Advanced Theory of Statistics, Volume 1*. Oxford University Press, New York.

Sturges, H.A. (1926). "The Choice of a Class Interval" *J. Amer. Statist. Assoc.* **21** 65–66.

Switzer, P. (1980). "Extension of Linear Discriminant Analysis for Statistical Classification of Remotely Sensed Satellite Imagery" *Mathematical Geology* **12** 367–376.

Tapia, R.A. (1971). "The Differentiation and Integration of Nonlinear Operators." In *Nonlinear Functional Analysis and Applications*, L.B. Rall (ed.), pp. 45–101. Academic Press, New York.

Tapia, R.A. and Thompson, J.R. (1978). *Nonparametric Probability Density Estimation*. John Hopkins University Press, Baltimore.

Tarter, M.E. and Kronmal, R.A. (1970). "On Multivariate Density Estimates Based on Orthogonal Expansions" *Ann. Math. Statist.* **41** 718–722.

Tarter, M.E. and Kronmal, R.A. (1976). "An Introduction to the Implementation and Theory of Nonparametric Density Estimation" *Amer. Statist.* **30** 105–112.

Tarter, M.E., Lock, M.D., and Mellin, C.C. (1990). *Curves: Background and Program Description*. Precision Data Group, Berkeley, CA.

Taylor, C.C. (1989). "Bootstrap Choice of the Smoothing Parameter in Kernel Density Estimation" *Biometrika* **76** 705–712.

Terrell, G.R. (1983). "The Multilinear Frequency Spline." Technical Report, Department of Math Sciences, Rice University.

Terrell, G.R. (1984). "Efficiency of Nonparametric Density Estimators." Technical Report, Department of Math Sciences, Rice University.

Terrell, G.R. (1985). "Projection Pursuit Via Multivariate Histograms." Technical Report 85–7, Department of Math Sciences, Rice University.

Terrell, G.R. (1990). "The Maximal Smoothing Principle in Density Estimation" *J. Amer. Statist. Assoc.* **85** 470–477.

Terrell, G.R. and Scott, D.W. (1980). "On Improving Convergence Rates for Nonnegative Kernel Density Estimators" *Ann. Statist.* **8** 1160–1163.

Terrell, G.R. and Scott, D.W. (1983). "Variable Window Density Estimates." Technical report presented at ASA meetings in Toronto.

Terrell, G.R., and Scott, D.W. (1985). "Oversmoothed Nonparametric Density Estimates" *J. Amer. Statist. Assoc.* **80** 209–214.

Terrell, G.R. and Scott, D.W. (1992). "Variable Kernel Density Estimation" *Ann. Statist.* **20** 1236–1265.

Thompson, J.R. and Tapia, R.A. (1990). *Nonparametric Function Estimation, Modeling, and Simulation.* SIAM, Philadephia.

Tibshirani, R.J. (1996). "Regression shrinkage and selection via the lasso" *J. Roy. Statist. Soc. B* **58** 267–288.

Tierney, L. (1990). *LISP-STAT.* John Wiley, New York.

Titterington, D.M., Murray, G.D., Murray, L.S., Spiegelhalter, D.J., Skene, A.M., Habbema, J.D.F., and Gelpke, G.J. (1981). "Comparison of Discrimination Techniques Applied to a Complex Data Set of Head Injured Patients" *J. Roy. Statist. Soc. A* **144** 145–175.

Titterington, D.M., Smith, A.F.M., and Makov, U.E. (1985). *Statistical Analysis of Finite Mixture Distributions.* John Wiley, New York.

Tufte, E.R. (1983). *The Visual Display of Quantitative Information.* Graphics Press, Cheshire, CT.

Tukey, J.W. (1977). *Exploratory Data Analysis.* Addison-Wesley, Reading, MA.

Tukey, P.A. and Tukey, J.W. (1981). "Graphical Display of Data Sets in 3 or More Dimensions." In *Interpreting Multivariate Data*, V. Barnett (ed.), pp. 187–275. John Wiley, Chichester, England.

Turner, D.W. and Tidmore, F.E. (1980). "FACES-A FORTRAN Program for Generating Chernoff-Type Faces on a Line Printer" *Amer. Statist.* **34** 187–187.

U.S. Bureau of the Census (1987). "U.S. Decennial Life Tables for 1979–81; Some Trends and Comparisons of United States Life Table Data: 1900–1981." Volume 1, Number 4, DHHS Pub No. PHS 87-1150-4.

Van Ryzin, J. (1973). "A Histogram Method of Density Estimation" *Comm. Statist.* **2** 493–506.

Wagner, H.M. (1959). "Linear Programming Techniques for Regression Analysis" *J. Amer. Statist. Assoc.* **54** 206–212.

Wahba, G. (1971). "A Polynomial Algorithm for Density Estimation" *Ann. Math. Statist.* **42** 1870–1886.

Wahba, G. (1977). "Optimal Smoothing of Density Estimates." In *Classification and Clustering*, J. Van Ryzin (ed.), pp. 423–458. Academic Press, New York.

Wahba, G. (1981). "Data-Based Optimal Smoothing of Orthogonal Series Density Estimates" *Ann. Statist.* **9** 146–156.

Wahba, G. (1990). *Spline Models for Observational Data.* SIAM, Philadephia.

Wahba, G. and Wold, S. (1975). "A Completely Automatic French Curve: Fitting Spline Functions by Cross-Validation" *Comm. Statist.* **4** 1–17.

Walter, G. and Blum, J.R. (1979). "Probability Density Estimation Using Delta Sequences" *Ann. Statist.* **7** 328–340.

Wand, M.P. (1994). "Fast computation of multivariate kernel estimators" *J. Comp. Graph. Statist.* **3** 433–445.

Wand, M.P. and Jones, M.C. (1991). "Comparison of Smoothing Parameterizations in Bivariate Density Estimation." *J. Amer. Statist. Assoc.* **88** 520–528.

Wand, M.P. and Jones, M.C. (1994). "Multivariate Plug-in Bandwidth Selection" *Comp. Statist.* **9** 97–116.

Wand, M.P. and Jones, M.C. (1995). *Kernel Smoothing*, CRC Press, London.

Wand, M.P., Marron, J.S., and Ruppert, D. (1991). "Transformations in Density Estimation (with Discussion)" *J. Amer. Statist. Assoc.* **86** 343–352.

Wang, F.T. (1990). "A New Method for Robust Nonparametric Regression." Ph.D. thesis, Department of Statistics, Rice University.

Wang, F.T. and Scott, D.W. (1994). "The L_1 Method for Robust Nonparametric Regression" *J. Amer. Statist. Assoc.* **89** 65–76.

Wang, P.C.C. (Ed.) (1978). *Graphical Representation of Multivariate Data*. Academic Press, New York.

Watson, G.S. (1964). "Smooth Regression Analysis" *Sankhyā Ser. A* **26** 359–372.

Watson, G.S. (1969). "Density Estimation by Orthogonal Series" *Ann. Math. Statist.* **40** 1496–1498.

Watson, G.S. (1985). *Statistics on Spheres*. John Wiley, New York.

Watson, G.S. and Leadbetter, M.R. (1963). "On the Estimation of the Probability Density I" *Ann. Math. Statist.* **34** 480–491.

Watson, G.S. and Leadbetter, M.R. (1964). "Hazard Analysis I" *Biometrika* **51** 175–184.

Weaver, C.S., Zollweg, J.E., and Malone, S.D. (1983). "Deep Earthquakes Beneath Mount St. Helens: Evidence for Magmatic Gas Transport?" *Science* **221** 1391–1394.

Wegman, E.J. (1970). "Maximum Likelihood Estimation of a Unimodal Density Function" *Ann. Statist.* **41** 457–471.

Wegman, E.J. (1990). "Hyperdimensional Data Analysis Using Parallel Coordinates" *J. Amer. Statist. Assoc.* **85** 664–675.

Wegman, E.J. and DePriest, D.J. (eds.) (1986). *Statistical Image Processing and Graphics*. Marcel Dekker, New York.

Weisberg, S. (1985). *Applied Linear Regression*. John Wiley, New York.

Wertz, W. (1978). *Statistical Density Estimation: A Survey*. Vandenhoeck and Ruprecht, Göttingen.

Wharton, S.W. (1983). "A Generalized Histogram Clustering Scheme for Multidimensional Image Data" *Pattern Recognition* **16** 193–199.

Woodroofe, M. (1970). "On Choosing a Delta-Sequence" *Ann. Math. Statist.* **41** 1665–1671.

Worton, B.J. (1989). "Optimal Smoothing Parameters for Multivariate Fixed and Adaptive Kernel Methods" *J. Statist. Comp. Simul.* **32** 45–57.

Wu, C.F.J. (1986). "Jackknife, bootstrap and other resampling methods in regression analysis (with discussion)." *Ann. Statist.* **14** 1261–1350.

Author Index

Multivariate Density Estimation, First Edition. David W. Scott.
© 2015 John Wiley & Sons, Inc. Published 2015 by John Wiley & Sons, Inc.

Subject Index

Multivariate Density Estimation, First Edition. David W. Scott.
© 2015 John Wiley & Sons, Inc. Published 2015 by John Wiley & Sons, Inc.

WILEY SERIES IN PROBABILITY AND STATISTICS

ESTABLISHED BY WALTER A. SHEWHAR AND SAMUEL S. WILKS

Editors: *David J. Balding, Noel A. C. Cressie, Garrett M. Fitzmaurice, Geof H. Givens, Harvey Goldstein, Geert Molenberghs, David W. Scott, Adrian F. M. Smith, Ruey S. Tsay, Sanford Weisberg*
Editors Emeriti: *J. Stuart Hunter, Iain M. Johnstone, Joseph B. Kadane, Jozef L. Teugels*

The *Wiley Series in Probability and Statistics* is well established and authoritative. It covers many topics of current research interest in both pure and applied statistics and probability theory. Written by leading statisticians and institutions, the titles span both state-of-the-art developments in the field and classical methods.

Reflecting the wide range of current research in statistics, the series encompasses applied, methodological and theoretical statistics, ranging from applications and new techniques made possible by advances in computerized practice to rigorous treatment of theoretical approaches.

This series provides essential and invaluable reading for all statisticians, whether in academia, industry, government, or research.

[†] ABRAHAM and LEDOLTER · Statistical Methods for Forecasting
AGRESTI · Analysis of Ordinal Categorical Data, *Second Edition*
AGRESTI · An Introduction to Categorical Data Analysis, *Second Edition*
AGRESTI · Categorical Data Analysis, *Third Edition*
AGRESTI · *Foundations of Linear and Generalized Linear Models*
ALSTON, MENGERSEN and PETTITT (editors) · Case Studies in Bayesian Statistical Modelling and Analysis
ALTMAN, GILL, and McDONALD · Numerical Issues in Statistical Computing for the Social Scientist
AMARATUNGA and CABRERA · Exploration and Analysis of DNA Microarray and Protein Array Data
AMARATUNGA, CABRERA, and SHKEDY · Exploration and Analysis of DNA Microarray and Other High-Dimensional Data, *Second Edition*
ANDĚL · Mathematics of Chance
ANDERSON · An Introduction to Multivariate Statistical Analysis, *Third Edition*
[*] ANDERSON · The Statistical Analysis of Time Series
ANDERSON, AUQUIER, HAUCK, OAKES, VANDAELE, and WEISBERG · Statistical Methods for Comparative Studies
ANDERSON and LOYNES · The Teaching of Practical Statistics
ARMITAGE and DAVID (editors) · Advances in Biometry
ARNOLD, BALAKRISHNAN, and NAGARAJA · Records
[*] ARTHANARI and DODGE · Mathematical Programming in Statistics
AUGUSTIN, COOLEN, DE COOMAN and TROFFAES (editors) · Introduction to Imprecise Probabilities
[*] BAILEY · The Elements of Stochastic Processes with Applications to the Natural Sciences
BAJORSKI · Statistics for Imaging, Optics, and Photonics
BALAKRISHNAN and KOUTRAS · Runs and Scans with Applications
BALAKRISHNAN and NG · Precedence-Type Tests and Applications
BARNETT · Comparative Statistical Inference, *Third Edition*
BARNETT · Environmental Statistics
BARNETT and LEWIS · Outliers in Statistical Data, *Third Edition*
BARTHOLOMEW, KNOTT, and MOUSTAKI · Latent Variable Models and Factor Analysis: A Unified Approach, *Third Edition*

BARTOSZYNSKI and NIEWIADOMSKA-BUGAJ · Probability and Statistical Inference, *Second Edition*

BASILEVSKY · Statistical Factor Analysis and Related Methods: Theory and Applications

BATES and WATTS · Nonlinear Regression Analysis and Its Applications

BECHHOFER, SANTNER, and GOLDSMAN · Design and Analysis of Experiments for Statistical Selection, Screening, and Multiple Comparisons

BEH and LOMBARDO · Correspondence Analysis: Theory, Practice and New Strategies

BEIRLANT, GOEGEBEUR, SEGERS, TEUGELS, and DE WAAL · Statistics of Extremes: Theory and Applications

BELSLEY · Conditioning Diagnostics: Collinearity and Weak Data in Regression

† BELSLEY, KUH, and WELSCH · Regression Diagnostics: Identifying Influential Data and Sources of Collinearity

BENDAT and PIERSOL · Random Data: Analysis and Measurement Procedures, *Fourth Edition*

BERNARDO and SMITH · Bayesian Theory

BHAT and MILLER · Elements of Applied Stochastic Processes, *Third Edition*

BHATTACHARYA and WAYMIRE · Stochastic Processes with Applications

BIEMER, GROVES, LYBERG, MATHIOWETZ, and SUDMAN · Measurement Errors in Surveys

BILLINGSLEY · Convergence of Probability Measures, *Second Edition*

BILLINGSLEY · Probability and Measure, *Anniversary Edition*

BIRKES and DODGE · Alternative Methods of Regression

BISGAARD and KULAHCI · Time Series Analysis and Forecasting by Example

BISWAS, DATTA, FINE, and SEGAL · Statistical Advances in the Biomedical Sciences: Clinical Trials, Epidemiology, Survival Analysis, and Bioinformatics

BLISCHKE and MURTHY (editors) · Case Studies in Reliability and Maintenance

BLISCHKE and MURTHY · Reliability: Modeling, Prediction, and Optimization

BLOOMFIELD · Fourier Analysis of Time Series: An Introduction, *Second Edition*

BOLLEN · Structural Equations with Latent Variables

BOLLEN and CURRAN · Latent Curve Models: A Structural Equation Perspective

BONNINI, CORAIN, MAROZZI and SALMASO · Nonparametric Hypothesis Testing: Rank and Permutation Methods with Applications in R

BOROVKOV · Ergodicity and Stability of Stochastic Processes

BOSQ and BLANKE · Inference and Prediction in Large Dimensions

BOULEAU · Numerical Methods for Stochastic Processes

* BOX and TIAO · Bayesian Inference in Statistical Analysis

BOX · Improving Almost Anything, *Revised Edition*

* BOX and DRAPER · Evolutionary Operation: A Statistical Method for Process Improvement

BOX and DRAPER · Response Surfaces, Mixtures, and Ridge Analyses, *Second Edition*

BOX, HUNTER, and HUNTER · Statistics for Experimenters: Design, Innovation, and Discovery, *Second Editon*

BOX, JENKINS, and REINSEL · Time Series Analysis: Forecasting and Control, *Fourth Edition*

BOX, LUCEÑO, and PANIAGUA-QUIÑONES · Statistical Control by Monitoring and Adjustment, *Second Edition*

* BROWN and HOLLANDER · Statistics: A Biomedical Introduction

CAIROLI and DALANG · Sequential Stochastic Optimization

CASTILLO, HADI, BALAKRISHNAN, and SARABIA · Extreme Value and Related Models with Applications in Engineering and Science

CHAN · Time Series: Applications to Finance with R and S-Plus®, *Second Edition*

CHARALAMBIDES · Combinatorial Methods in Discrete Distributions

RUBINSTEIN, RIDDER, and VAISMAN · Fast Sequential Monte Carlo Methods for Counting and Optimization

RYAN · Modern Engineering Statistics

RYAN · Modern Experimental Design

RYAN · Modern Regression Methods, *Second Edition*

RYAN · Sample Size Determination and Power

RYAN · Statistical Methods for Quality Improvement, *Third Edition*

SALEH · Theory of Preliminary Test and Stein-Type Estimation with Applications

SALTELLI, CHAN, and SCOTT (editors) · Sensitivity Analysis

SCHERER · Batch Effects and Noise in Microarray Experiments: Sources and Solutions

* SCHEFFE · The Analysis of Variance

SCHIMEK · Smoothing and Regression: Approaches, Computation, and Application

SCHOTT · Matrix Analysis for Statistics, *Second Edition*

SCHOUTENS · Levy Processes in Finance: Pricing Financial Derivatives

SCOTT · Multivariate Density Estimation: Theory, Practice, and Visualization, Second Edition

* SEARLE · Linear Models

† SEARLE · Linear Models for Unbalanced Data

† SEARLE · Matrix Algebra Useful for Statistics

† SEARLE, CASELLA, and McCULLOCH · Variance Components

SEARLE and WILLETT · Matrix Algebra for Applied Economics

SEBER · A Matrix Handbook For Statisticians

† SEBER · Multivariate Observations

SEBER and LEE · Linear Regression Analysis, *Second Edition*

† SEBER and WILD · Nonlinear Regression

SENNOTT · Stochastic Dynamic Programming and the Control of Queueing Systems

* SERFLING · Approximation Theorems of Mathematical Statistics

SHAFER and VOVK · Probability and Finance: It's Only a Game!

SHERMAN · Spatial Statistics and Spatio-Temporal Data: Covariance Functions and Directional Properties

SILVAPULLE and SEN · Constrained Statistical Inference: Inequality, Order, and Shape Restrictions

SINGPURWALLA · Reliability and Risk: A Bayesian Perspective

SMALL and MCLEISH · Hilbert Space Methods in Probability and Statistical Inference

SRIVASTAVA · Methods of Multivariate Statistics

STAPLETON · Linear Statistical Models, *Second Edition*

STAPLETON · Models for Probability and Statistical Inference: Theory and Applications

STAUDTE and SHEATHER · Robust Estimation and Testing

STOYAN · Counterexamples in Probability, *Second Edition*

STOYAN and STOYAN · Fractals, Random Shapes and Point Fields: Methods of Geometrical Statistics

STREET and BURGESS · The Construction of Optimal Stated Choice Experiments: Theory and Methods

STYAN · The Collected Papers of T. W. Anderson: 1943–1985

SUTTON, ABRAMS, JONES, SHELDON, and SONG · Methods for Meta-Analysis in Medical Research

TAKEZAWA · Introduction to Nonparametric Regression

TAMHANE · Statistical Analysis of Designed Experiments: Theory and Applications

TANAKA · Time Series Analysis: Nonstationary and Noninvertible Distribution Theory

THOMPSON · Empirical Model Building: Data, Models, and Reality, *Second Edition*

THOMPSON · Sampling, *Third Edition*

THOMPSON · Simulation: A Modeler's Approach

THOMPSON and SEBER · Adaptive Sampling

THOMPSON, WILLIAMS, and FINDLAY · Models for Investors in Real World Markets

TIERNEY · LISP-STAT: An Object-Oriented Environment for Statistical Computing and Dynamic Graphics

TROFFAES and DE COOMAN · Lower Previsions

TSAY · Analysis of Financial Time Series, *Third Edition*

TSAY · An Introduction to Analysis of Financial Data with R

TSAY · Multivariate Time Series Analysis: With R and Financial Applications

UPTON and FINGLETON · Spatial Data Analysis by Example, Volume II: Categorical and Directional Data

† VAN BELLE · Statistical Rules of Thumb, *Second Edition*

VAN BELLE, FISHER, HEAGERTY, and LUMLEY · Biostatistics: A Methodology for the Health Sciences, *Second Edition*

VESTRUP · The Theory of Measures and Integration

VIDAKOVIC · Statistical Modeling by Wavelets

VIERTL · Statistical Methods for Fuzzy Data

VINOD and REAGLE · Preparing for the Worst: Incorporating Downside Risk in Stock Market Investments

WALLER and GOTWAY · Applied Spatial Statistics for Public Health Data

WEISBERG · Applied Linear Regression, *Fourth Edition*

WEISBERG · Bias and Causation: Models and Judgment for Valid Comparisons

WELSH · Aspects of Statistical Inference

WESTFALL and YOUNG · Resampling-Based Multiple Testing: Examples and Methods for p-Value Adjustment

* WHITTAKER · Graphical Models in Applied Multivariate Statistics

WINKER · Optimization Heuristics in Economics: Applications of Threshold Accepting

WOODWORTH · Biostatistics: A Bayesian Introduction

WOOLSON and CLARKE · Statistical Methods for the Analysis of Biomedical Data, *Second Edition*

WU and HAMADA · Experiments: Planning, Analysis, and Parameter Design Optimization, *Second Edition*

WU and ZHANG · Nonparametric Regression Methods for Longitudinal Data Analysis

YAKIR · Extremes in Random Fields

YIN · Clinical Trial Design: Bayesian and Frequentist Adaptive Methods

YOUNG, VALERO-MORA, and FRIENDLY · Visual Statistics: Seeing Data with Dynamic Interactive Graphics

ZACKS · Examples and Problems in Mathematical Statistics

ZACKS · Stage-Wise Adaptive Designs

* ZELLNER · An Introduction to Bayesian Inference in Econometrics

ZELTERMAN · Discrete Distributions—Applications in the Health Sciences

ZHOU, OBUCHOWSKI, and MCCLISH · Statistical Methods in Diagnostic Medicine, *Second Edition*

---

*Now available in a lower priced paperback edition in the Wiley Classics Library.

†Now available in a lower priced paperback edition in the Wiley–Interscience Paperback Series.